The Difference Between God and Larry Ellison

ALSO BY MIKE WILSON

Right on the Edge of Crazy:
On Tour with the U.S. Downhill Ski Team

The Difference Between God and Larry Ellison*

With a New Epilogue

Mike Wilson

*God Doesn't Think He's Larry Ellison

HARPER

NEW YORK • LONDON • TORONTO • SYDNEY

A hardcover edition of this book was published in 1997 by William Morrow.
A paperback edition of this book was published in 1998 by Quill.

HarperCollins books may be purchased for educational, business, or sales promotional use. For information please write: Special Markets Department, HarperCollins Publishers Inc., 10 East 53rd Street, New York, NY 10022.

First HarperBusiness paperback edition published 2003

Designed by Laura Hammond Hough

Library of Congress Cataloging-in-Publication Data:
Wilson, Mike.
The difference between God and Larry Ellison : inside Oracle Corporation / by Mike Wilson.
p. cm.
ISBN 0-688-16353-X
1. Ellison, Larry. 2, Oracle Corporation—History. 3. Computer software industry—United States—History. 4. Businesspeople—United States—Biography. I. Title.
HD9696.C62E57648 1997
338.7'610053'092—dc21 97-20100
CIP

ISBN 0-06-000876-8 (pbk.)

10 11 12/ RRD / 20 19 18 17 16 15 14 13 12

For the Wilsons
Carol, Dyami, Lena, and Kirby

And the Taylors
Phil, Joanne, Emily, Ben, and baby Jess

You're a reporter. You want to know what I think about Charlie Kane. Well, I suppose he had some private sort of greatness, but he kept it to himself. He never gave himself away. He never gave anything away. He just left you a tip.

He had a generous mind. I don't suppose anybody ever had so many opinions. But he never believed in anything except Charlie Kane, he never had a conviction except Charlie Kane in his life. I suppose he died without one. It must have been pretty unpleasant. Of course, a lot of us check out without having any special convictions about death. But we do know what we're leaving. At least we believe in something.

—From Orson Welles's 1941 film *Citizen Kane*

The Difference Between God and Larry Ellison

One

LARRY ELLISON WALKED DOWN THE LONG HALLWAY, HIS SNEAKERS chirping quietly as he approached the living room. The hallway was narrower at his end than at mine, so Ellison seemed to grow larger as he moved toward me, an optical illusion that would have pleased him. He wore white gym shorts and a white T-shirt, and his youthful face was still glowing red from his afternoon workout.

I had arrived thirty minutes earlier, at the appointed time. Klaus, a member of Ellison's domestic staff, had ushered me inside, and Maria, the maid, had given me a soft drink.

"Hi, sorry I'm late," Ellison said, extending an outsize hand. Embroidered onto his shirt was the word SAYONARA, the name of his yacht. His mouth was saying hello, but his shirt was saying goodbye. The contradiction seemed appropriate: As the entire world of high technology knew, Larry Ellison was a hard man to pin down.

We chatted for a moment. Then—*sayonara*—he excused himself and disappeared up the spiral stairway to take a shower.

After he had rotated out of sight, I had a few more minutes to look around. Standing in the vast living room of Ellison's San Francisco home was like being transported into the pages of *Architectural Digest.* A couple of black lacquer stereo speakers stood like monoliths near the fireplace, a Steinway grand piano stood three-legged in the corner, and on the coffee table a pair of wire-frame glasses lay on top of a stack of art history books. The house had many other impressive features that could not be seen from the living room. A computer in the closet controlled more than five thousand lights, all of which could be dimmed or brightened according to the time of day or Ellison's mood. In the entertainment area an eight-thousand-dollar video projector produced a razor-sharp picture even when the room was not fully darkened. (Someone had told me that NASA used the same kind of projector at Mission Control.) The house also had a courtyard patio made of bronze tiles, a unique handmade snooker table in the game room, and two shiny stainless steel garage doors that looked great except that they showed fingerprints.

Probably the most remarkable thing about the house was the view—"the best view money can buy," the designer had told me.[1] I couldn't quarrel. Through the broad living-room windows I could see Alcatraz Island, Sausalito, the Golden Gate Bridge—the great, misty San Francisco panorama.

And this was just Ellison's entertainment house, his "pied-à-terre," the home where he gave parties and did business whenever he happened to be in the city. His main residence, a ten-thousand-square-foot Japanese-style house with a large garden and an authentic teahouse, was down the peninsula in Atherton. And he was currently building a new Japanese estate on twenty-three acres in Woodside, a wooded village in the Silicon Valley hills. The projected cost: forty million dollars.

He could afford it. Only days before this interview, my first with Ellison, *Forbes* magazine had estimated his net worth at six billion dollars. According to the magazine, that made Ellison the fifth-richest person in the United States. Some of the people on the Forbes 400 didn't like having their fortunes discussed publicly; as I was about to see, Ellison was not one of them. He enjoyed the attention.

Soon he reappeared, wet-headed from the shower and dressed casually in slacks and a black, short-sleeve polo shirt. He led me to a table in the far corner of the room and offered me a seat. The table had place settings for two. After a moment the staff arrived with plates of fresh greens, the first course of what turned out to be a three-course lunch. Ellison, a man of legendary appetites, drank about a quart of carrot juice with his meal.

I began by asking him to talk about his childhood in Chicago. He did, briefly. Then—seamlessly and effortlessly—he segued into a speech about Bill Clinton's failed attempt several years earlier to remake the American health care system. Ellison believed that the President should have left health care alone and focused on education.

I have added punctuation to make the following paragraph readable, but when Larry Ellison was speaking, there were no commas or hyphens. He talked so fast that he mutilated some sentences and even some words. For example, when he said the word "graduating," it sounded like *gradjing*. He ate "uate."

"People get on airplanes all over the world and fly to the United States to get health care. And we have an educational system that would be an embarrassment to a third world country," he said, overstating the case, as he often did. "No one flies here to send their kids to the seventh- and eighth-grade public schools in San Francisco or New York City or Chicago. I just think Bill Clinton was wrong. . . . First of all, health care is not a government monopoly. Education is. So government taking responsibility for fixing this problem is absolutely appropriate. Government taking responsibility for fixing health care, which isn't broken—People get on airplanes and fly here. Don't get me wrong. I think that there are access problems with our health care system. The very poor can't get at it. Interestingly enough, one of the big groups who don't get to have health care are people just graduating"—*gradjing*—"from college. . . . So a lot of these statistics can be very misleading. Some of the people who don't have access really are not at risk. There are others—the very poor—who really are at risk, and they don't have access, so we have to do something about that. But I don't think you revamp the entire health care system, which delivers incredible quality health care, the best health care in the world, you don't revamp that system of private-public cooperation that exists right now and turn it into a public monopoly."

Around the time of my interview with Ellison the business pages were full of articles about search engines, computer programs designed to help people find information on the World Wide Web. That was the way Ellison's mind worked. He was like a search engine gone haywire. If you asked him for information about Chicago in the 1950s, he told you about the Clinton presidency.

I tried again. I asked Ellison how he had seen his adult life when he was a kid. What had he thought was going to happen to him?

"You mean, did I anticipate being the fifth-wealthiest person in the United States in 1996? No," he said. "I mean, this is all kind of surreal. I don't even believe it now. Not only did I not believe it when I was fourteen, but when I look around, I say, this must be something out of a dream." Coming from almost anyone else, the phrase "something out of a dream" would have sounded hackneyed, but Ellison, an utterly self-made man, meant it. A dozen years after he accumulated his first million, his good fortune still amazed him.

Then the conversation shifted again, and Ellison began to talk about some of the bizarre and farfetched stories that had been published about him. For example, the *Wall Street Journal* once said that he used to sign his checks in green because green was the color of money. Ridiculous, he said: "My favorite color happens to be green, but not because it's the color of money." He also mentioned a 1991 article in a technology industry magazine called *Upside* that talked about the way Ellison had bought this very house. As Ellison remembered the article, it said he and his then girlfriend drove past the house at 2:00 A.M. and liked it. Ellison woke up the owner, bought the house for four million dollars in cash (he had the money in the trunk), then took his girlfriend inside and had sex with her.

Five years after the article appeared, Ellison was still astonished that such an outlandish story had been published. "Four million dollars for sex?" he said. "I mean, excuse me?" He said he bought the house from the estate of the owners, who were dead at the time.

Later I looked up the *Upside* magazine article. The actual anecdote wasn't half as juicy as the one Ellison had told me. It said Ellison and his girlfriend were driving home from the symphony when they saw the house. "[He] rang the doorbell and offered to write a check on the spot. The owner, in his bathrobe, was shocked and initially refused, but over the ensuing months Ellison managed to buy the house anyway." *Upside* attributed the story to anonymous sources, who said they'd heard it from Ellison. There was no mention in the article of wads of cash and not a whisper of celebratory sex.

This was classic Larry Ellison: He had embellished the story even while denying it. If he was going to be the subject of a rumor, then by God, he wanted it to be an interesting one.

If Hollywood ever wants to update the legendary film *Citizen Kane,* it might consider Lawrence Joseph Ellison for the lead character. Ellison is the Charles Foster Kane of the technological age. He is bright, brash, optimistic, and immensely appealing, yet somehow incomplete. Like the movie character, he wants desperately to be loved, even idolized, but love does not come easily to him or from him. Also like Kane, Larry Ellison is oversize, a myth of his own making. Only a big screen could contain such a man.

The fictional Kane created his own legend by publishing newspapers (Orson Welles's movie was based on the life of William Randolph Hearst). Ellison did the same through the mystery and promise of high technology. Born to an unwed mother in 1944, he was adopted by relatives and raised in a middle-class Jewish home on the South Side of Chicago. He went to college in Illinois and California but, to his embarrassment, never got a degree. All his life, Ellison said, his adoptive father told him that he would never amount to anything. In 1966 he headed to California, where he attempted to prove his father wrong.

He worked in the computer industry for several years but never had a job that suited what he saw as his superior intellectual gifts. In 1977 the thirty-two-year-old Ellison went into business for himself, writing computer programs on a contract basis. About that time the International Business Machines Corporation, which loomed like a colossus over the industry, announced an idea for a new kind of business software. Ellison, seeing an opportunity, borrowed the idea and turned it into a product before IBM did.

What a product it was. While Microsoft founder Bill Gates was fulfilling his vision of a computer in every home, Larry Ellison's company, Oracle Corporation, was fomenting a revolution at the office. In the mid-1970s computer databases could do a few things extremely well. They could, for example, keep track of thousands of customer accounts, updating them after every new order or payment. What those databases could not do, or could not do fast enough, was answer unanticipated questions, the kind businesspeople often asked before they made decisions. If a company wanted to know which of its products were selling best in Dubuque, or which regional office did the most business in August, or how many of its female employees had more than five years of service, the database of yesteryear could not produce a fast answer.

Ellison's database software could. Oracle's so-called relational databases gave businesses and government agencies something they desperately needed: quick and easy access to information. When people saw what this software could do for them—when they understood that knowledge really was power—they lined up to buy it. By the mid-1990s Oracle software brought order to people's lives in ways they were not even aware of. Anyone who made a hotel reservation, bought or sold stock, ordered something from a catalog, rented a video, or used a credit card came into contact with software made by Oracle or its competitors.

Larry Ellison's company, the market leader in relational database software, qualified as one of the extraordinary success stories in California's Silicon Valley—or for that matter, in American business. Oracle Corporation doubled its sales in eleven of its first twelve years, mutating from 4 employees and a few hundred thousand dollars in revenue in its first year to 4,148 employees and $583 million in sales in 1989. By the mid-1990s Oracle no longer cloned itself once a year—no company could sustain that kind of growth—but it still grew at a robust pace. Though few outside the computer industry knew it, Oracle was the second-largest independent software company in the world, behind Microsoft. ("Independent" meant a company that did not also sell computers, like IBM or Digital Equipment Corporation.) It employed 23,113 people in the United States and fifty-five other countries. In fiscal 1996 the company averaged more than $1 billion in sales each quarter, for a total of $4.2 billion. Oracle's profit was $603 million.

Investors adored the company. If a person had sunk $10,000

into Oracle stock on the day the company went public in 1986 and had kept the stock for the next ten years, those shares would have been worth more than $790,000.[2]

The stockholder who benefited most from Oracle's performance was Larry Ellison, exactly what he intended. Ellison started the company because he wanted to be a boss, and he stayed in control throughout his tenure at Oracle, always holding on to enough stock that his power and authority could never be seriously challenged. In 1996 he still owned about 23 percent of the stock in Oracle, worth about six billion dollars. He had amassed this almost unimaginable fortune in less than two decades, beginning with an investment of twelve hundred dollars.

Ellison was not one to bury his money in the backyard; he spent it gleefully. He drove Ferraris until they blew up, then switched to convertible Bentleys and Acura NSX sports cars. When he became interested in sailing, he commissioned a seventy-eight-foot maxi-yacht, *Sayonara,* in which he and his crew won Australia's Sydney to Hobart yacht race in 1995. Ellison collected fine art, not just to cover the walls but to complement his knowledge and love of history. And when his thirteen-year-old son, David, went mad for the action film *Independence Day* in the summer of 1996, Ellison arranged for the movie studio to send him a copy. Then he had his limousine pick up David and his friends and drive them to a private screening. Let other fathers take their children to the movies; Ellison took the movies to his children.

Ellison also spent lavishly on airplanes. He flew a Cessna Citation jet, a Lancair kit plane called the Dreamcatcher, and several others and was considering buying a used Soviet MiG fighter jet. He spent weekends flying with his son. David and a flight instructor would climb into the Dreamcatcher, and Larry would get into one of the other planes. Then father and son would stage mock dogfights over the Pacific Ocean, swerving and swooping and diving as if the controls were joysticks and the sky a big video screen. Why settle for virtual reality when you can have the real thing?

As Ellison became rich, many of his employees did too. In the early years, at least, Ellison was generous with stock grants, and the more people he made rich, the more glorious and powerful he felt. One early Oracle employee—a woman who now spends her days relaxing in her homes in California and Hawaii—said, "Larry is very self-centered, where everything revolves around Larry, but he's not selfish."[3] In the mid-1980s Ellison asked one of his assistants to keep a list of the Oracle employees who had become millionaires. In those days there were twenty or thirty; a decade later the number was surely in the hundreds.

One of the first half dozen employees of Oracle was a programmer named Stuart Feigin. When he joined the company in 1978, Feigin received a large amount of stock, much of which he kept. In the mid-1990s he bought a personal finance program called Quicken to help him manage his money. There was a problem: Quicken could handle only seven digits to the left of the decimal point.

How did Ellison create so much prosperity? Certainly he had a generous mind. He could converse freely about almost anything: the Holocaust, poetry, education, architecture, and the future of high technology. He was well versed in the world's religions from Judaism to Buddhism, but to him there was no power greater than the human mind. One of Ellison's interests was molecular biology; he once spent a two-week "vacation" in a university laboratory, assisting scientists in transferring genetic information from one bacterial strain to another. "Things he doesn't know he picks up very, very quickly. As a result, we have very compressed conversations," said his friend Joshua Lederberg, who invited Ellison to the lab. "He is one of the most agile, insightful minds I have ever met." Lederberg, it is worth mentioning, has met some pretty agile minds. He is a Nobel laureate.

When hiring help, Ellison valued intelligence more than experience and maturity; he often looked for unruly geniuses instead of solid, steady workers. People who went into job interviews expecting to talk about software engineering often ended up discussing other things. Jenny Overstreet, who became Ellison's longtime assistant, remembered talking with him about Thomas Jefferson and the Louisiana Purchase. A man applying for a job in corporate finance chatted with Ellison about thirteenth-century Italy.[4] Once, while interviewing a woman who had worked on the Alaska pipeline, Ellison started an argument about the best way to build an ice bridge. He liked people who argued back; he wanted employees who were as sure as he was that they were right.

Ellison insisted that his recruiters hire only the finest, and cockiest, new college graduates. "When they were recruiting from universities, they'd ask people, 'Are you the smartest person you know?' And then if they said yes, they'd hire them. If they said no, they'd say, 'Who is?' And they'd go hire that guy instead," Oracle engineer Roger Bamford said. "I don't know if you got the smartest people that way, but you definitely got the most arrogant." Ellison's swaggering, combative style became a part of the company's identity. This arrogant culture had a lot to do with Oracle's success. But it also explained why Oracle's competitors—and quite a few of Oracle's customers—despised the company and distrusted its founder. As Stuart Feigin put it, "If he hadn't made me rich, I'd probably hate him, because he's obnoxious. He's not nice to people."

While Ellison demanded absolute loyalty, he did not always return it; the people he liked best were the ones who were doing something for him. The people he hired were all geniuses until the day they resigned, when, in Ellison's view, they became idiots or worse. Several times people were shocked to be let go just before their last stock options would have vested. Among those who lost a million dollars' worth of stock was an executive who was leaving the company to become an Episcopal priest.[5] But Ellison's charms were such that even the aggrieved pastor said he still liked him. As a friend of Charles Foster Kane's says in *Citizen Kane,* "[It's] not that Charlie was ever brutal. He just did brutal things."

The people around Ellison knew that he was neither all good nor all bad. He was capable of chilling selfishness and inspiring generosity. He could dazzle people with his insights and madden them with his lies. He was a fundamentally shy man who could delight audiences with his colorful speeches. Though he was known for his healthy ego, he often seemed deeply insecure. Over time many people learned to accept Ellison's contradictory nature. A business associate[6] once said Ellison "uses people and spits them out," but this person did not see that as a fatal character flaw. Ellison had too many other good qualities to commend him.

The ambiguous Ellison helped build an ambiguous industry, the software business. Software (the word was first used about 1959) has only a brief history as a separately packaged commodity. In the 1950s IBM and its competitors threw in programs for "free" every time they sold a computer, in the same way that Topps gave away baseball cards when it sold gum. This was known in the computer industry as bundling. "There was no concept . . . of software being a salable item: Most manufacturers saw applications programs simply as a way of selling hardware," according to one book about the history of computing.[7] The 1960s were boom years for contracting companies, which wrote custom programs for organizations that could not write software themselves. Then everything changed. In 1969 IBM, under threat of antitrust action, "unbundled"—that is, it began selling computer programs separately. About the same time computers became much faster and more powerful, able to do more complex work. The demand for packaged software boomed. For the first time entrepreneurs formed independent software companies that wrote programs and then sold them over and over again, like blue jeans or cans of corn. In 1970 sales of packaged computer programs amounted to only seventy million dollars a year.[8]

The people who started software companies in the 1970s had a lot to say about how business would be done in the fledgling industry. After all, this industry had almost no history, no éminences grises, no widely accepted ethical standards, and no time-honored way of doing business.

Into this ethical void strode Larry Ellison, a man who misled people about his past, had precious little experience in business, and was prepared to do whatever it took to achieve success. One of his contemporaries was the ruthless and paranoid Bill Gates, who was to lord it over the personal computer market the way John D. Rockefeller ruled the oil-refining business at the turn of the century.

There was a stark ideological difference between these software entrepreneurs and some of the industry's elder statesmen. In 1939

Bill Hewlett and David Packard, who grew up during the Great Depression, founded the Hewlett-Packard Company with certain basic values in mind. They believed their company should provide opportunity and security to employees, contribute to the betterment of society, build first-rate products, satisfy customers, and make money. Their way of doing business eventually became known as the HP Way. "We thought that if we could get everybody to agree on what our objectives were and to understand what we were trying to do, then we could turn them loose and they would move in a common direction," David Packard wrote.[9]

Larry Ellison's vision was narrower. The Oracle Way, to the extent that such a thing existed, was simply to win. How that goal was achieved was secondary. As a former Oracle board member put it, Ellison established no "magnetic north"—no common direction, no sense of how things would or would not be done.[10] Sometimes Oracle succeeded by writing good programs, delivering them on time, and making sure they worked as promised. But Ellison was also known to make grand claims for his products only to deliver them months or years late or not at all. When he did deliver them, they sometimes did not work the way they'd been advertised. A lot of corporate information managers made their careers—or ruined them—by believing the things Ellison said.

"No one wants to go to court, so I'll just give you my own description of Mr. Ellison," one columnist wrote. "Smarmy."[11]

By the mid-1990s people in the computer industry agreed that most of Oracle's products were excellent. Yet the company found it difficult to shake the dubious reputation it had acquired in its early years. In 1996 a joke memo made its way around the high-tech world by electronic mail. It asked the reader to imagine what toasters would be like if they were made by various companies. For example, if Xerox made toasters, the slices would get lighter and lighter as you went along. If IBM made toasters, the company would make just one big toaster, and people would have to submit bread for overnight toasting. And what if Oracle made toasters? "They would claim that their toaster was compatible with all brands and styles of bread, but when you got it home, you would discover the Bagel Engine was still in development, the Croissant Extension was three years away, and indeed the whole appliance was just blowing smoke."

Did reputation matter? Oracle's bottom line suggested that it didn't. Yet others wondered what the bottom line might have been if in the early days Ellison had been less smug and more concerned about customer satisfaction. Gary Kennedy, a sales executive who helped build Oracle early on, believed that Oracle's competitors existed largely because "the marketplace wants and needs an alternative to Larry."

Oracle was hardly the only software company that tended toward exaggeration. The entire industry seemed propelled by it. Bill Gates's Microsoft was well known for making grand announcements only to deliver mediocre products months after they were promised. The best known of those, Windows 95, was widely seen as a poor imitation of Apple's Macintosh operating system, which preceded it by a decade.

In 1996 computer industry columnist Dan Gillmor imagined what he figured must have been the Silicon Valley motto: "What's acceptable is what you can get away with." He wrote: "Too many tech companies don't launch products anymore. They launch announcements, intending to ship the products by a certain date—often knowing that the statements are optimistic at best. Stock prices rise, options vest and everyone is happy. They're getting away with it, after all."

Larry Ellison did not invent this way of doing business, but he certainly helped perfect it. He helped make the software business what it is today: viciously competitive, sporadically honest, shamelessly hyperbolic, and fabulously profitable.

If Ellison's business life was tumultuous, his personal life was sometimes even more so. He had been married and divorced three times, the last breakup occurring in 1986. After that he dated—indeed became well known for it—a series of much younger women, most of whom worked at Oracle. These relationships caused a lot of consternation at Oracle, where, it was noted, the chief executive officer invited a lot of women to dinner but never appointed one to the board of directors. One employee/girlfriend made sensational headlines when she sued Ellison, claiming that he had angrily fired her after she refused to have sex with him. The same woman made news again when it was revealed that she'd made up the whole story to frame him. She was eventually convicted of a felony, with Ellison testifying as the key witness.

Ellison's former wives hoped, for the sake of his girlfriends, that he would stay single, but he never stopped hoping that he would someday have a happy marriage. "I wish I was married. . . . It's nice to go through life with shared experiences," he once told an interviewer.[12] One night he called his third and final wife, Barbara Boothe Ellison, the mother of his two children, and said he was lonely. "He told me, 'You're never alone, because you have the kids. You always have a body. I'm here all by myself in the house,' " she said.

Though he longed for close attachments, many people found Ellison difficult to know. He thought of Oracle as his family, but if so, it was not an especially close family. Ellison, who usually worked at home, went to the office so infrequently that some employees referred to his appearances there as "Elvis sightings." When programmer Kirk Bradley celebrated his fifteenth anniversary with the company in 1996, Ellison popped in for the celebration. Of the fifty or so Oracle employees present, fewer than a half dozen had ever seen Ellison. He had few close friends, and people who had worked with him for years said they hardly knew him. Like Charles Foster Kane, he did not give much away.

Yet in many ways Ellison's life was not at all like the movie character's. While the elderly Kane grew pathetic and died profoundly alone, Ellison appeared to grow more relaxed, more comfortable with himself. Once a year, at cherry blossom time, he served dinner to a dozen or so carefully chosen guests at his Atherton house; among those attending on one occasion were Apple computer founder Steve Jobs, a close friend of Ellison's, and retired football star Joe Montana. Josh Lederberg, the Nobel laureate, was a frequent houseguest in Atherton. Ellison cultivated a friendship and business partnership with Michael Milken, the junk bond king; he was more interested in Milken's efforts in cancer research than in his past as a federal prisoner. And once Ellison dined with President Bill Clinton at the San Francisco home of California Senator Dianne Feinstein. As one Oracle engineer put it, "I think Larry has really arrived."[13]

By the mid-1990s Oracle had clobbered most of its direct competition; giving it 30 percent of the market for relational databases. Having slain Ingres, having mortally wounded Sybase, and having taken a sizable lead over Informix, Ellison needed a new enemy. There was only one software company left to vanquish, and it was a worthy opponent; Microsoft, the industry juggernaut, the IBM of the 1990s, a company that was once portrayed on the cover of *The New York Times Magazine* as a thousand-pound gorilla. Becoming larger than Microsoft would take some doing: Bill Gates's company did $8.6 billion in business in fiscal 1996.

In the industry Ellison was known as "the other billionaire," the one who was not Bill Gates, the boy wonder of the personal computer era who was not coincidentally the richest man in the world. (*Forbes* estimated Gates's fortune at $18.5 billion.) That Ellison was not Gates—not as rich, not as famous, not as sought after—may have been his greatest frustration. He attributed Gates's success not to intelligence or business acumen but to IBM's decision to use the Microsoft operating system in its first personal computer. Once, when an interviewer suggested that Gates was brilliant, Ellison bristled: "Bill Gates, brilliant? Really? . . . I think Bill is a very bright guy, but I don't think his strength is in his intellect at all. It's in his relentlessness."[14] This was not a great compliment, coming from Ellison, who was relentless in his pursuit of Gates. "He wants to get Gates. Larry told me that in no uncertain terms," one business associate said.[15]

For Ellison, the competition with Gates was not about money; they both had plenty of that. "If I pick up another billion this coming year, it has no meaning in terms of being able to buy something that I wanted that I couldn't buy last year. It doesn't change my life in any way," he once said.[16] No, the competition was about status, about who was the greatest in the pantheon of technological gods. Clearly Gates was; he had a best-selling book, his face was on major magazine covers almost weekly, and his public appearances drew overflow crowds. If Gates had mentioned offhandedly that he saw a tremendous market opportunity in Mozambique, a thousand software developers would have rushed out to learn Swahili. Ellison wanted people to listen to him and idolize him, the way they did Gates.

Ellison finally dreamed up something to get at Gates or at least to distract him. His strategy was inspired by the emergence of the Internet and the World Wide Web as the next big thing in computing. In a series of speeches Ellison laid out his vision for a new kind of computer, a five-hundred-dollar box that people could use to send electronic mail and crawl around the World Wide Web. His premise, a valid one, was that personal computers were ridiculously expensive and complex. If a PC broke, only an experienced computer scientist could fix it, and only a hopeless geek would try. With a five-hundred-dollar Internet box—Ellison called it a network computer (NC)—people would not store their data on a hard disk but would send it across the network to a big computer called a server. They could retrieve it from the server anytime they wanted. The price and ease of use of the NC would democratize computing, Ellison said. Every schoolchild would have one.

Oracle was not interested in going into the hardware business and had no plans to build network computers. Instead it hoped to make money by selling the software that would make the NCs communicate with the servers. Best of all, the NC would not depend on Microsoft software to make it work. If you wanted to use one, you wouldn't have to pay Microsoft eighty or ninety dollars for a copy of Windows 95. As an Oracle executive put it, Ellison wanted "to stick it to Bill."[17]

The question, of course, was whether Ellison's vision for the future of personal computing would succeed in toppling Bill Gates. The computer press didn't think so. Several writers noted that Microsoft ended fiscal 1996 with seven billion dollars in the bank, enough cash to withstand almost any competitive move. Perhaps the larger question was whether any customer would really want to buy a less powerful personal computer. Christopher Barr, a columnist for c/net, an Internet news service, called Ellison's idea "the $500 rip-off," and said, "I don't think it can be done. Hey, Larry, the emperor has no clothes!" Another industry columnist wrote, "The day Larry Ellison uses his NC to compose confidential Oracle memos—and stores them out in the wild on the Net—is the day I'll drop my curmudgeonly predisposition and certify that all Microsoft products are 100% absolutely bug-free." This same writer concluded, "In a few years, those NCs will make great doorstops."[18]

Maybe so, but a lot of companies believed in them anyway. In the summer of 1996 several major companies, including Netscape, IBM, Apple, and Sun Microsystems, agreed on a technical standard for the cheap computers. Soon after, several hardware companies began producing prototype versions—all because Ellison thought doing so would be a good idea. It would be years, probably, before the world knew whether he was a visionary or—as the joke said—just blowing smoke.

Ellison continued to sell the NC—at least the notion of the NC—at every opportunity. And though he seemed to believe in the idea, seemed to talk about it with real conviction, there was always a possibility that he would abandon the NC and move on to something else, something that would serve him better. Like Charles Foster Kane, Ellison had only one real conviction, and it did not have anything to do with computers.

Two

LARRY ELLISON WOULD STOP AT ALMOST NOTHING IN BUSINESS OR IN his personal life. This had a lot to do with his success. He referred to a Microsoft product as "Stalinist" because it worked on only one operating system, while Oracle's worked on many different ones. He repeatedly promised new products by a certain time only to release them late. When other companies complained that he did it to forestall competition, he ridiculed them and their products. Ellison sprang a prenuptial agreement on his third wife a couple of hours before their wedding. When a former girlfriend embarrassed Ellison with a trumped-up lawsuit, he didn't rest until he got even.

But before all that—before the money and the power and the growing fame—came the ego, which had a life of its own. People who knew Ellison as a boy had a sense of what kind of man he'd be. He was capable of almost anything, they said, meaning it mostly as a compliment. But that's not the same as saying they believed in him.

Lawrence Joseph Ellison was born on August 17, 1944, on the Lower East Side of Manhattan. His birth was not entirely joyous. His mother, Florence, was nineteen years old and unmarried, his father nowhere to be found. (Years later Larry Ellison would not talk about his father at all.) Florence tried for a while to bring up the baby herself. But when Larry was nine months old, he got pneumonia and almost died. "She thought it would be a good idea if she gave me up," he said. "She couldn't work and care for a child. It was difficult."

Florence sent the child to the affluent North Side of Chicago to live with her aunt Lillian Ellison and Lillian's husband, Louis. The Ellisons adopted him and gave him their name. Larry Ellison did not know until he was twelve years old that he was adopted and was not aware until he was an adult that his adoptive mother was a blood relative.

That Larry was given up by his mother was a tender subject for him. "Is there anyone for whom that wouldn't be a tender subject?" he said. "I mean, it's not a traditional—it certainly isn't Ozzie and Harriet. I think it's very difficult to be that different, to wonder if you belong to your own family. I think it's difficult for a child to deal with." Ellison once told a friend that he dealt with it partly by trying hard to succeed in life.[1] From the beginning he felt he had something to prove.

Ellison's adoptive father was the most influential, though not the most beloved, figure in his childhood. Louis Ellison was among the many thousands of oppressed Russian Jews who fled their country in the early years of the twentieth century. "It was right out of *Fiddler on the Roof,*" Larry said. Louis was smuggled out of the Crimea in the back of a hay cart in 1905 and placed on a steamer on the Black Sea. He carried only two possessions: a gold locket bearing a Cyrillic inscription and a brass samovar, used to boil water for tea. After stops at several foreign ports he arrived in New York, where he traded in his hard-to-pronounce Russian name for a new one: Ellison, after Ellis Island.

Like many of his fellow immigrants from Russia, Louis moved on to Chicago, where he married and had a son and a daughter. According to Larry Ellison, he was a big success. As a young man he put together enough money to make down payments on some apartment buildings, then leveraged those properties to buy more. Larry Ellison said his adoptive father was "a millionaire, Russian immigrant, championship tennis player, candidate for Congress." According to Larry, Louis Ellison's tenants stopped paying rent to him when the Great Depression began, and as a result, Louis didn't have the money he needed to make the mortgage payments on the buildings. "In one year he lost all of his money, his wife, and the election," Ellison said.

It was hard to know how much of that story was true, how much was family lore, and how much, if any, was invented by Larry Ellison. If Louis Ellison had run for Congress in Chicago, his candidacy probably would have been mentioned in the city's newspapers, but I found no clippings bearing his name. Nor was there anything in the Illinois State Archives to suggest that he campaigned for state legislature. If he was well known in Chicago for his fortune or his forehand, there was no way to prove it. The Chicago Historical Society, which kept track of prominent people, had no file on him.

This much is clear: If Louis Ellison ever achieved anything, his greatness was no longer apparent when Larry was in his twenties. "His dad was an accountant, and he was just the quietest little guy. I cannot imagine that he ran for Congress unless they were absolutely desperate for candidates," said Adda Quinn, Larry Ellison's first wife.

Louis's second wife, Lillian, worked as a bookkeeper at a dairy near the family home on the North Side. Larry, who attended nearby Eugene Fields Grammar School, was the only child Lillian raised; Louis's children from his first marriage were grown. "She was always wonderful," Ellison said. Unlike most other mothers in the neighborhood, she was not around to give Larry lunch during his break from school, but she did arrange for another family to feed him.

Larry was a religious skeptic from the beginning. The Ellisons, who were Jewish, attended synagogue regularly—"and dragged me along," he said. "While I think I'm religious in one sense, the particular dogmas of Judaism are not dogmas I subscribe to. I don't believe that they're real. They're interesting stories, they're wonderful mythology, and I certainly respect people who believe that these are literally true, but I don't. . . . I see no evidence for this stuff."

To please his parents, Ellison tried to study the Torah, to no avail. "I couldn't make myself do it. . . . I lost interest. My mind wandered in four seconds. It was an impossibility," he said.

The argument over religion culminated when, at age thirteen, Larry refused to be bar mitzvah. "He saw no point in going through the sham of a religious ceremony just to make his parents happy," first wife Adda Quinn said.

When Larry was a sophomore in high school, his family relocated to the South Side, to a neighborhood called South Shore. There they lived next door to Doris Linn, Louis Ellison's daughter by his first wife, and her husband, David Linn. Doris Linn was about twenty years older than Ellison.

Of all the vague or cloudy issues in Ellison's life, none was more obscure, years later, than how he spent his teenage years. The confusion was usually created by Ellison himself. Time after time he gave people the impression that he grew up in poverty and that he used his wits to survive on the mean streets. "I didn't know how bad the neighborhood was until I left," Ellison told the *Wall Street Journal* in 1989.[2] Other publications got the same idea, referring to Ellison's neighborhood variously as "rough," "notoriously rough," and "tough." *USA Today* once said Ellison "grew up poor in a tenement building."[3]

What Ellison said about his past depended on when you talked to him. In a 1994 conversation with the San Francisco *Examiner,* he said, "We didn't have money, but we weren't poor." Two years later, in an interview for this book, he told me, "We had no idea that we were poor."

He went on, "It really wasn't dangerous growing up. I mean, sure, there were fights, but no one got shot, rarely did people get shot. There was a peculiar month in my neighborhood when Richard Speck killed nine nurses.[4] The Blackstone Rangers were just coming up, and they killed four or five people in the neighborhood. We went into one of our favorite places, and all of the windows were shot out. But that was the exception. That was not the rule."

Ellison's older sister, Doris, had a joke about the way he described his childhood: Every time she read an article about Larry, she said, his old neighborhood got worse and worse.[5] As Ellison surely knew, his success was much more impressive if you believed he had overcome adversity to achieve it.

The truth was that Ellison did not grow up in a tenement, his family was not poor, and his neighborhood was not rough—at least not while he was there. South Shore, along the shore of Lake Michigan, was once one of the South Side's most desirable neighborhoods. In the 1920s Protestants from England and Sweden bought fancy houses along the water and held frequent lavish parties at the South Shore Country Club. Later the population boomed as Irish- and German-American families settled in the area. They were followed by large numbers of Jewish immigrants from Russia and Germany, who moved into newly built apartment houses a few blocks inland from the lake. The Jewish families, excluded from the country club, soon formed their own social network. By the late 1950s the neighborhood was not just stable but strong: People worked as college professors and shoe store owners and lawyers and so on. And the community valued education: According to one of Ellison's friends,[6] most students from South Shore High School went on to college. The writer Stanley Elkin, who attended in the 1940s, was one of South Shore High's noted successes.

South Shore later went through the same sort of seismic change that affected many American cities in the 1960s: the arrival of large numbers of poor or lower-middle-class African-American families, followed by massive white flight.

But Ellison was never really touched by the changes in his neighborhood. He and his parents lived in an apartment building at Clyde Street and Eighty-second. The Ellisons' building was called a four-flat because it had four apartments, two on each floor. They lived upstairs, in a two-bedroom, one-bath flat. Certainly the Ellison home was modest, but it was not seedy or run-down. "It was a pretty nice way to grow up," Ellison's friend Dennis Coleman said. Richard Speck and the Blackstone Rangers were never really a threat.

At South Shore High School, Larry Ellison was not exactly a big man on campus; people hardly knew he existed. He joined no school-sponsored clubs, played no varsity sports, apparently did not distinguish himself in any way. "He was very quiet, very withdrawn, not at all in the mainstream," said Sheila Maydet Gutterman, the valedictorian of Ellison's class. Ellison did not stand out academically either: South Shore High had an honors program, but he was not in it. The problem apparently was that Ellison read the books he wanted to read instead of the ones he was told to read. "I never accepted conventional wisdom. This got me in a lot of trouble. It served me well later in life, but it got me in terrible trouble in a school system that tries to get you to conform," he said.

Though Ellison spent a lot of time reading, he also did a lot of goofing around. Even as a teenager he was the kind of person whom other people followed—when they probably shouldn't have. Sometimes he and his buddies would go to the local Laundromat, climb into the big Speed Queen dryers, and go "laundernating," leaving the dryer door open to avoid asphyxiation. One time Ellison and his friends happened upon a new house that was being built in the neighborhood and, just for kicks, tried to topple one of the walls with their bare hands. They nearly did. "Those were the kinds of things you did with Larry," Dennis Coleman said.

Though he always had a few good friends, Ellison always did things his own way. While other boys let their fathers cut their hair, young Ellison had his hair done by a professional, a fellow on Seventy-first Street named Baltimore George. While other kids listened to rock music, Ellison preferred show tunes and standards. One of his favorites, Coleman recalled, was *An Enchanted Evening with Earl Wrightson.* Ellison was full of surprises. For a while he answered his home telephone by saying, "Russian embassy—Boronov here."

One of Ellison's passions was basketball. He never played on the school team—he was unwilling to yield to the authority of a coach—but he spent hours shooting baskets at the YMCA. While playing, Ellison would refer to himself grandly as the Kid, offering a running analysis of his own dribbling and shooting skills.

About the only organization Ellison joined in high school was a Jewish fraternity called Tau Omicron Mu, whose members were known as the Tommies. The Tommies hosted dances, threw parties, and—probably Ellison's favorite part—competed in sports against other fraternities. He played basketball, football, and sixteen-inch softball for the Tommies.

He also sang for them. Each year the eight or nine fraternities on the South Side got together in a hotel ballroom for a competition known as the Sing, in which the fraternities wrote new lyrics to popular tunes and sang them while marching. Although most of the other kids' families attended, Ellison's apparently never did. His friends thought that Ellison's relationship with his parents was distant at best. "I don't think Larry did anything with his parents," boyhood friend Rick Rosenfield said.

Apparently the only thing Ellison and his father did was disagree. To hear Ellison tell it, his adoptive father was "pretty much the ultimate conformist," an accusation that was never leveled against Ellison. "My father was not rational," he said. "My father believed that if the government said something, the government was always right. And if the police arrested someone, the person was always guilty. But that [attitude] is not uncommon to a lot of immigrants who are just thrilled to be in America. . . . Clearly the American police were always right, and the teachers were always right. So I just heard this all the time." He set a powerful example for Larry, one that Larry never wanted to follow.

Louis Ellison didn't have much respect for Larry either. Once, during a basketball game, Larry got confused and accidentally scored a basket for the other team, a blunder that was mentioned in a newspaper story about the game. According to Ellison, his father kept the clipping just so he could embarrass him with it.

Ellison said his father never missed a chance to tell him he would never amount to anything. "Oh, it was a powerful motivation. I think my dad had a wonderful effect on me. If fire doesn't destroy you, you're tempered by it. Thanks, Dad," Ellison said. "I'm not sure I would recommend that everyone raise their children like this. There's got to be a better way. But it certainly worked."

The tension between Ellison and his father was apparent to Ellison's friends. "He hated his father. It wasn't a pleasant homelife for him at all," Dennis Coleman said. "Larry was basically the kind of guy who would say 'fuck you' to anybody, even his father."

After a while Ellison stopped saying even that much to his father. As Ellison got older, "I would try to hold my tongue. . . . I wasn't going to change him and win him over to my side, have him see it my way. So why should I constantly make these arguments that do nothing but upset him? That wasn't in my interest to do."

If there was a dad Ellison respected, it was his friend Dennis's. Like Ellison, the elder Coleman was an outspoken skeptic: He had changed the family name from Goldman to Coleman because he believed that being identified as Jewish would hold him back, and he once told Dennis that he would rather see him in a jail cell than a synagogue. But mostly Ellison admired the elder Coleman for his intellect. Harold M. Coleman was an industrial research chemist and college professor who seemed to know a little bit about everything. Ellison asked him about everything from the theory of relativity to the salaries of airline pilots. The fact that Harold Coleman had received his doctorate from the University of Chicago impressed Ellison no end; he always called Dennis's father Dr. Coleman, even though the other kids called him mister.

"Talk about your slave to reason," Ellison said. "With Dr. Coleman, there was empirical evidence followed by deductive reasoning, and that was it." Dr. Coleman had an admirer in Larry Ellison, but the reverse was not always true. One day, Dennis said, Ellison showed up at the Coleman home with a bagful of new clothes from Marshall Field, the Chicago department store. He bragged about having returned his worn-out sweaters and shirts, claiming there was something wrong with them, and having received brand-new clothes in exchange. "He had recycled his clothes," Dennis Coleman said. (Said Ellison: "That could have happened. I just don't remember it.") Dennis thoroughly enjoyed the prank—"He was smart enough to figure out, 'Hey, there's an angle' "—but his father was aghast. "But, Larry," Dennis remembered his father's saying, "that's dishonest."

The other important influence on Ellison was David Linn, the husband of Ellison's older sister, Doris. Linn attended law school at the University of Chicago and later became a professor at Northwestern University. He eventually became a judge, as did his son, James. Larry Ellison named his first child after David Linn. "He was a wonderful role model," Ellison said. "He was the most conspicuously successful adult in my life. And as such he was a beacon of strength that I looked up to." Though Ellison admired his brother-in-law, he did not spend as much time with him as he would have liked. "David worked all the time," Ellison said.

By most accounts, including his own, Larry Ellison felt that a lot of things were missing from his life. His birth parents were God only knew where (and God only knew who); his adoptive father was a nonentity; high school rewarded conformity and punished free thinkers. Ellison's material life also left a lot to be desired. His friend Rick Rosenfield believed that Ellison "was not highly proud" of his family's middling economic status; Ellison hardly ever invited his buddies to their apartment, and he often complained about his father's car, a two-tone 1956 Dodge with a push-button transmission. "That was very unsatisfactory to Larry, because it had no acceleration. Larry was very pedal to the metal," Rosenfield said. Ellison eventually saved enough money from his summer jobs as a lifeguard to buy himself a sportier car, a Sunbeam Alpine convertible. But he was never very happy with the humdrum facts of his life.

So he changed them. Beginning when he was a child, and continuing into his days in the Forbes 400, Ellison lived partly in a world of his own invention. Even as a teenager he was an engaging storyteller, a raconteur. When reality was not interesting enough for him, he simply made up delightful and often plausible details as he went along. His stories all had certain things in common: They were funny, they glorified Larry Ellison, and unless you had the authority to issue subpoenas, they were damned near impossible to disprove. They were also mostly benign. Ellison was neither cynical nor mean-spirited; he was always unflaggingly positive and optimistic, and his fractionally true stories reflected those personality traits. (He wasn't going to be smothered by the dreary circumstances of his life; he was going to leap over them.) He tended to see the world as he wanted it to be, rather than as it was. "Larry didn't fabricate," Rick Rosenfield said. "Larry just truly believed in everything he said." ("I would disagree with that," Dennis Coleman said. "I think Larry knew how full of shit he was.") Ellison's tendency to say what he wished was true, instead of what was actually true, became a defining issue in his business life. Everyone who knew him agreed that he was a wishful talker. Those who liked him called him a visionary. Those who didn't called him something else.

Ellison graduated from South Shore High School in 1962 and enrolled at the University of Illinois in Champaign-Urbana that September. He arrived on campus on a three-wheel Harley motorcycle that he had bought cheap at a police auction. Even the motorcycle was the subject of a wonderful and not very likely story.

"A fraternity brother of mine—the largest person in the fraternity, unfortunately—borrowed it one weekend while I was gone and rammed it into a tree," Ellison said. "Then he bought it from me—but at a reduced price because as he pointed out, it was damaged."

As in high school, Ellison did well in the areas he enjoyed and did a lot of reading on his own. But he was unwilling to study a subject just because the university required it or complete an assignment just because a teacher said he had to. The irony of this was that Ellison was hoping to become a doctor. He may have been an iconoclast, but he definitely wasn't a realist.

Ellison never got a degree from the University of Illinois. According to the university registrar's office, he left Champaign-Urbana in May 1964, after his sophomore year. The question was why. The school would not say. All I knew was what *Business Week* magazine had said in its May 15, 1995, cover story about Ellison: "Records show he was dismissed in June, 1964, for failing to maintain a C average—after skipping final exams two semesters in a row." When I asked if *Business Week* had given the right reason for his departure from Champaign-Urbana, he laughed, but not joyfully; the sound was more like a cough, as if he were using his lungs to expel an unpleasant thought from his body.

"Well, my mother died during finals week," he said, almost in a whisper. "I just left without taking finals. I never went back."

"How old were you?" I said.

Ellison drew a deep breath. "It was at the end of my sophomore year," he said, then paused for a long time. "So. I mean, literally, it happened the first day of finals week. She was dying of cancer. I didn't, uh—boy, this is a lot of personal stuff." Pause. "So it's funny—my family told me that she would want me to stay there and take finals. Give me a break." He inhaled deeply again, then let the air out through his mouth. "I really didn't know she was dying. You know, it's interesting, what you—I'm not a dumb person. She was at Billings Hospital at the University of Chicago, which is a cancer research hospital. And I just never . . . never . . . looked closely. Denial is an amazing thing. You know—repression 'r' us."

"What kind of cancer did she have?"

"Kidneys."

"That must have been awful," I said.

"It was awful. I cried a lot. I just left and never went back. I never liked Champaign-Urbana anyway."

The conversation wandered for a few minutes. I asked Ellison how it was possible for him not to know that his mother was dying of cancer.

"Exactly," he said. "I knew she was sick. I had no idea that it was terminal. I just didn't expect it. I think it was just repression. I was a pretty alert, aware child. . . . I didn't miss much, actually. And then this. It's an interesting—you know? It's interesting."

I asked him if his family had kept the truth about his mother's illness from him.

"No. I don't blame anyone but myself for not seeing," he said.

A few minutes later I asked Ellison if he had shared the pain of his mother's death with his closest friends: high school pal Dennis Coleman, best friend Errol Getner, college fraternity brother Rick Rosenfield.

"No way," he said sharply, putting up his hands, as if to fend off the idea. "No way."

Hearing Ellison's answer, I understood why none of his friends remembered anything about his mother's death. Coleman did not remember Lillian Ellison at all and could not say when she died. Rosenfield said he did not know why his friend had dropped out of school. No wonder their memories were so vague: Ellison's mother had died, and even though these were his closest friends—in some ways the best friends he would ever have—he had not told them about it.

When I asked Ellison why, he laughed that same rueful laugh again. "I think you only share things with friends when you come to terms with them yourself. It's very hard. I was too dazed to even be able to express myself. The English language is only useful when you've translated feelings into words. There were no words. There was nothing to share."

After another summer as a lifeguard Ellison enrolled at the University of Chicago, which he found far more stimulating than the University of Illinois. "The whole culture was a much more intellectual culture; it was more about learning, and it was much more accepting of diversity," he said.

Despite his failure at Illinois, Ellison had not given up on the idea of becoming a doctor. Indeed, according to one friend, Ellison gave people the impression that he was well on his way.

One day Ellison paid a visit to his old high school pal Dennis Coleman. According to Coleman, Ellison showed him an acceptance letter to the University of Southern California School of Medicine. The letter was typed on USC stationery, but "it didn't look entirely genuine to me," Coleman said. He thought that Ellison might have talked his way into med school or that his friend was just trying to impress him. "Larry was capable of anything," he said. Ellison said Coleman must have been mistaken about seeing an acceptance letter. But he did remember applying to medical school in California, a place he had always wanted to live.

Ellison was not planning to go to California alone. He hoped to be accompanied by his longtime girlfriend, who was attending the University of Michigan. After he mailed off his medical school applications, he drove to Michigan to propose marriage.

"We had talked about getting married for a long time, but you know, when confronted with the actual question, she said, 'I just can't give you an answer, I just can't give you an answer.' I said, 'Well, I can understand. . . . This is the single most important decision you ever make in your life.' She said, 'No, that's not it.' 'Excuse me? What is it?' She said, 'Well, I'm dating somebody.' I said, 'Excuse me?'

"I asked her, 'How long have you been dating this guy. She said, 'Nine months.' Nine months! Nine *months*? That's when things snapped. Nine months. Excuse me? Nine months."

About the time Ellison abandoned the idea of becoming a doctor, he learned to program a computer. For physics class Ellison was required to program an IBM 1401 computer, the first IBM model to be equipped with transistors instead of vacuum tubes. He said he parlayed this skill into a part-time job as a programmer for the university. He had no plans to make his career in high technology; programming was just a way to make some money.

Although Ellison said he was studying at the University of Chicago, he was actually spending a lot of his time on the Northwestern University campus in Evanston, where Dennis Coleman was going to school. Coleman introduced Ellison to Steve Abramowitz, a New Yorker also studying at Northwestern. Coleman, Ellison, and Abramowitz were inseparable during the year or so they were all in Evanston; three decades later Ellison's friends remembered those days as some of the happiest in their lives. "I learned what it was to really almost love two people in a relationship—just care so much for them that you had to be sarcastic so there would be some distance, to make it safe," Abramowitz said. He and Coleman enjoyed Ellison so much that it never occurred to them to ask how he was getting through the University of Chicago if he rarely went to class—indeed rarely went to Chicago.

Ellison could often be seen tooling around greater Chicago in his little Sunbeam Alpine, a convertible sports car with a tiny rear seat. One night Ellison, Coleman, and their girlfriends double-dated.

Ellison and his girlfriend were in front, and his friends were squished into the back. They were speeding along the Outer Drive, the scenic highway that runs along the water on the South Side, when Ellison saw blue lights in his rearview. Ellison started talking almost before the police officer reached the driver's side door. "On the spur of the moment out of his mouth came that he was a resident at the University of Chicago Medical School," Coleman said. Coleman and the women were dumbstruck. "Larry said he'd just gotten a phone call, and they wanted him to witness a craniotomy—whatever that was—at Michael Reese Hospital. He said that was why he was speeding. He was so convincing that the policeman wanted to give him an escort to the hospital." Ellison, who did not receive a speeding ticket, politely turned down the escort. To Coleman, the moment illustrated "just how quickly this guy could think on his feet."

Another time he and Coleman picked up a couple of young women someplace in Chicago. The women spoke to each other only in French and said in halting English that they were French exchange students. Ellison and Coleman, titillated by the thought of being with these exotic young women, took them out and bought them pizza, only to have their giggling companions admit at the end of the evening that they were American high school girls who happened to be proficient in French.

Ellison was generally self-centered—he was, Abramowitz said, "always inside his own head"—but he was also capable of showing great concern and tenderness for his friends. "He wasn't a nurturing person in that he would ask how your day was, or whatever. But if you told him that this was really a bad day, and you needed to take a walk and talk about something, he'd be right there," Abramowitz said. Thirty years later Abramowitz still remembered the time his girlfriend broke off their relationship. Abramowitz, heartsick and desperate, came up with a plan to win her back: He would buy her a puppy, proving his sensitivity and his love for her. Ellison told Abramowitz that he would help him pick out a puppy if that was really what he wanted to do. But he advised against it, explaining that if the young woman did not want him, she would not want his dog either. She will be angry if you do this, Ellison said. Puppies are a lot of work. She'll feel that you're harassing her. Abramowitz gave her a puppy anyway, and sure enough, the young woman was furious. She kept the dog for a few days, then gave it away. After Abramowitz got over the hurt, he realized how perceptive Ellison had been and how hard Ellison had tried to spare his feelings. Even when Ellison was very young, Abramowitz said, "There was a way in which he was older than us."

There were also ways in which he was more sensitive and vulnerable than his friends. Abramowitz, who went on to become a clinical psychologist, thought of Ellison even then as a "very complex" person. During Abramowitz's years in Evanston he drove a two- or three-year-old Corvette that his father had bought him. Ellison, the middle-class kid, was entranced by this car. "Even though he had the Sunbeam, he would ask me if he could drive the Corvette," Abramowitz said. "I suspect that he just really liked being in that higher-powered, higher-status car, although he had a very nice car himself." Ellison eventually traded in the Sunbeam for an aqua blue Ford Thunderbird convertible, the kind with a hardtop that retracted into the body. Coleman believed that Ellison had an unending thirst for attention; he seemed to fear that people would not like him if he did not constantly wow them. "On some sort of fundamental level he felt inadequate," Coleman said. "I didn't understand it. Larry was a very adequate guy."

Ellison's relationship with the two Northwestern men was intense but fleeting; after he left the Chicago area, his friends rarely heard from him again. Still, Ellison did not forget them. When Abramowitz called Oracle in the early 1990s, he got right through to Ellison, and the two had a nice talk. And in 1996, three decades after Coleman and Ellison had last been close, Coleman sent an E-mail saying hello and telling Ellison that he was married and had two children. Ellison wrote back right away, congratulating Coleman for "getting and staying married." Ellison said he would "love to get together, or short of that, chat on the phone."[7] Coleman wasn't sure it would happen, but he hoped Ellison could find the time.

"You have to understand," Coleman said, "I loved the guy. You always had a good time with Larry."

* * *

I asked Larry Ellison why he never received a diploma from the University of Chicago. "I think there were lots of reasons, none more compelling than the French comprehensive exam. . . . Once I had learned everything that was on the menu, I saw no other reason to learn French," he said.

While it may have been true that Ellison was no whiz at foreign languages, there was a much simpler reason why he didn't get a college degree. According to the university, he was actually registered there for only one semester. The story about the language requirement was entertaining—once he had mastered the menu, he didn't need any more French—but it had little to do with Ellison's failure to graduate.

That Ellison was still telling such a story was surprising. When he made his excuse about the French comprehensive exam, he had long since achieved astonishing success in business and was considered by many to be a genius and even a visionary. Who cared if he had not gotten it together to finish college?

Larry Ellison cared; that was who. He cared so much that for many years he led people to believe he not only had finished college but had distinguished himself academically. "I've been in meetings where Larry has told people he has an advanced degree from the University of Chicago," said Gary Kennedy, a former top executive at Oracle. Ellison was consistent, according to Kennedy: He always said the degree was in physics. Ellison also made the claim publicly or let his public relations people do so. A February 28, 1988, article in *The New York Times,* "Moving Up Fast in the Software Sweepstakes," said confidently that Ellison held "B.S. and M.S. degrees in physics from the University of Chicago." Other publications made the same error. Ellison said he regretted not correcting the errors when he read them, but denied that he ever lied about his academic credentials.

Ellison often said that the twentieth-century figure he admired most was Winston Churchill. "I think, perhaps, he saved Western civilization, and he had incredible courage to stand alone and support unbelievably unpopular issues," Ellison once said.[8] He had a lot in common with Churchill: Both were mediocre students; both desperately sought the approval of their fathers, to no avail (the difference, according to Ellison, was that "Winston Churchill adored his father"); and both were witty, insatiably curious, and charming when it suited them. Despite Ellison's constant public posturing, he was, also like Churchill, fundamentally shy. Reading about Churchill reassured him that "even these gods have moments of insecurity," a friend said.[9]

He shared at least one other trait with Churchill: Both men were masterful manipulators of public opinion who were motivated largely by self-interest. In 1898 the young Churchill wrote his mother, "I do not care so much for the principles I advocate as for the impression which my words produce & the reputation they give me. This sounds very terrible. But you must remember that we do not live in the days of Great Causes."[10] Ellison's story about his college career was Churchillian in that sense. During his business life Ellison was to promote an endless list of ideas, from relational technology to massively parallel computing to interactive television. But his greatest cause, always, was Larry Ellison.

Ellison left his hometown for good in the summer of 1966. Still California dreaming, he went to Berkeley, which he thought of as "the cultural nucleus of what was going on in the sixties." He was no radical, but he always wanted to seek out interesting people and ideas. "I had originally planned to go to Southern California and be a doc, so I decided, 'I'll go to Northern California and figure out what I'll be when I get there.' " He rolled into the San Francisco Bay Area in his aqua blue Ford Thunderbird.

Ellison needed money to pay his rent, so he went to an employment agency looking for a job as a programmer. "I was broke," he said. "I went into the employment agency, and they said, 'Would you like to work [as a job counselor] here?' 'Well, when can I start?' 'How about now?' 'I'll take it!' "

At this agency Ellison began the first real romance of his new, independent life. "We met, and there was just an instant attraction," Adda Quinn said. A recent graduate of San Jose State University, where she had majored in Chinese history, Quinn was now studying secondary education at Berkeley and working at the employment agency to pay the bills.

She thought Ellison was electrifying. He was tall and gangly but not at all awkward, and though his nose was badly bent (he had broken it several times playing basketball), he was extremely magnetic. There was something about the way he carried himself. Most attractive of all was his mind. His attention span was about three seconds long; that was all the time it took him to master a new idea. He was intensely active, and he always set the agenda. He and Quinn went where he wanted to go, saw the things he wanted to see, learned the things he wanted to learn. She didn't mind; she had fun. Sometimes Ellison had to break dates because he did not have enough money to put gas in the car, much less take Quinn anywhere. (Ellison said he was so poor that he subsisted on dime packages of Kraft macaroni and cheese.) After he had earned some money, he would always reappear, ready to lead Quinn on another adventure. The guy was a one-man amusement park, Larryland. Within a couple of months Ellison and Quinn decided to marry.

"I hardly knew him," Quinn said. "I agreed to marry him because he was the most fascinating man I'd ever met in my life. I knew I would never be bored." When I read that quotation to Ellison, he quipped, "I think she was better off bored."

After they set a date, Ellison wrote a letter to Dennis Coleman and Steve Abramowitz, his friends back in Evanston. He did not invite them to the wedding but told them where they could send gifts. "It was just a 'fuck you' sort of a letter, which was really uncalled for," Coleman said. Abramowitz felt differently about it: "It was just someone telling his friends that he had made it." Said Ellison: "I certainly hadn't made it. Maybe I felt I'd had a good week."

Ellison and Quinn were married on January 23, 1967, Quinn's twenty-second birthday. No one from Ellison's family attended.

Ellison told her that he had finished college in Illinois and was attending graduate school at Berkeley. She was under the impression that he was enrolled at Berkeley during the first couple of years of their marriage. The truth was that he was enrolled there only in the summer of 1966, the summer he arrived in California. If he attended classes after that—and he apparently did—he was just sitting in. "I paid for a lot of textbooks, so I know that he was in a lot of classes," Quinn said.

Over the next few years Ellison did computer-related work for a series of big corporations: Wellsco Data Systems, Fireman's Fund Insurance Company, and others. In most of the jobs he was a systems programmer for IBM mainframes, meaning that he hung tapes, backed up data, and generally kept the machines running. The work was monotonous and unchallenging; Ellison passed a lot of hours reading books. He always worked nights and weekends so he could go to school—or to Yosemite—during the day.

Ellison and Quinn lived in an unremarkable one-bedroom apartment on Rock Ridge Avenue in Oakland. They hung tinfoil on the windows so Ellison could sleep after his night shifts, and they often rode the bus because the T-bird was not as dependable as it was good-looking. The only new piece of furniture they owned was a bed. When the bed frame broke after a few months, Quinn was almost too embarrassed to bring it back to the store. "I was just dying nine deaths. Here we were these newlyweds, and we had broken our bed," she said.

The late 1960s were turbulent times at many American college campuses, but especially at Berkeley. Ellison, no hippie, kept his hair relatively short and once refused to wear beads to a party. "I said, 'The beads look ridiculous. Someday I'm going to have a child. And that child is going to see a picture of me, and I'm going to have to explain myself. Why am I wearing beads? I won't be able to convince him of anything if I'm wearing beads. So I'm not wearing beads.' " Ellison said he tried marijuana only once, and even then by accident: He ate some pot-filled chocolate-chip cookies at a party and got "absolutely blitzed." His participation in politics was also purely recreational. He distributed some flyers after President Nixon announced the invasion of Cambodia and was devastated when Robert F. Kennedy was assassinated.[11] But compared with some other people at Berkeley, he was "not political," Quinn said.

An exception was when Israel went to battle against several Arab nations during the Six-Day War of 1967. According to Quinn, Ellison was "an ardent Zionist. When the war was on, Larry took off work and stayed glued to the television for a week." He named his cat Yitzhak, after the Israeli defense chief and future prime minister Yitzhak Rabin. "When the cat died, he took off two weeks in mourning. He was nonfunctional," Quinn said.

In 1970 Ellison and Quinn came up with the money for a down payment on a thirty-thousand-dollar house on Mountain Gate Way in Oakland. Ellison poured all his energy into fixing up the place. That was the beginning of his interest in home improvement, an interest he retained long after he had earned his first billion.

Quinn knew that Ellison was a perfectionist, but she saw that trait most clearly one day when he was hanging wallpaper. "He got the wallpaper just a tiny bit crooked on the wall. And he lost his temper and ripped it off the wall. And it was expensive. It was expensive wallpaper. And he wadded it all up and threw it on the floor," she said. "All he had to do was shift it around a little on the wall. But no. That's just a perfect example of his perfectionism. He will not settle for less than what he thinks something should be." Said Ellison, "I was trying to match the patterns exactly. But the printing itself wasn't very good; the patterns were slightly off. So I went nuts."

As a result of his perfectionism, Ellison was extremely hard on himself, Quinn said. "He kind of had a mental image of where he should be and what he should be, and he was not able to attain it."

After Ellison and Quinn had moved into the new house, Louis Ellison came to California to live with them. He was very old now and needed someone to take care of him. Quinn enjoyed her father-in-law for the most part; she would help him make borscht and listen to his stories about the old days. Ellison was less enthused. "Larry became more distant, and really didn't like to just come home and relax because he knew he had to deal with his dad," Quinn said. "He wanted to be a success for Louis, but he felt that no matter what he did, he wasn't."

When I asked Ellison about those days, he sighed heavily, paused, and said, "We all seek détente with our parents," adding that he tried to achieve it with his father. He never did. Louis Ellison lived with his son and daughter-in-law for a couple of years, then moved to a nursing home in Oakland, where he died.

If Ellison had any future at all, Adda Quinn couldn't see it. During the seven years of their marriage he bounced from job to job, sometimes taking a cut in pay when he made the change. Even so, Ellison spent money lavishly. He bought a thousand-dollar bicycle, treated himself to lunch in nice restaurants, and went to Beverly Hills to have his nose fixed by a plastic surgeon. "He couldn't go to just anybody to have this nose job. He had to go to Beverly Hills. . . . He had champagne tastes on a beer budget," Quinn said. He also went out and borrowed three thousand dollars to buy a thirty-four-foot sailboat even though he was still making monthly payments on a smaller boat. At the time he and Quinn were earning about sixteen hundred dollars a month, combined.

Ellison never worried about how the bills would be paid. "The financial stresses were Adda's. They weren't mine," he said. "She had a program, you know? Get promoted, have an ambition, go do something. That's just the way the world is. And I was perfectly happy to do a little writing, play the guitar. I adored going to Yosemite. . . . I valued my time more than [I did] the pursuit of money." He admitted that he was "cavalier" about spending money and that his attitude drove Quinn half crazy.

In 1974 she decided to leave him. She could no longer stand to watch him flounder. Besides, she thought that marriage was "not a good venue for him." Yes, he was energetic and exciting, but also uncompromising. She told him she wanted to pay off the debts and separate.

He did not take it well. "He didn't want to let me go. I don't know whether that was ego or what. But he kept me awake for three days straight," she said. "Finally I just said, 'There's no doubt in my mind that I'm going. But we've been together so long I want to be sure that we salvage something from this relationship. So I'll go to a psychiatrist with you and help you work through what's happening to you.' "

Quinn believes something happened to Ellison during the visit to the therapist that gave him a clear vision for his life, one that seemed to come out of nowhere.

"He said to me, 'If you stay with me, I will become a millionaire and you can have anything you want.' " She did not know where that idea came from. He had never said anything like it before. "There was never any clue. Believe me. Not a clue." In Quinn's view, Ellison "made a commitment to himself that he was not going to be a failure. That was the turning point of his life." Still, she did not take him seriously at the time. "I said, 'Well, why don't you just go on and become a millionaire and make yourself happy? Because you're never going to make anybody else happy until you're happy.' " That was the end, after seven years of marriage. She left him the house, the furniture, everything; she just wanted to get out. She was developing an ulcer.

When I asked Ellison about the appointment with the therapist, he laughed. "Oh, that idiot counselor. I remember that," he said.

"I remember that I was very curious about human nature. I talked about this feeling called love. I said, 'I don't understand. I don't understand love. I don't understand people bonding to one another. What is it? What's going on inside my head? What's going on inside my heart?' It was a genuine question. I wanted an explanation," Ellison said. Love wasn't something he could just feel; he had to analyze it, break it down, discover the source code of the human heart.

What about his promise to Quinn that he would someday become a millionaire?

"I actually do remember saying something like that," he said. "I remember at that point saying [to myself], 'God, everyone seems to be very concerned that I'm not going to know how to make money.' . . . There was that very consistent point of view throughout most of my life that success wasn't going to come my way. So I think it was probably in that context that I said something."

Ellison and Quinn remained friends; she never lost her fascination with him. Years later he did her many kindnesses. When Quinn's parents became seriously ill, he bought them a house so she could care for them without worrying about money, as she had always done when she and Ellison were together. When her second husband was recovering from cancer and needed something to do, he gave him a job for a year—at a six-figure salary. Another time he gave her his Mercedes 500SEC instead of trading it in because, he told her, "You've never had a nice car." He had refused to let her buy a car when they were married; Quinn believed that the gift was his way of making amends. She appreciated it—and was glad she got the car, not him.

"He is extremely intense. I was married to him for seven years, and by the time I left, I was worn out," she said. "And I'm a fairly multiphasic, high-energy person with a lot of diverse interests and a Type A personality. Goal-oriented. And I was worn out. He's beyond anything I've ever experienced. And I'm sure that's what accounts for his enormous success now. He has incredible intelligence, and he applies it with incredible intensity. And that intensity does not let up.

"People say, 'Gee, don't you wish you were still married to him?' I say, 'No, I'm perfectly happy with the relationship I have with him now because it's not going to ruin my health.' "

Three

THE MAN ACROSS THE HALL TALKED CONSTANTLY. THAT IS WHAT Stuart Feigin remembers. The foaming white water of words, the rushing river of noise. Larry Ellison just never quit.

The year was 1973, and Feigin had just landed a job at the Amdahl Corporation, a high technology start-up in Sunnyvale, California. Amdahl hoped to take customers away from IBM, the industry giant, by building mainframe computers that were exactly like IBM's, only faster. Feigin, a young programmer, had been hired to write software for the new company's first machine, the 470 V/6. It was a good job, but Feigin rarely got a chance to do it because his new colleague would not . . . stop . . . talking. Ellison, who had experience with IBM mainframes, had signed on to teach Amdahl's engineers about the new machine. But mostly he just talked. All day long. He was a talking doll with a pull string that never reached his back. He talked to anyone and everyone: bosses, secretaries, telephone callers, visitors, delivery people, copy machine technicians, maintenance guys, and passersby.

But mostly he talked to Stuart Feigin, who was always within talking distance. There was, Feigin said, "an aura about him," a powerful sense of possibility. There was also something unsettling about Ellison, something vaguely dangerous, a sense that you didn't know what was going to happen. "He was the kind of person you would like to follow," Feigin said.

The way he looked was part of it. Ellison walked into the Amdahl building as if he felt everyone there had been awaiting his arrival. You couldn't help noticing him when he was around. And he had enormous hands. To someone like Feigin, who was short and plain, Ellison looked like a god; with those hands and all that wonderful height it seemed he could accomplish anything. What Feigin was seeing was the embodiment of charisma, the characteristic that Ellison soon began to exploit so that everyone around him would do his will.

And the talk! The talk was interesting, or at least most of it was. Ellison talked about books, music, basketball, real estate, presidential politics, Israeli national security, automobiles, the stock market, God, technology, and the price of gasoline. Sometimes he even talked about work, but as far as Feigin could tell, he never did any.

If you disagreed with something he said, all the better. He liked it when people argued with him because arguing was talking with a purpose. Ellison did not care what the argument was about. He could make a point with born-again conviction, then return to the office the next day and vigorously defend the opposite position. He did this earnestly, without embarrassment or irony. The argument was his true love, the idea itself just someone he dated.

The subject he liked best was himself. He was forever telling people "how wonderful he was, how smart he was, and how rich he was going to be," Feigin said. Ellison's visit to the counselor may not have been "the turning point in his life," as Adda Quinn said, but clearly something about him had changed. If Ellison had ever had a weak or uncertain moment in his life, Feigin would not have known it. Somehow Ellison had escaped the tight clammy grip of everyday fear and doubt.

Then Ellison lost his job. Amdahl was behind schedule and running out of money, so it laid off nonessential personnel, and if anybody fitted that description, it was Ellison. "He was mightily annoyed," Feigin said. Didn't they see what a mistake they had made?

After that Feigin occasionally met Ellison for lunch. Ellison always showed up forty-five minutes late, an hour late, an hour and a quarter late. It apparently never occurred to him to apologize. The world ran on Larry time. Besides, apologizing was not in his nature. Does the queen ant thank the workers for building her nest? No. She expects them to do it. When the bill came, Feigin always paid.

A couple of years later Feigin received a call from Ellison, who was going to start his own software company, though it was not entirely clear to Feigin what the company would do. "Come with me," Ellison told Feigin. "I'll make you rich."

Feigin hesitated. He was earning thirty-five thousand dollars a year at Amdahl, good money for a young man just a couple of years out of school. The job was challenging and reasonably secure; he was loathe to give it up. Besides, Ellison was all talk. The guy couldn't manage a lunch date. How was he going to manage a company?

Feigin said no. "I told myself, 'This guy is a complete flake, and he doesn't have a chance of succeeding.' . . . I was half right."

Feigin eventually joined Ellison's new company but did not receive as much stock as he would have if he had been one of the founders. He later estimated that his mistake cost him a couple of billion dollars.

For Ellison, success was still far in the future. After his layoff from Amdahl he got another programming job at a Silicon Valley technology concern. This time it was Ampex, an audio and video equipment company in Sunnyvale. He did not last long there either, but his time at the company was important nonetheless. At Ampex Ellison met two men who would change his life and accompany him on the first part of his multibillion-dollar journey.

Soon after he joined the company, Ellison went to work for a wry and unsentimental man named Bob Miner. "I thought that my manager, the manager they assigned me to, was not technically competent. So I refused to work for him," Ellison said. "So I said, 'I'll work for Bob. He's the best guy. I'll work for him.' " Thus began a collaboration that would become one of the longest-lasting and most lucrative in the history of Silicon Valley. If Larry Ellison was to be the brains behind Oracle Corporation, Bob Miner was surely the company's heart.

Soon Ellison met another man who would become enormously important in his life. Ellison was talking with Miner one day when he happened to mention the name of his former wife. Just then a young man poked his head into Miner's office and said to Ellison, "Did you say Adda?"

"Yes," Ellison said.

"Adda who?"

"Adda Quinn," he said. "I was married to her."

"Adda Quinn was my lab partner in high school biology class," the man said. She did not like dissecting frogs, so he did the cutting and she drew diagrams of the parts.

The man was Edward A. Oates. A career programmer, Ed had operated IBM mainframes in the Army, then worked for several years at Singer Business Machines, an IBM competitor. When Singer began having serious problems, Oates had jumped to Ampex. The day he met Larry Ellison was his second day on the job. Later he joined Ellison and Miner as a cofounder of what became Oracle Corporation, and though circumstances forced him out of the company in its infancy, he later returned to make a significant contribution.

That was still some time away. At Ampex, Ellison, Miner, Oates, and the rest of the crew were trying to solve a long-standing problem in the world of high technology. In those days there was no affordable and efficient way to store and retrieve enormous amounts of digital information. Disk drives could retrieve information quickly but did not have much capacity; a roomful of them would have been needed to create a large database. Traditional magnetic tape could store lots of information, but you had to search through miles of tape to find what you wanted.

Ampex had come up with a possible solution. The company was working on a way to store a terabit—a trillion bits—of data using videotape. The Ampex system could search a videotape at a thousand inches a second and rewind it at two thousand inches a second. By the standards of the 1990s, such a system would have been considered hopelessly poky and inefficient. For example, in 1996 Oracle and Sun Microsystems put together a 5.5-terabyte system—a database about fifty times as large as the one Ampex envisioned. But in its day the Ampex terabit memory system was, Ellison said, "astounding, astounding technology." The whole project was funded by the Central Intelligence Agency, which was always looking for more efficient ways to manage information. The CIA had a code name for the project: Oracle. Ellison and Miner later gave that name to their first software product and then to their company.

It was the job of Ellison, Miner, and Ed Oates, among others, to write the software that would make the Ampex system work. And it did work—most of the time. "But most of the time isn't good enough. It had to work all the time," Ellison said. That was the trick: making the terabit memory system work all the time.

As a boss Bob Miner was not exactly a taskmaster; he and Ellison spent a lot of the workday playing chess and eating lunch. "Bob and Larry would disappear every afternoon at three to go play tennis," Ed Oates said. "One of the differences between Bob and Larry was that between tennis matches Bob was writing code and implementing things, while Larry was mostly talking about becoming rich and famous. The normal Larry stuff."

Ellison said he was also writing code—and writing it well. "I thought I was a very good programmer. I thought I was a *really* good programmer. And I don't think that many people could program as well as I could," he said. Erik Salbu, a supervisor at Ampex at the time, agreed that Ellison had talent. Still, Ellison probably wasn't cut out to be a programmer; he wouldn't have enjoyed the monkish, antisocial lifestyle. A software engineer named Ellen Ullman once described programming as "an illness, a fever, an obsession. . . . When you are programming, you must not let your mind wander. As the human-world knowledge tumbles about in your head, you must keep typing, typing. You must not be interrupted. Any break in your concentration causes you to lose a line here or there. Some bit comes, then—oh no, it's leaving, please come back. But it may not come back. You may lose it. You will create a bug and there's nothing you can do about it. . . . A real programmer wants to stay close to the machine."[1] Ellison may have liked machines, but he probably would not have liked being tied to one. "I find it hard to believe that [Larry] was ever a true programming addict," Stuart Feigin said. "The true programming addict finds the program itself more interesting than the customer, his needs, or the money. Larry loses on the last point."

Eventually Ellison left his programming job for a position in sales and marketing. "In a lot of ways it was a more natural role for him," Ed Oates said. "Bob could put his arms around the details of a complex software project or technology project and really understand the bits and pieces on the inside. Larry could do that, but he didn't really want to. He wanted to put his arms around the technology and understand its implications. 'What does this mean for the product and the company and the world?' Larry was looking at the horizon. Bob was steering the ship." And so it would be, for two decades to come.

During their brief time together at Ampex, Miner, Ellison, and the equally independent-minded Ed Oates spent a lot of lunch hours discussing what they considered the shortcomings of the corporate world. "We all agreed companies tend to pay people for seniority and for being good guys," Oates said. They also thought that "the people who make the actual significant contributions to products rarely get rewarded very well," Oates said. Not surprisingly the three men considered themselves among the overworked and undercompensated. "This sounds terrible," Ellison said, "but I'll say it anyway. We used to joke that we did a hundred forty percent of the work, and the other people did negative forty percent of the work."

It became clear after a while that the terabit memory system was never going to work. Ellison, Miner, and Oates blamed the Ampex management for the failure. They talked a lot about what they would do differently if they were in charge. Perhaps inevitably that led to talk about going into business together. Ellison in particular was enthusiastic about this. "I thought I was a fairly good business guy. And then I'd see these guys making decisions that I didn't understand, that I didn't think were rational. And therefore I lost confidence in them," he said. "I thought I was better technically than they were, and I thought I was a better business guy than they were. I know it's not terribly modest to say this. But yeah, I thought I had better judgment not only about technology but also about markets than the people who were my bosses. They served as inspirations: If they can run companies, I'll try. I'll give this a shot."

Ellison and Oates soon left for different companies, while Miner stayed at Ampex. But it was not long before they joined up again—with Miner as the glue that held them together.

Robert N. Miner arrived in a newly warring world on December 23, 1941, in Cicero, Illinois. His parents had immigrated years earlier from the part of northern Iran that once formed the heart of the Assyrian empire. His father was a hotel clerk and busboy, his mother a homemaker. Bob, the youngest of five children, was not an only child, but he grew up with an only child's sense of independence. When he was twelve or thirteen, he told his devout Presbyterian mother that he did not believe in God and would no longer go to church with her. His religious skepticism was one of the many traits he shared with his future business partner.

Like Ellison, Miner attended the University of Illinois in Champaign; unlike Ellison, he graduated, receiving bachelor's degrees in math and philosophy. To fulfill his draft requirement, he joined the Public Health Service, which put him to work writing custom computer programs for researchers at the National Institutes of Health. He had never written code before but got the idea quickly and found that he enjoyed the work. That was the beginning of Miner's long, lucrative career as a programmer.

When he finished his stint with the Public Health Service, Miner briefly worked on database technology at IBM, then jumped to Applied Data Research, one of the early software consulting companies. He was in his mid-twenties when the Computer Sciences Corporation asked him to go to Europe to build operating systems for its customers there. Why not? He was unmarried and looking for adventure. He spent three years working in Amsterdam, Paris, and London, and everywhere he went, other computer professionals deferred to him. They assumed that he knew everything there was to know about high technology simply because he was an American. "Americans were kind of like the Japanese [are] today—you could do no wrong. Just by being an American, you were suddenly a lot smarter than you had been when you were in the states," he said.[2]

Miner may have been smart, but he was also frustrated. Every project he worked on eventually failed, either because it was mismanaged or because it had little chance of working to begin with. The failures were not his fault, but the fact remained that he was going nowhere in his career.

Still, his time in Europe had its joys. Miner, less fluent in French than in FORTRAN, was attending a language school in Paris when he met Mary MacInnes, an Englishwoman working in Paris. Miner was not the sort of man most women would have considered devastating: He was short and round, with dark eyes and a perpetual Fred Flintstone shadow. Still, he had qualities that people found immensely attractive. He was extremely bright, with a quick, clever, self-effacing sense of humor. Once a business associate ribbed Miner by saying, "You're small, aren't you?" "I didn't mean to be," Miner responded.[3] Many people were drawn to this man who was no more and no less than what he appeared to be. Certainly Mary MacInnes was. She and Bob Miner were married in 1969, and their first child, Nicola, was born the following year.

The plainspoken American soon moved his family to Washington, D.C., where Miner went to work for Informatics, another early player in the custom software business. Before long Informatics asked Miner to move to California to help Ampex with the terabit memory project. Miner, who liked San Francisco and knew Mary would too, quickly said yes. For a while Miner was Informatics's man at Ampex. But when Ampex asked him to become a manager in its programming department, he jumped at the chance. Soon after, Larry Ellison hired him as his boss.

Ellison and Miner could not have known it when they met in the mid-1970s, but they were about to become part of a great tradition of high technology partnerships. About the same time they formed the company that became Oracle, two other legendary computer industry companies were taking shape. One was Microsoft, the other Apple Computer. The three companies had widely different cultures, products, and ideals, but they all had the same formula for success: Each was founded by a visionary and technically competent entrepreneur with the help of a head-down programming wizard.

Bill Gates had Paul Allen, Steve Jobs had Steve Wozniak—and Larry Ellison had Bob Miner.

Ellison and Miner could hardly have been more different. Miner was almost the anti-Larry. While Ellison was always performing, always seeking attention, Miner kept his own counsel. Even when Oracle became a story worthy of the cover of *Fortune* magazine, he rarely gave interviews. Wealth did not change either man. Ellison was brash when he was poor, and Miner remained humble after he got rich. While Ellison used his money to buy one fabulous house after another, Miner and his family continued to live in the same San Francisco Victorian that he had bought when he had nothing. While Ellison sought out all the best things in life, Miner had the simplest of tastes. According to his friend Roger Bamford, Miner "wouldn't touch any 'weird' food, which was anything remotely ethnic." When he found something good to eat, he ate just a little bit. When he was eating something that wasn't so good, "He'd eat a whole lot of it because it took more to get to the level of pleasure that he was expecting."

Miner was so humble that his daughter Nicola had no idea that she was a rich kid; she found out when she read her father's name in the Forbes 400. "My dad didn't want money to change him, and he was very concerned that it would change us. He was scared about the kind of people that we'd end up being if we had a lot of money," she said. "The only things that he would spend money on were Porsches—he had one at a time—and later we bought a winery in Napa." While Ellison sacrificed everything—love, family, friendships—for his business, Miner insisted on being home each night at six-thirty to have dinner with his wife and kids. He was, one colleague said, "the most unaffected near billionaire you could ever want to meet."[4]

Nor did he tolerate affectation in others. Miner enjoyed few things more than puncturing overinflated egos. "Every time Bob smelled any cant or hypocrisy or just plain fakery, his response was always the same: 'Give me a bleeping break,' And the bleep started with an F," one friend remembered.[5] Many Oracle employees eventually heard those words or a variation of them. Just before Christmas in 1989 Oracle executive John Luongo—who, it is important to know, employed three nannies to care for his two children—sent Miner an E-mail message objecting to the placement of a Christmas tree in the Oracle lobby: "Bob, while I welcome the addition of some color to the lobby, I think it is inappropriate for the company to have a Christmas tree in the lobby of the building. On my staff alone, I have a number of Buddhists, Hindus, Moslems and Jews who, I believe, would find this affirmation of a uniquely Christian symbol inappropriate (if not offensive) in a company that claims to have a global perspective. I'd appreciate it if you would have it removed."

Miner's reply: "John, this pathetic attempt at reaffirming your liberal roots is pure and utter twaddle. Christmas is part of our heritage and as secular as Thanksgiving. I copied Larry on this to see if the Jewish part of him is offended. I represent the non-believers in the world and have no problem with Christmas trees. Furthermore, anyone with three or more nannies cannot speak for the common man."

Of course few people were more vainglorious than the man who was to become Miner's chief business partner. Miner knew Larry Ellison was almost always putting on a show, but it was a show Miner enjoyed. He once told Ellison, "Larry, the only reason you don't like Bill Gates is he does it better than you." By "it," Miner meant the way Gates persuaded people to do his will. Another time, after they had started Oracle, Ellison heard about a Texas company that was offering fighter jet rides. Each ride cost a few thousand dollars. Ellison sent an E-mail message asking Miner if he wanted to come along. If it is possible to harrumph in an electronic mail message, Miner did so. "Are you crazy? That's a lot of bread for an upset stomach," he wrote. "You obviously have far more money than you should. It's things like this that caused the French Revolution." One longtime Oracle employee said of Miner, "I think he was amused at Larry's conduct, almost like you'd be amused at a teenager's conduct if you were an adult. . . . As long as they weren't going to hurt themselves, you'd tolerate enormous things."[6]

On the few occasions when Miner did not enjoy Ellison's act, he put a stop to it. Early on he got fed up with Ellison's habit of showing up late for appointments. (People expecting to meet with Ellison could expect him to show up sometime between thirty minutes late and not at all.) "He said to Larry, 'The next time you're late, I'm not going to be here.' And the next time Larry was late, he wasn't there. Larry was always on time for my dad after that," Nicola Miner said.

While Miner was amused by Ellison, he also respected him. Some people thought Ellison was just an egomaniac, or just a performer, or just an exaggerator. Miner never made that error. Miner could see that he was also smart and shrewd and fiercely determined. It would have been a mistake to take Ellison too seriously, but you couldn't disregard him. Ellison may not always have been completely honest with others, but he never shaded the truth with his partner. Bruce Scott, Oracle's fourth employee, once asked Miner if Larry had always been loyal to him. "He looked at me like he was just stunned that I would even ask him that," Scott said. "He said, 'Well, yeah, sure he was. He did things that really annoyed me; he never was on time for anybody. But he was always loyal to me.' " Nicola Miner added, "He really liked Larry. He just thought that people reacted badly to Larry because Larry had a lot of personal insecurities. But he thought that Larry was deep down just a really good person."

Ellison thought Miner was a good person too. He admired Miner's generosity and respected his technical ability, a compliment he paid to few people. "In the realm of technology, I was always willing to listen to what Bob thought," he said. "Especially when the company was small, his good nature made him a natural leader. People not only respected Bob, they had affection for him."

Ellison was sizzle, and Miner was steak. In the coming years Ellison peddled the product and Miner built it—always in that order. Ellison the Performer recruited good people into the company, and Miner the Regular Guy got them to stay. The two men complemented each other well. Even more than that, they needed each other. Bob Miner never could have created Oracle Corporation by himself. He couldn't have sold software the way Ellison did; he was too inward, too modest, too *honest.* Nor was he willing to make the sacrifices Ellison made. For Ellison, Oracle was a holy mission; for Miner, it was always just a job. Ellison needed Miner too. Few people liked, respected, and enjoyed Larry Ellison more than Miner did. And nobody would have put up with him for as long.

Four

HUMAN BEINGS HAVE BEEN TRYING FOR CENTURIES TO FIND EFFICIENT ways to store and manage data. Before computers that was often painstaking work. The U.S. census of 1880 is a good example.[1] First, cities and towns were divided into small districts for counting. Then census workers went out and asked the age, sex, and ethnicity of each person in each district. They recorded this information and gave it to a tally clerk, who transferred it to a large tally sheet divided into columns and rows. Each box in the resulting grid represented people of a certain age, sex, and ethnicity. The clerk read the forms and placed check marks inside the boxes. Another clerk counted the check marks, and eventually the government learned something about the population. This consumed a lot of human energy, not to mention a lot of paper.

Computers changed all that. The technological age enabled organizations (and later individuals) to store information digitally, in databases. By the end of the 1960s, the decade when computers came into widespread use, two kinds of database had emerged as the most viable. Both did certain things extremely well, and both left something to be desired.

The first was the hierarchical database. To understand how such a database works, imagine a company that supplies buttons to a clothing manufacturer. The button company's database would hold lots of information about that customer. The main record would contain the customer's name, address, and so on. Attached to that would be a series of orders the customer had placed. If you drew a picture of a hierarchical database, it would resemble a corporate organizational chart, with the customer at the top and the orders arranged below. This database was easy to use, as long as you searched it from the top down. If the clothing company changed its July order from fifty thousand buttons to seventy thousand buttons, the computer could make the change quickly. It would simply start at the top of the hierarchy of records and work its way down.

The trouble started when you wanted information that was hidden somewhere in the middle of the hierarchy. Suppose the company was filling orders for a hundred clothing manufacturers instead of just one. And imagine that the button maker wanted to know in June how many alabaster buttons it would have to ship in August. (The price of alabaster was at a peak, so the company wanted to buy as little as possible.) Because of the rigidity of the hierarchical database, the computer could not go directly to the records that contain information about alabaster buttons. Instead it would have to examine every order placed by every company, slowly pulling out the desired information along the way. It might take until September for the button company to get its information about the August shipment.

Here's another way to look at it. Suppose that you were at someone's home for a dinner party, a house you had never visited before. When the host told you to help yourself to a cold beer, would you search every room until you found one? Would you open all the kitchen cabinets and rummage through the sock drawers? Of course not. You'd look in the refrigerator. Well, a hierarchical database would search the whole house.

All that searching takes time, and that explains why businesspeople had trouble with hierarchical databases. When such databases became popular in the 1960s, the cost of running computers was extremely high, making some database searches prohibitively expensive. If you ran your business on a hierarchical database, there was some information you simply could not afford to have.

And what if the button company diversified into the zipper business? Because of their rigid structure, hierarchical databases were extremely hard to modify. Grafting zipper information onto a button database was like trying to find a good place to paint a third eyeball on the Mona Lisa.

A second kind of data structure, called a network database, solved some of these problems. With the network structure, you could see data from many different points of view, not just top down. If you looked at the records for a clothing manufacturer, you would be directed to information about the types of buttons it had ordered. And if you checked the data about buttons, you would be pointed to all the companies that had ordered them. Everything was connected to something else. A picture of the network structure would look like a spider's web: intricate but functional. Adding new records to the database was as easy as drawing a new line from one end of the web to the other.

The flaw of this database was its cat's cradle complexity. To find your way through it, you had to know a lot about how the network was structured. (Yes, the beer is in the refrigerator, but it is stored in baby bottles, and the only way to the kitchen is through the garage.) Most businesspeople were not experts in database structure, so they had to depend on programmers to untangle the thicket of information. This gave enormous power to the stoop-backed, sun-deprived nerds in the computer room. That was nice for the nerds but not for management. Besides, writing special programs to get information from the database took a long time—years sometimes.

Then came the idea that launched a new wing of the software industry, the idea that eventually made Larry Ellison a billionaire. In June 1970 an IBM researcher named Edgar H. "Ted" Codd published an article describing a new kind of database. The article, "A Relational Model of Data for Large Shared Data Banks," was pretty dense stuff. Twenty-five years later an IBM database expert said of the paper, "A couple of us from the Systems Department had tried to read it—couldn't make heads nor tails out of it. At least back then, it seemed like a very badly written paper—some industrial motivation, and then right into the math."[2]

The theory's solid algebraic foundation was what made it so powerful. Codd's idea had a certain elegant simplicity. He proposed to organize data in tables, just as the census clerks had done with their tally sheets almost a century earlier. His theory was that useful information could be gleaned simply by putting the tables together. Suppose that the button company created a table containing the name, address, and identifying number of each of its hundred customers. Let's say it then created a table to keep track of orders. That table would contain the number and type of buttons ordered, the date the order was placed, and the identifying number of the customer who needed the buttons. Another table might contain information about buttons: their size, color, cost, and price. That makes three tables: customers, orders, and buttons.

In Codd's theory, you could combine the orders and buttons tables to find out how many alabaster buttons were awaiting shipment at any given time. Or you could combine the customers and orders tables to find out when a certain customer had placed its most recent order. If you created a few more tables, you could find out how many customers in New York had not yet paid their July invoices. Or you could discover which buttons were selling fastest west of the Mississippi. The possibilities were endless.

Best of all, the structure of the database was a nonissue. Hierarchical and network databases differed in many ways, but they had one important thing in common: Their structure determined how quickly and easily you could get information out of them. That was not so with a relational database. The tables in the database would be connected logically but not physically. If the button company wanted to go into the zipper business, it could simply create a table of its zipper customers. And if the company had customers ordering both buttons and zippers, asking the computer to display the contents of a couple of tables would show it.

People who manage information liked to say that data are not worth much until they are organized in a meaningful way. More than any idea that had come before it, Ted Codd's relational theory held the promise of turning raw data into something worthwhile: *information.*[3] Database technology wasn't sexy or exciting—unless you were a businessperson interested in running your company much more efficiently and profitably.

The trouble with the relational model was that it was slow. It took computers a long time to put tables together. And the larger the tables, the longer thc process would take, and the more time and money would be wasted. A lot of people doubted that the relational model would ever be commercially viable. "The conventional wisdom was that it was really kind of an interesting idea, but it could never be practical," Bob Miner said.[4] Still, there were many who believed that faster, cheaper computers would eventually make relational databases workable. A database textbook published in the late 1970s acknowledged that relational technology was still far from perfect but added, "Many experts contend that the relational model holds the greatest promise for future development."[5]

In 1970 there was still a long, long way to go. Codd's theory was brilliant—almost twenty-five years later a database magazine honored him with an article titled "So Help Us Codd"—but still only a theory. His paper discussed how a relational database system might work but did not really say how it *would* work. Now someone had to develop a way to mine the relational database for information. As always, IBM was among the first to take up the challenge, and Larry Ellison would be forever grateful that it did.

One day in 1976 Larry Ellison went to talk to a couple about buying a used Mercedes and ended up getting married again. "We were chatting away, and finally this woman says, 'You know, I have the perfect girl for you.' I'm thinking, *This is a really bad idea,*" Ellison said, "Normally with a used car a woman doesn't come along with it. I said, 'That's interesting, that's nice.' I was being polite. She said, 'Charlie'—or whatever her husband's name was—'don't you think Nancy would be perfect for Larry?' Charlie said, 'Well, I don't know.' She said, 'Don't you think Nancy's the best-looking girl you've ever seen in your life?' He said, 'Yeah, that's probably true.' *Now* I'm paying a lot of attention."

Ellison got a phone number. The young woman's name was Nancy Elizabeth Wheeler, and she was from what Ellison described as "an aristocratic family" from Louisville, Kentucky. The daughter of a successful businessman, she was now attending Stanford University. "We met for lunch and started dating," said Ellison. "She was very funny, very smart, very athletic, very outgoing, and very sheltered. We got married rather quickly."

He also bought the car.

When Ellison saw that Ampex's terabit memory project was not going to work, he quit the company for a job at a small technology firm called Precision Instrument Company. The company gave him a good raise and—perhaps more important to him—the title of vice-president of systems development. For the first time in his working life he would not be just another drone. "Larry liked that. He was definitely in charge of something," Ed Oates said.

Precision Instrument had been founded in 1957 as a manufacturer of audiotape recorders. But for several years it had been trying to break into the world of computers with a device that could store and retrieve large amounts of data. The company hoped to sell the device to big government agencies that dealt in information: the Census Bureau, the Social Security Administration, intelligence agencies, and so on. The PI 180, as the machine was known, was straight out of *The Jetsons*. It was seven feet long, two feet wide, and five feet high. (Imagine several tall filing cabinets standing side by side.) Within the contraption were twelve canisters, each about three feet long. Inside the canisters were ten long, thin strips of Mylar, each one plated in rhodium. When the computer operator typed a command, a mechanical arm would withdraw one of the strips and guide it onto a drum that was spinning at a high speed. A laser would then write data onto the rhodium coating or—if the data had already been written—read the information and display it on a computer screen.

That was the theory, anyway. What actually happened was another matter. In testing the device, the company's engineers found that the tiniest speck of dust could cause the PI 180 to misread data. If so much as an eyelash or dust mote got between the laser and the rhodium-plated strip, it would make a mistake. You would still get the information—minus a few dozen words or numbers. Clearly, customers would not take kindly to this. What was worse, the mechanical arm sometimes loaded strips onto the drum at a slightly screwy angle. When this happened, the whirling drum would sandpaper away the rhodium coating, erasing all the data. In the database business, losing data was a mortal sin. The Census Bureau was not going to be happy about losing track of the population of, say, Illinois. Those problems and others eventually doomed the PI 180 as a data storage device.

About the time Larry Ellison joined the company, Precision Instrument came up with a clever idea: It decided to market the PI 180 as a replacement for microfilm and microfilm readers. Now, instead of storing data, the device would store *pictures of data.* The difference was critical. A dust mote could ruin a table of census information, but it would not do much harm to a photograph. Let's say you used the PI 180 to read a page from an old newspaper. If the device made an error, it might give you a pockmarked picture of the page, but at least it would not rearrange the words themselves. Precision Instrument could make a persuasive case that the PI 180 would be easier to use than a microfilm reader. Instead of scrolling through endless reels of microfilm in search of a certain piece of information, a user would simply type in a request and—poof!—the information would appear. Again, that was the theory. The reality turned out to be something else entirely.

Before it could bring the device to market, Precision Instrument needed one more thing: software. The company didn't have the computer programs it needed to make the machine work. Nor did it have a staff of programmers who knew how to write them. All Precision Instrument had was a young vice-president named Larry Ellison. Talented as he thought he was, he couldn't possibly write all the software by himself.

Seeing no alternative, Precision Instrument began soliciting bids from outside programmers. Years later there was some disagreement about the size of the bids that came in. According to Ed Oates, one company said it could do the work for $700,000, but Ellison believed the lowest bid was $2.3 million. Whatever the other bids were, Ellison thought they were far too high, and he could not have been happier about it. This was his big opportunity.

Ellison called Bob Miner and Ed Oates, his former colleagues at Ampex, and suggested that the three of them form a company and put in a bid. If they got the contract, Miner and Oates would write the software while Ellison supervised the project from within Precision Instrument. Eventually he would give up his beloved vice-presidency and join them in the new company. After the three men finished the Precision Instrument job, they would continue on in business. Beyond that, Ellison's plan wasn't completely baked. Maybe the new company would become a contracting house, and maybe it would develop a product. Ellison wasn't sure. But somehow they all would become rich.

Ellison possessed no vision of the future, no great plan to conquer the software industry. His sole motivation was to be his own boss. "I think I was driven to do this because I knew I could never really survive inside a conventional corporation," he said. "I was not suited to being able to work my way up the corporate ladder." He had the same problems in business that he had experienced in school: "If people asked me to do things that didn't make sense, I just couldn't do [them]. I couldn't start my own school, but I could start my own company."

Miner, who was still hanging on at Ampex, went right along with Ellison's scheme. Miner, who was suffering through the collapse of the Ampex terabit memory project, believed that forming a new company would be a chance "to do something right."[6] Besides, his salary would be "the same or better than his salary at other jobs," Ellison said, "and if this didn't work out, he'd just get another job. He was employable, so there was no risk to him." Bob's wife, Mary, wasn't so confident. The Miners had two small children (a third would come soon) and had just bought a Victorian house in San Francisco. What if Bob wound up unemployed? What would the family do then? "My mom was very, very worried about the money," Nicola Miner said.

Ed Oates, who was working at Memorex, also had a wife, kids, and a mortgage. But like Miner, he felt sure he could get another job if the new company failed. In less than a year at Memorex, Oates had received three job offers. "If it all went belly-up in a year, it really didn't make any difference," he said.

In the summer of 1977 Ellison, Miner, and Oates put together a bid on the Precision Instrument contract. They arrived at a figure of four hundred thousand dollars, which they believed would be enough to write the software, hire an extra programmer (they already had one in mind), and fund their new company for a while after the project was finished. The three business partners had every confidence that they could do the job for what they were bidding. "We had big egos. We thought we were good programmers," Oates said. That kind of self-confidence—some would say arrogance—became a distinguishing characteristic of the new company.

Precision Instrument was skeptical at first. "They were very concerned. They weren't going to give us the contract because our price was too low," Ellison said. Maybe that was true. If so, Precision Instrument eventually put aside its concerns and awarded the contract to Ellison and the others. In June 1977 the three programmers officially founded their new company, with Miner as president and Oates as vice-president. Ellison, the mastermind, remained an employee of Precision Instrument, with responsibility for overseeing the contract. But he made it clear to his employers that he would soon resign to join the new venture. Years later Ellison enjoyed noting that he was not even the first employee of the company he founded.

The partners had some fun naming the new company. Among the names they joked about using were Nero Systems ("We Fiddle While You Burn"), Uranus Systems ("The logo would be a sphincter with products emerging," Oates said), and Intergalactic Tytanic Octopus. Eventually the partners settled on something straightforward: Software Development Laboratories, Inc. Oates said they liked the name because it recalled that of System Development Corporation, the first great American software contracting company. In the company logo the letters *SDL* had vertical stripes cutting through them. echoing the horizontal stripes in the logo of IBM.

The founders authorized the issuance of a hundred thousand

shares of stock. Ellison bought sixty percent of the shares for twelve hundred dollars, or two cents a share. His partners each bought twenty percent of the company, paying four hundred dollars each for their shares. Why did Ellison get the majority of the stock? "It was my idea to do this," Ellison said. Oates agreed. "Larry was the prime mover behind this thing. There was no question about the fact that Larry was pushing this idea a lot harder than either Bob or I would have pushed it. He had more chutzpah than the two of us combined. One and a half times as much. So he got the chutzpah bonus. Bob and I both recognized that we were going to build the software, but this organization was going to be a success because of Larry's chutzpah."

Software Development Laboratories soon received a fifty-thousand-dollar advance from Precision Instrument and began its work. The first person to write a line of code for the new company was a lightning-fast young programmer named Bruce Scott, whom Bob Miner had hired as soon as the details of the contract were finalized. (Scott had worked for Miner at Ampex in the days before Ellison got there.) Scott worked for a couple of weeks in his home office, using a pencil to write code in longhand. Then SDL leased some space in the Precision Instrument building in Santa Clara. The company's first official day in business was August 1, 1977.

Bruce Scott's first experience with Larry Ellison was an eye-opener. On one of his first days at SDL, Scott was trying to connect SDL's computer terminals to the Precision Instrument computer. There was a problem: A Sheetrock wall stood between the SDL offices and the computer room. Scott said, "Larry, we need to hook up these terminals. How are we going to hook them up?" "I'll show you how," Ellison replied. He grabbed a hammer and smashed a hole through the wall. Bruce Scott came to believe that Ellison's entire business philosophy could be summed up in that single act. "Find a way or make one. Just do it," Scott said.

As determined as he was, Ellison was also a lot of fun. One day he realized that SDL did not have a sign announcing its existence to the world. The lack of a sign hardly mattered; the company had only one customer, and the customer was in the next room. Still, Ellison wanted a sign. He rounded up his coworkers (all three), hustled them into his used Mercedes, and drove them into San Jose to get one. "The whole company went out and bought this plastic sign," Scott said. When they got back to the office, Ellison glued the letters *SDL* onto the sign and planted the sign in the grass.

Another time Larry and his wife, Nancy, took Bob Miner and Bruce Scott to the ballet in San Francisco. Nancy, who appreciated high culture, apparently enjoyed the performance. Her companions behaved like adolescents, smirking and giggling and cracking jokes. There was a part of the dance when the performers rolled over one another. "Bob and Larry and I are sitting next to each other and just laughing. It's a ballet, you know? Most people don't usually laugh at a ballet. We were laughing uncontrollably," Scott said. At times like those SDL did not feel like a serious technology company. This was Our Gang, and Larry Ellison was Spanky.

Yet only a few miles from the SDL offices in Santa Clara a group of engineers was doing work that would inspire Ellison and his colleagues to get very serious indeed.

In the mid-1970s IBM engineers began migrating to the IBM Research Laboratory in San Jose to work on various database projects. Eventually they were assembled into a team of about forty and given the challenge of creating a functioning relational database system based on Ted Codd's theory. Part of the job would be to come up with a computer language that could be used to ask questions of the database. Codd was not a part of the forty-member group. "I think he may have wanted to maintain a certain distance from it in case we didn't get it right," group member Don Chamberlin said.

Early on the bosses instructed the team members to come up with a name for the project. One engineer suggested Relational User Friendly Universal System, or RUFUS, which happened to be the name of the engineer's dog. That idea was rejected. Several members of the group had once worked on an IBM project called System A. They still liked the sound of that—it had a certain mystery to it—so they suggested System R, for "relational." That did it.

The system R group began experimenting with a language called SQUARE, which Don Chamberlin and others had developed a couple of years earlier. But SQUARE, which stood for Specifying Queries as Relational Expressions, was, Chamberlin said, far from ideal. "First of all, you couldn't type it on a keyboard because it had a lot of funny subscripts in it," he said. Any language that could not be typed on a keyboard was not going to be helpful to people who worked with keyboards. The System R group then developed a simpler, yet more functional language based on plain English. They called it Structured English Query Language, or SEQUEL. They later learned that SEQUEL was the registered trademark of a British aircraft company, so Chamberlin changed the acronym to SQL. Some people pronounced the name "sequel"; others said "ess-cue-ell."

Either way you said it, SQL was a breakthrough. With SQL, a user could interact with the database in wonderful ways. Early on the System R group set up a database for an imaginary company. It included all employee names and salaries. Using SQL, the engineers gave the command "Find the names and salaries of employees who earn more money than their managers." The names came up. Later the engineers told the database to give a 10 percent raise to every employee who earned less than twenty-five thousand dollars. It did so. This was exactly the sort of thing that businesses needed. Still, SQL was not perfect: In early tests the database interpreted the command about salaries to mean that it should keep giving raises until every employee earned at least twenty-five thousand dollars. Even so, it had great potential.

The System R people could still not afford to become complacent. They had competition, and it was coming from just up the road. About that time a group of professors at the University of California, Berkeley, were also trying to turn Ted Codd's theory into a working relational database. Their project, funded by the U.S. government, was called Ingres. Ingres was eventually marketed as a commercial product, providing vigorous competition for Larry Ellison and bringing out some of his fiercest and most arrogant competitive behavior. In the mid-1970s the Berkeley group was working on a query language called . . . well, called Query Language, or QUEL. The IBM people knew all about the Berkeley bunch, and vice versa. The relationship between the groups was sometimes strained. The Ingres people accused IBM of swiping its ideas, and IBM accused Ingres of the same thing. "We came to the conclusion that the best thing was not to talk to each other," said System R engineer Jim Gray.

While the System R and Berkeley groups were busy building databases and not talking to each other, Larry Ellison, Bob Miner, and Ed Oates were considering their options. (By now Ellison had resigned from Precision Instrument and taken over as president of SDL.) One possibility was that they would continue to write software on a contract basis. There were several advantages to contract work. For one, they would get to write lots of different kinds of programs; every day would be a new challenge. Contractors could often write the fun part of a program and leave the dreary cleanup work to the customer. But contract work also had a major disadvantage. In that kind of business, Bob Miner said, "you're either working or you're trying to get work."[7] He and the others did not want to spend the rest of their careers scrounging around for their next contract.

The three partners decided that the best idea was to write a single program and sell it over and over again. As hardware technology improved and the demand for useful software intensified, more and more companies were going into business writing ready-to-use computer programs. The packaged software business was about to take off, and Ellison and company wanted to be along for the ride. "We just figured that a product-oriented business would be more profitable and more satisfying," Miner said. Besides, Ellison, Miner, and Oates were tired of working on failed projects. This time, as Miner said, they wanted to do something right.

But what? The answer was supplied by the engineers at the IBM San Jose Research Laboratory. The System R people might not have been talking to the Berkeley professors, but they were not exactly keeping their work a secret. Quite the contrary. They were publishing papers that explained how System R worked. The first, called "System R: Relational Approach to Database Management," appeared in June of that year in the technical journal *Transactions on Database Systems.* The researchers also published articles in the *IBM Systems Journal* and gave papers at industry conferences. A lot of people in the computer industry got their first real understanding of relational technology by reading those papers.

Ed Oates was one of them. He had been interested in the possibilities of relational technology ever since Codd's first paper and had followed its development closely. For a guy like Oates, the *IBM Systems Journal* was light reading. "We all knew that relational was the trick. We all knew especially that network and hierarchical databases were not it. Those were old technology," he said. He studied the System R papers and gave Miner and Ellison a copy of the document that described the SQL user language. "It was a very terse specification," Miner said. "It was maybe six pages of text describing the language, and then a couple of pages of formal but incomplete language specification."[8] Still, the papers offered a pretty good idea of how to build a working relational database management system. Ellison understood that he could not go wrong if he built a system just like IBM's. "Larry was the strongest IBM follower, and believer that IBM could do no wrong, among us," Oates said.

That cinched it. Ellison and his colleagues now knew what kind of product they would build: a relational database system.

The guys from SDL were not academics, not mathematical theorists. They probably could not have developed the SQL language on their own. But once IBM did so, they knew what to do. Building a database management system based on System R would be like constructing a model airplane from a kit—an extremely complicated kit. "We were basically implementers," Bruce Scott said.

After SDL finished "a significant amount"[9] of work on the Precision Instrument contract, Scott and Miner began writing the code for the relational database product. (Ed Oates also contributed some code to the first version.) They did the work on a Digital Equipment Corporation PDP-11 minicomputer for the simple reason that it was the only machine they had access to. The computer, which was owned by Precision Instrument, was the same one they had used to write the software for the mass storage device.

Scott and Miner, though far from expert in database systems, needed only a few months to complete the first version of the database. "We were good. Seriously, we were good," Scott said. "There's a bell curve of programmers, and there are the ones who are way out on the edge who are better than most. And when they're better, they're really better." Ellison apparently agreed that Scott was good. After a couple of months, he gave Scott about 4 percent of the company—simply handed the shares to him unconditionally. There was no schedule for vesting; Scott could have left that day owning 4 percent of SDL. In later years Ellison became much more conservative about bestowing stock.

The SDL guys named the finished product Oracle, after the CIA project they all had worked on at Ampex. They thought that Oracle, the source of wisdom, was a fitting name for software that answered users' most difficult questions. It also seemed appropriate that a borrowed idea would have a borrowed name.

Miner and Scott may have been good, but that didn't mean their database system was. The first version of Oracle could be used to answer a simple relational query, but that was about all. It could not be counted on to store and manage large amounts of information and therefore could not be marketed as a working product. Version 1 "was really just a toy," according to Robert Brandt, an early employee. SDL never sold it to anyone. It was going to take time—a long time—to make Oracle work the way it was supposed to.

SDL's work on Oracle raised some eyebrows at Precision Instrument. In order for SDL to keep receiving payments on the contract, its software had to pass certain tests along the way. Precision Instrument hired a consultant, Irv Tjomsland, to perform the tests and report on the progress. Tjomsland said SDL always did just enough work to get paid. But he believed that Ellison and the others devoted most of their energy to the relational database project. He didn't think that was fair to Precision Instrument. "If you're being paid to do a job, you've got to get it done," he said.

But the SDL people said they *did* get the job done. Though it was true that Scott and Miner spent a lot of time creating Oracle, they did so only after they had finished writing the programs for Precision Instrument, Scott said. The SDL software "had the usual amount of bugs, but it was certainly passing these independent tests," said Oates, who was overseeing the contract.

Software was not the problem, Oates said. Hardware was. Sometime during the first year of the SDL contract, Precision Instrument changed its name to the more technical sounding Omex. Unfortunately the name change did nothing for the performance of the mass storage device. The PI 180 still did not work. According to Oates, the SDL software worked perfectly when you used it to retrieve an image from a floppy disk. But the PI 180 didn't use floppy disks; it used rhodium-plated strips. No matter how many times Oates tried, the PI 180 would not deliver what the software had asked for. The software talked, but the hardware did not listen.

Finally the men from SDL decided they needed the final twenty-five-thousand-dollar payment on the contract. But they could not test the reliability of their software unless the Omex contraption could be made to work. The two sides made an agreement: Omex would keep the PI 180 working for seven eight-hour days, with no more than two days of downtime. SDL would test the software while the machine was running. But the Omex engineers could not make it happen. Within three days the company violated the agreement.

"I was trying to debug stuff. You'd say, 'Load strip,' and the machine would smoke," Oates said. Sometimes, instead of loading a strip, the mechanical arm would retrieve an entire canister and toss it on the floor. Omex eventually went broke.

Larry Ellison and company did not suffer; they had already been paid. On SDL's first anniversary Ellison, Oates, Miner, and Scott had a little celebration in their offices in Santa Clara. Somebody brought in a cake with a candle in the shape of the number 1, and the four men posed for a picture. Oates, Scott, and Miner were grinning so broadly that they appeared to be almost laughing. The tall, almost gawky Ellison towered above them all, his sunglasses dangling from his shirt pocket, a knowing half-smile on his lips. The Omex deal had been "enormously successful" for SDL, Miner said.[10]

Even after expenses Ellison had enough money left to go after a fortune.

Which raises a question: Why did IBM, the very seat of American corporate power, the titan of the computer industry, give away what turned out to be a multibillion-dollar idea? One reason was that IBM wanted to remain the leader in the world of high technology lest another company take its place. The only way to do that was to plant new ideas and give them enough light to grow. "You don't set standards by keeping them to yourself," said Frank King, who supervised the System R group. IBM wanted to set the agenda for the future, and it saw relational technology as part of that agenda.

Another reason had to do with the culture of the IBM Research Division. The company had been researching new technologies since the days of old Tom Watson, who founded IBM in 1924. By the 1970s IBM was spending 10 percent of its revenue on research and development, accounting for 10 percent of all private research done in the United States.[11] Not all the ideas coming out of the Research Division were practical or even close. People in the industry often referred to it as the place "where rubber meets the sky." Still, there was no question that IBM employed some of the finest computer scientists in the world.

Ted Codd, the father of the relational database, is a good example. While writing his landmark paper on relational theory, Codd held the exalted title of IBM fellow, something like occupying an endowed chair in computer science at the world's finest technical university. Even though he worked for an American corporate colossus, Ted Codd was basically an academic.

The people who developed System R were academics too. Certainly they all wanted to build a marketable product—they knew where their next meal was coming from—but the System R people were mainly interested in ideas, not products. Trained as academics, they believed strongly in the principle of sharing knowledge, so there was never really any doubt that they would publish what they knew. Besides, publishing papers and giving speeches made them famous, to the extent that anyone can become famous doing something so arcane. Why should the Berkeley guys get all the glory? Of *course* the IBM people passed on their knowledge. "That was how we got our renown," said Mike Blasgen, a senior member of the System R group.

Twenty years later, at a reunion, members of the group debated the wisdom of having published the System R papers. One said he believed that "publishing everything was a big mistake." Blasgen had a different view. Publishing created the momentum that was needed for the technology to catch on. "If we had not published those papers, [relational database technology] would have failed. The reason it would have failed is that IBM would have ignored it," he said. Such was the power of IBM. Without Big Blue backing the new technology, not many people would have wanted it, no matter how charming and persuasive Larry Ellison was.

If Blasgen and his colleagues had a regret, it was that IBM did not turn System R into a product sooner than it did. How slow was IBM? The System R group had its relational database up and running around 1977, but IBM did not introduce a commercial product until February 1982. Larry Ellison's nimble and opportunistic little company had a relational database product on the market *before IBM managed to move System R from the Research Division to development.* While IBM moseyed along, Ellison was out there gobbling up market share.

Why did IBM move so slowly? As many, many commentators noted, the company was like a nation unto itself, and a very cautious nation at that. It had a massive, richly layered bureaucracy, with committees reporting on committees reporting on committees. Nothing could get done without endless review and re-review. According to a former IBM programmer, the company once did a study of the way things were done. "What they found is that it would take at least nine months to ship an empty box," he said.[12]

Besides, unlike Ellison's fledgling company, IBM had something to lose: money. For years IBM had been selling a hierarchical database product called IMS, which stood, straightforwardly enough, for Information Management System. IMS ran on big, mainframe computers, the kind that crunched numbers for banks and insurance companies and so on. In the mid-1970s IBM was making a whole lot of money on IMS. According to Mike Blasgen, about fifteen hundred customers were paying IBM fifteen hundred dollars a month to use the software. (Blasgen was not sure of the figures, but he was sure of his point.) Some people in the company believed that selling this new technology would hurt sales of the tried-and-true hierarchical product.

There was another major concern within IBM. A lot of people—executives, programmers, salespeople, technical support people—had built their careers on IMS. Indeed the engineers at IBM's Santa Teresa Research Laboratory were still making improvements in IMS when their colleagues in San Jose published the first of the System R papers. According to longtime industry consultant Jeffrey Tash, the IMS loyalists "fought like crazy" to keep the relational product from going to market. They did not want IBM to sell anything that might replace the product they had built.

At first IBM tried to resolve the conflict between IMS and System R by creating a database that was somehow both relational and hierarchical. "The problem with programs that try to do both is they end up doing neither," Blasgen said. The schizophrenic IBM database was like that. Finally, in February 1982, IBM introduced its first relational product, SQL/DS. One of the main features of SQL/DS, from IBM's point of view, was that it posed no threat to IMS. The product—really just a souped-up version of System R—ran only on certain small mainframes, while IMS was engineered strictly for big iron. IBM did not introduce its industrial-strength database, DB2, until 1985. By then Larry Ellison was a multimillionaire.

Indeed Ellison could not have dreamed of a more amiable and helpful competitor than IBM. Think of the marketing of relational technology as a race, with Ellison and IBM as two of the main entrants. IBM taught Ellison to walk, bought him a pair of track shoes, trained him as a sprinter, and then gave him a big head start. How could he lose?

An interviewer once asked Ellison to comment on another fateful IBM decision: to use Microsoft's MS-DOS operating system in the IBM personal computer. Ellison, no fan of Microsoft or its products, called the decision "the single worst mistake in the history of enterprise on earth." (The Indians who sold Manhattan to the Dutch for twenty-four dollars might beg to differ.) But if IBM's choice of Microsoft was, as Ellison said, "a hundred-billion-dollar mistake," its decision to publish the System R papers, and its failure to market quickly a relational product, certainly ranked right up there. By the mid-1990s the relational database industry was generating billions of dollars a year. IBM was getting a significant share of that money, but not as much as it might have. "IBM could have had all of the relational database market if they'd been more aggressive," said Michael Stonebraker, one of the directors of Berkeley's Ingres project.

But then, it is not unusual for the inventors of new technologies to miss out on some of the resulting riches. About the time that IBM was doing unintentional favors for Larry Ellison, the Xerox Corporation did a similar kindness for an equally brash—and equally motivated—young entrepreneur. In 1979 Xerox's Palo Alto Research Center invited Apple Computer's Steve Jobs to have a peek at some technology it had developed. This technology completely transformed the face of the computer: What once had been an inscrutable machine now looked like a desktop, with little pictures, or icons, representing programs and documents. To read a document, a person merely touched a device called a mouse. Xerox—which was run by "copier heads," according to Jobs[13]—did nothing with the graphical user interface. Jobs incorporated it into the Apple Macintosh, the first (and some would say last) great personal computer. By adopting Xerox's idea, Steve Jobs changed an industry, much as Larry Ellison did by adopting IBM's. It really wasn't much of a coincidence that Jobs and Ellison later became friends.

What did IBM have to say about all that? When the System R group held its reunion in 1995, Mike Blasgen brought along a cartoon that pretty well summed up System R's contribution to the relational database market. In the drawing a beaver and a rabbit are standing next to the Hoover Dam, having a chat. The beaver is saying to the rabbit, "I didn't actually build it, but it was based on my idea." Said Blasgen: "This little beaver is System R."

Later I asked Blasgen how it felt to give away such a lucrative idea. "I feel bad that IBM didn't capture this market, and I feel bad for myself that the work didn't personally enrich me more than it did. I wish that I'd figured out how to get a billion dollars out of it," he said. "But I don't feel anger. I'm sort of proud that the work we did has led to success for some people. And I hope that they are the deserving ones." He said he was happy for them "even if they're not."

A lot of people summed up Larry Ellison's success by saying he was in the right place at the right time. One early employee of Oracle[14] said of Ellison and his partners, "They didn't have a great idea; they found a great idea." That was true. But so what? Lots of people in the computer industry read the System R papers, but only Ellison seized on the opportunity to build an actual database product. Only Ellison took an idea and used it as the foundation for a huge corporation. Yes, he was in the right place at the right time. But as he told me, "I don't know of any place or any time where there aren't great possibilities." Yes, IBM gave him the idea. But it did not give him six billion dollars. He made himself rich through ceaseless work, brilliant strategy, unrelenting optimism, and ruthless determination.

Larry Ellison achieved the first success of his life by doing what no one else could or probably would do. He did it by being himself.

Five

ABOUT THE TIME LARRY ELLISON WAS CELEBRATING THE FIRST ANNIversary of the new business, he also suffered a loss. On July 7, 1978, Nancy Elizabeth Wheeler Ellison left him after only eighteen months of marriage and filed divorce papers at the San Mateo County Courthouse a week later.

When Ellison needed someone to talk to about the breakup, he turned to an old and trusted friend, his first wife, Adda Quinn. "He spent, I'm not kidding, days on the telephone with me, talking to me about why he wasn't successful, what was it in him that was preventing him from having a successful relationship," she said. When Quinn left him, only his ego had seemed hurt. But "with Nancy it was just total pain," Quinn said.

The divorce was so uncomplicated that Ellison did not even hire a lawyer. He simply read the papers prepared by Nancy's attorney and signed them. He kept his house in Woodside and the one he was building in Orinda. He agreed to pay Nancy five hundred dollars a month for the first year after the divorce, four hundred a month the second year, and so on, until he was paying just one hundred a month. He also paid two thousand dollars in fees to Nancy's lawyer. Nancy took back her maiden name. The couple had no children.

Larry and Nancy had only one other asset to dispose of. Ellison had founded Software Development Laboratories about six months after he and Nancy were married. She was at his side throughout the first year, when he had often worked long hours overseeing the first version of Oracle. Larry had given some of his original sixty thousand shares to Bruce Scott, but he probably still owned at least fifty thousand. Nancy could have made a reasonable claim to half of Larry's stock in the company.

But at the time SDL had no functional product and few, if any, assets. It wasn't really a company; it was an idea for a company. And the chances of its succeeding seemed remote. Larry Ellison had never had any kind of lasting success—not in school, in business, or in his personal life. Why would he have one now? Besides, Nancy came from an affluent family. What did she need with Larry's all but worthless stock?

Nancy Wheeler sold her interest in Software Development Laboratories to Larry Ellison for five hundred dollars. Had she taken half the stock and held on to it for the next two decades, it would have been worth several billion.

"Nancy's family was less concerned that she was selling the Oracle stock than the fact that I was getting a truck," Ellison said. "I got this white panel truck that had belonged to her dad before we got married. That was the thing that seemed to really bug them the most. Now I think they'd let me keep the truck."

With the Precision Instrument/Omex work finally done, it was time for Larry Ellison's little company to move on. On December 1, 1978, Software Development Laboratories left the Omex building in Santa Clara and moved into a suite of offices at 3000 Sand Hill Road in Menlo Park, in the heart of Silicon Valley. The office complex was an odd choice because it was populated almost exclusively by stockbrokers, insurance underwriters, and venture capitalists. There wasn't another start-up company in the whole development, though some of the venture capitalists occasionally lent space to start-ups. Very few start-ups chose Sand Hill Road because they could not afford the rent. Ellison, who still had "champagne tastes on a beer budget," as his first wife put it, didn't mind spending the money. SDL was flush with cash from the Omex deal. Besides, he liked the idea of being near all those successful people and thought the address would give the company much-needed legitimacy. What was more, the offices were close to Interstate 280, the scenic highway that winds through the green hills of the valley. That was convenient for Bob Miner, who commuted from San Francisco every day. Ellison took about three thousand square feet at 3000 Sand Hill Road—way too much space for a company with five guys and no marketable product. As always, he had big plans.

Soon after changing its location, the company also changed its name. Ellison, Miner, and the others no longer liked Software Development Laboratories because it wasn't specific. Now that the partners were creating a product, they wanted a name that explained what they were doing. They chose Relational Software Inc., or RSI, though the product was still called Oracle. "They didn't want to be a development laboratory anymore," one early employee said. "They wanted to do a software package and sell it like a doughnut—you know, sell the same thing over and over."[1]

But first they had to sell it once. Their first customer was the Central Intelligence Agency. The CIA did not do any intelligence work to find RSI; it found the company by accident. In the late 1970s some people from the CIA made a trip to IBM to see a working relational database. Why the agency would be interested was no secret. It was in the business of gathering, storing, and analyzing information—about governments, about weapons, about God knew what. Relational technology promised to make data analysis a lot faster and easier. The agency—and a lot of other American intelligence agencies—had been following its development since the publication of Ted Codd's paper.

The CIA people were so impressed with the IBM demonstration—so enthralled with the idea that the average user could actually get information out of a database—that they decided to get a relational database system for themselves. But IBM was not ready to make a sale, so the CIA began looking elsewhere. The person in charge of finding a commercial relational database was Dave Roberts, who worked in research and development at the agency.

Roberts heard about a little company called Relational Software

Inc. that was working on just the product the CIA wanted, and when he called, he was surprised to hear a familiar voice on the line, Bob Miner. Roberts had been Miner's supervisor a few years earlier, in the Washington office of Informatics. That was when Informatics was helping Ampex with its terabit memory project for—coincidentally—the CIA. Later Miner moved to California, and Roberts took a job with the CIA, and the two lost touch. Roberts had no idea that Miner was working there at RSI. But when Miner told him that he had followed the IBM specification for SQL, Roberts's ears perked up. Eventually the agency gave RSI a contract to build a relational database management system. "The CIA became customer number one for Oracle," Roberts said.

The agency was surprised to hear that RSI's product was named Oracle, the name Ampex had used when working on the terabit memory project for the CIA. "They said, 'You have a lot of nerve coming in here with that name,'" Ellison remembered.

The CIA project was hardly top secret, but there was nonetheless a mystery about it. A couple of RSI people got low-level security clearances, and when the agency made payments on the contract, it sent the checks in plain white envelopes with no return address.[2] Eventually word got around 3000 Sand Hill Road that RSI had something to do with the CIA. Ellison's old colleague from Amdahl, Stuart Feigin, joined the company about that time. He could only imagine what people must have thought they were doing. "Here we were, five guys in beat-up cars who had too much office space and appeared to do nothing," he said.

But the appearances were misleading. The CIA required RSI to make Oracle work on an IBM mainframe and on a Digital Equipment Corporation minicomputer called a VAX. Those two machines had entirely different operating systems, so they understood different kinds of computer code. About the same time Ellison sold a copy of Oracle to Navy Intelligence in San Diego. The Navy wanted to make the software run on the VAX or on another operating system called Unix. The tally so far: two customers, three operating systems. This created enormous problems for RSI. Bob Miner and Bruce Scott had written Oracle to run on a DEC PDP-11, and only a PDP-11. But already customers were insisting that it run on everything from a toaster oven to a dishwasher. "We had to scramble," Miner said.[3]

Soon after RSI signed the CIA deal, Larry Ellison set out to buy a mainframe computer that could be used to write the IBM version of Oracle. Stuart Feigin, an IBM expert, was going to do most of the work. But Ellison quickly ran into a roadblock: IBM had a waiting list for the machines that was several months long.

The CIA took care of that. Because RSI was a defense contractor (though a minor one), it was given priority. RSI's 4331 arrived soon after Ellison put in his order. "I'm not sure how Larry ever managed to pay for it," Feigin said.

RSI wheeled the new machine into the computer room—and then pretty much forgot about it. "It looked like a big meat locker. It just sat there and didn't do anything," one employee remembered.[4] The IBM 4331 became "the most expensive coffee table in Silicon Valley," another said.[5] Even so, the machine served a purpose. Soon after it arrived at 3000 Sand Hill Road, a man named Bob Preger interviewed for a job as a salesman. When he saw the 4331, he assumed that RSI was writing software for IBM mainframes. That was enough to convince him that RSI was a serious company, so he took the job. "What I didn't realize at the time was that the machine was just sitting there. It wasn't hooked up. Nobody had even turned it on," Preger said.

The reason RSI didn't use the 4331 was that it was busy trying to please its customers by making Oracle work on other operating systems. This was a lot like trying to wedge a Ford carburetor onto a Chevy engine. Ellison and Miner knew that jury-rigging Oracle in that way would not work in the long term. If Oracle were to be a success, they would have to rewrite it in a language that lots of computers could understand. They chose a new language called C for a couple of reasons. First, they knew they could get C compilers for many different kinds of machine. (A compiler takes a programming language such as C and translates it into something that the computer can understand.) Second, they knew that C compilers were inexpensive. Even though RSI had done well so far, it had to be careful with money.

Miner and Scott wrote Version 3 of Oracle entirely in C. Without really setting out to do so, they ended up creating something remarkable: software that could run on more than one kind of computer. In the past most packaged software programs had been written with specific operating systems in mind. For example, a company called Cullinane Corporation (later Cullinet) was making millions writing network database software specifically for IBM mainframes. Oracle was different. Thanks to the tough demands of the CIA, the Navy, and other customers, RSI produced software that was *portable.* "We had the first portable piece of software of any real size. And it was forced upon us, really. It was just a matter of necessity," Bob Miner said.[6]

"We'd sell anything anyone wanted to buy," Ellison said. "So if they wanted it on this platform, we said, 'Fine, we'll move it to that platform.' It became very clear that there was no dominant minicomputer platform, and we were a database for minicomputers at that time. There were several different operating systems and computers, and we had to survive. We had to make sure that our product ran in a variety of environments. The only way we could do that was to make it portable."

Larry Ellison may not have conceived the idea of writing portable software, but the way he exploited it was pure genius. In years to come he promoted Oracle as a one-size-fits-all computer program. It doesn't matter what kind of computer you have, he told customers; Oracle will work on it. For a long time he liked to say that Oracle software was "promiscuous" because it would work "with anybody."[7] This was a powerful selling point. Big corporations and government agencies often used more than one kind of computer, as the CIA did. The idea that they could run one kind of database software on many different machines was exciting; no longer would they have to waste time and money writing new programs for each machine. However, as time went by, some of Ellison's customers found out the hard way that his claims of portability were a little exaggerated; Oracle did not work as well on some machines as it did on others. Still, Ellison's marketing idea—his promiscuous software—served him well, and the portability feature became a major part of Oracle's success.

Early in *The Soul of a New Machine,* Tracy Kidder's classic 1981 book about the design and construction of a minicomputer, Tom West, an engineer for Data General Corporation, travels to a city "somewhere in America" to look secretly at a minicomputer built by the Digital Equipment Corporation, Data General's archrival. According to Kidder, West "walked into a building, just as though he belonged there, went down a hallway, and let himself quietly into a windowless room." There he saw a technician working on a DEC minicomputer.

Kidder writes:

> Although West's purposes were not illegal, they were sly, and he had no intention of embarrassing the friend who had given him permission to visit this room. If the technician had asked West to identify himself, West would not have lied, and he wouldn't have answered the question either. But the moment went by. The technician didn't inquire. West stood around and watched him work, and in a little while, the technician packed up his tools and left.
>
> Then West closed the door, went back across the room to the computer, which was now all but fully assembled, and began to take it apart.

What Tom West saw inside that box was a disaster for Data General, which was trying to compete with DEC. But it was wonderful, life-affirming news for Relational Software Inc. West was looking at the future; what he saw in that room, in a building somewhere in America, was the VAX.

If Larry Ellison owed part of his success to IBM (and he certainly did), he owed just as much to Digital Equipment Corporation.

IBM gave Ellison the idea for a relational database system, but DEC gave Ellison a place to put it.

The creation of the VAX, or Virtual Address Extension, was the culmination of a decade-old trend in computing. In 1965 DEC, founded by Ken Olsen, introduced a product that had a seismic effect on the world of computing. The PDP-8 was small enough to fit on the top of a lab bench,[8] yet it was powerful enough to quickly perform complex calculations. DEC was to sell fifty thousand PDP-8s during the machine's fifteen-year life span.

The PDP-8's importance went way beyond DEC's bottom line. The machine and its successors (one of which was the PDP-11, the machine that Bob Miner and Bruce Scott had used to write the first version of Oracle) were largely responsible for creating the minicomputer revolution, a paradigm shift in the world of technology. Though IBM amassed huge profits from the sale of giant water-cooled mainframes, DEC, Data General, and other minicomputer makers satisfied a growing demand for smaller, less expensive, yet still useful machines. Companies installed these machines in individual departments, which used them for typesetting, design, word processing, and so on. From that moment on, computing power no longer belonged exclusively to a few technicians in white lab coats. As Glenn Rifkin and George Harrar write in *The Ultimate Entrepreneur,* their book about Olsen, "DEC had brought computers out of the hands of the data processing high priests and now there were legions waiting to get at them in corporate settings."

In October 1977—only weeks after Ellison's little company was founded—DEC introduced the VAX. This computer, with its Dr. Seuss–like name, eventually gained mythic status; *The Soul of a New Machine* was all about Data General's efforts to build "the answer to VAX." But there was no easy answer to the VAX. It was designed to do everything the PDP machines could do, and more. The trouble with the PDP computer was that it had too little memory for big jobs. The VAX solved the problem by adding memory—without changing the fundamental way the machine worked. That meant the VAX could run PDP programs, a strong selling point indeed. It was as if every VAX had a trusty old PDP living inside it. Stuart Feigin described the VAX as "a PDP with thyroid problems." This was why it was somewhat easier to move the PDP-11 version of Oracle to the VAX than it was to move it to, say, an IBM machine. In an industry that valued innovation above almost everything else, DEC built a machine meant to last for years. (That was by itself an innovation.) The strategy of reliability and durability paid off. DEC shipped its first VAX in 1977, its twenty-five thousandth in 1984, and its hundred thousandth in 1987.[9]

Of course computer hardware is worthless without software. "But like most computer makers of the time, DEC paid little attention to software," Rifkin and Harrar say. "The company supplied the hardware and suggested uses it could be put to. Application software was left to the customers themselves or third-party developers to devise."

Larry Ellison was a third party. In 1979 he hired a DEC operating systems expert named Robert Brandt to develop Oracle on the VAX. But if Brandt, who did not have any experience with the VAX, was going to do this, he needed a VAX to work on. RSI could not afford to buy one (it had already invested heavily in the IBM "coffee table" mainframe), so Ellison, Mr. Find-a-Way-or-Make-One, got permission to use a VAX that belonged to the University of California, Berkeley. This was a real coup. For one thing, Brandt soon managed to get a look at the VAX's source code, which was licensed exclusively to the university. Once he knew exactly how the machine worked, his job was considerably easier.

The Berkeley deal was also clever because Berkeley was the birthplace of Ingres, the relational database system that became Oracle's chief competitor. "So here we are, we're working right under the nose of Professor Michael Stonebraker," Brandt said, referring to the chief architect of Ingres. "Of course he didn't know we were up there." And the guys from RSI did not tell him. About the time Ellison worked his deal to use the Berkeley VAX, Stonebraker created a new company so he could sell Ingres commercially. Stonebraker called the company Relational Technology Inc., or RTI—a name so similar to RSI that people routinely confused the two companies. Once RSI—Ellison's company—received a check that was meant for RTI. "We had to talk Larry out of keeping it," Brandt said. Ellison wasn't going to deposit it, just keep it. "He was wondering how badly the lack of cash flow would hurt them [RTI]."

Brandt eventually managed to make Oracle work on the VAX, thus becoming the first person to move the software successfully from one operating system to another. There was, however, a problem: The VAX version of Oracle ran even more slowly than the old PDP-11, which already wasn't too swift. This was not much of a marketing point. To solve the problem, Brandt would have to fiddle around with some of the code in the VAX operating system, and Berkeley wasn't going to go for that. Again Ellison made a spectacular deal. Somehow he heard about a company that had bought a VAX but had no place to put it. "Larry said, 'Have I got a deal for you. We'll let you keep it in our computer room if you'll let us have free use of it,'" Brandt said. The owners of the VAX agreed. Their employees would still be able to connect to the computer using phone lines. It seemed impossible that anything could go wrong.

Well, every time Brandt tinkered with the computer, the system crashed. The people connected to the VAX by phone couldn't work because their screens would freeze. Sometimes they got a message saying, "Please log on." This was a bad sign indeed, because they thought they already were logged on. Work was lost. Time was wasted. Not RSI's time, the other company's.

"We'd get this phone call: 'What happened?'" Brandt said. "For a while Larry would say, 'Oh, we must have had a power glitch.' He would come up with some excuse. They got to be suspicious one day when it happened three times in an hour."

Finally one of the VAX's owners called Ellison and told him he could not have system-level access to the computer anymore. According to Brandt, Ellison said, "We're moving the VAX onto the sidewalk right now, so you'd better come get it. . . . We can't use the machine. It's taking up room and electricity, and we want it out of our building." It was raining at the time. The owners quickly changed their minds, and Brandt was allowed to crash the computer for ten minutes every hour. Thus RSI developed its first workable version of Oracle for the VAX. When the owners finally found a place for the computer, they ripped it out of the RSI building in a hurry.

"They were so happy to get us out of the picture that they could hardly stand it," Brandt said.

Later Ellison borrowed money to buy a VAX for RSI. The investment turned out to be worthwhile. In 1984 Oracle sold three quarters of its software licenses to DEC minicomputer users, mostly VAX customers.[10] In 1986 Oracle sold about $50 million in new licenses, about 60 percent to VAX users. Two years later new licenses totaled $205 million, with more than half coming from VAX shops. By 1992 Unix, developed by Bell Laboratories in the 1970s, had become the operating system of choice at most corporations, and Oracle's financial statement reflected that trend. But even then VAX customers accounted for 20 percent of Oracle's $709 million in sales. All told, Larry Ellison sold several hundred million dollars' worth of software to VAX users.

During the 1980s and into the 1990s thousands of government agencies and corporate departments bought VAXes, and Larry Ellison did his best to see that every VAX ran a copy of Oracle.

Like a lot of people who worked at Oracle in the early years, Stuart Feigin never really did the actual job he had been hired to do. He was supposed to develop software for the IBM 4331 but wound up doing lots of other things instead. One was keeping the books, and Feigin, the first person in the company who could really be called a details guy, was not really qualified; he took the job on because nobody else was doing it. He also tried to keep a few bucks in the petty cash box "in case we needed to buy stamps." Keeping a few bucks on hand was not easy: According to Feigin, Ellison was forever raiding petty cash for lunch money.

Feigin did more than manage the company's assets; for a while he generated most of them. When times got tight, he took on consulting jobs to keep RSI going. Once Tandem borrowed him because it needed his expertise in IBM computer systems. "We charged them an outrageous consulting fee. And every time they agreed to it, we raised the price," Feigin said. Later National Semiconductor asked him to do some technical writing. RSI charged the company a thousand dollars a page, a high price considering Feigin finished each page in about ten minutes. Not long after he joined the company, Ellison rewarded him with a big chunk of stock.

Feigin was not the bookkeeper for long. One day a young college graduate walked into the RSI offices and said he was looking for work as a bookkeeper. Feigin got Ellison to hire the kid on the spot—a big mistake. The kid kept making disastrous errors; he would write $450 when he meant $540. Finally, RSI had to let him go. The young man had what was, for a bookkeeper, a dread affliction. He was dyslexic.

Don Lucas was the man who arranged a loan for Larry Ellison to buy a VAX. "The computer was worth quite a bit, so it wasn't a lot of risk," he said. But it wasn't the computer that Lucas wanted a piece of. It was Larry Ellison.

Lucas could not help noticing Ellison, Miner, and the other young men who worked in the offices on the floor below his. They were working when he arrived each morning and working when he left each night. He did not know what they were doing, but he figured it had something to do with computers; he could see their computer terminals glowing through the office windows. One day he poked his head into their offices and introduced himself. Ellison did not know it at the time, but Lucas was just what he needed.

Donald L. Lucas was a Stanford-educated businessman who had gotten rich by investing in new companies. His career as a venture capitalist stretched back to 1959, when he helped start National Semiconductor. He later served on the boards of such companies as Control Data, HBO & Company, ICOT Corporation, Kahler Realty Corporation, Tracor, Inc., and many others. He was a seasoned businessman, a veteran of the boardroom wars, a backslapping man's man, and a conservative—the kind of guy a lot of start-up companies could use.

Ellison liked Lucas but had no use for venture capitalists. To him, accepting venture capital was a Mephistophelian bargain: You got the money you needed to start a company, but you had to give up too much control in exchange. If things went badly, you might lose the company. Even in those days Silicon Valley echoed with stories of entrepreneurs who took venture money only to be booted out by money-hungry backers. In Silicon Valley venture capitalists were known by another name: vulture capitalists.[11]

Still, it was not long before Ellison found that he needed Lucas. Ed Oates was going through a rough patch. His marriage was unraveling, and his work was suffering. His diminished performance rankled Ellison, who expected everyone in the company to be as dedicated as he was. "I think there was a period there when Larry soured on having me as part of the founding trio of the company. And I don't begrudge him that," Oates said. He understood why Ellison was disappointed in him.

Finally Oates decided to leave. His plan was to sell his stock back to the company and give half the proceeds to his wife as part of the divorce settlement. To do this, he needed to know the fair market value of the stock. Oates and the others didn't want to set the value arbitrarily; if they underestimated it, Oates's wife—and, for that matter, the Internal Revenue Service—might demand further payment. The company needed an impartial valuation. Besides, knowing the fair stock price would come in handy when the firm applied for loans or sought investors. That was when Don Lucas, an expert in the financial dealings of high-tech companies, was hired to do the report.

At the time RSI had an expensive lease, considerable debt, an untested product, a president who had never run anything more complicated than a used Mercedes, and a heck of a time making payroll. Lucas set its value at a modest hundred thousand dollars, meaning that Ed Oates's 20 percent share was worth twenty thousand. The company gave Oates ten thousand dollars in cash, which he promptly turned over to his wife, and a ten-thousand-dollar promissory note. And though Ellison may not have been sorry to see Oates leave, he was also kind to him: He lent Oates forty thousand dollars to buy a house.

With the stock deal done, the company lost a founder but gained a business adviser, whom the company badly needed. "Yeah, we knew we had to sell things and whatnot," Ed Oates said. "But we were still under the misapprehension that a computer program and a product are the same thing. A lot of programmers believe that when they finish writing a line of code and do some testing, they have a product. But they don't have a product. They've done the first twenty percent. The thing still has to be packaged; it has to be marketed; it has to be sold; it has to be maintained. You have to have a support organization behind it. We sort of knew that instinctively, but we really didn't have the business acumen to run a rapidly growing company."

Lucas soon saw that. Early on Ellison asked Lucas to serve on his board of directors. It was clear to Lucas that Ellison did not know what a board chairman should do and that he did not understand the role of the compensation and audit committees and so on. Soon Lucas became chairman of the board, and he urged Ellison to instill some internal business discipline. Early in 1980 the company hired its first accountant, Roy Bukstein, whom Ellison had met at a wedding. This was a major move toward legitimacy: No longer would technical wizards or dyslexics be in charge of the checkbook.

Lucas believed that RSI had the potential to become a big public company, one that could someday make a lot of money, including some for him. So he did what he could to make Ellison's dream come true. Once he even lent Ellison tens of thousands of dollars to pay off his mortgages so he could concentrate on business.

According to Mike Seashols, who became RSI's head of sales, Lucas "could see that in Ellison he had an entrepreneur, he had a bright guy. No one had given Larry a chance, and that's what Don did. Larry would not have been successful without Don. [Lucas told

Ellison], 'This is a home run, Larry. This isn't a boutique company. You are going to take this to a new realm.' And that allowed Larry to go execute."

The only thing Lucas did not give Ellison was venture capital, which Ellison did not want anyway. There was never any doubt that RSI belonged to Larry. When the company went public in 1986, Lucas owned only 1.9 percent of the common stock, and Ellison retained 39 percent.[12] Lucas shaped the company but never controlled it.

The same can be said of Lucas's influence on Ellison: He shaped him but never controlled him. Lucas was determined to turn the young, cocky entrepreneur into a presentable businessman, someone who would inspire confidence in people who were about to unload forty-eight thousand dollars for a license to run Oracle on a VAX. This was like trying to transform a peacock into an owl. As one early employee put it, Ellison had "a lot of rough edges."[13] He habitually said things that were provocative or demonstrably untrue, he was always extravagantly and remorselessly late, and he never tried to hide his belief that most people he met were stupid.

And those were just a few of Ellison's charms. Ellison was driving a Mazda RX-7, which Lucas did not think projected the proper conservative image. (A two-seater, it also happened to be unsuitable for taking, say, a couple of potential customers to lunch.) Sartorially Ellison was in a regrettable cowboy phase; RSI's resident visionary was regularly attired in designer jeans and cowboy boots with holes in the soles. Even Bob Miner's grade school–age daughter Nicola thought he looked silly. Finally Ellison, who was between wives, lived, well, unmonastically.

"Larry had somewhat tumultuous affairs with women. And Lucas understood that when you come out with a prospectus, you can't have this kind of stuff in there," salesman Bob Preger recalled. "If you have a guy who is the driving force of the company and you look at his personal life and you find out he's shacking up with different women every couple months or something—well, if I was an underwriter, I guess I'd be concerned."

Lucas, the Henry Higgins of venture capital, started with the easy stuff. "I bought him a new pair of boots," he said. Ellison eventually began the transition to business suits. Lucas also arranged for Ellison to drive a two-door Mercedes coupe. This was Larry's first real Mercedes. While Lucas was at it, he persuaded Bob Miner to get rid of his Fiat and buy a BMW. "If he ever collided with anything, he's history," Lucas said. (Miner was not always so quick to do as he was told. Once Lucas urged him to sell his house in San Francisco and buy a place closer to the office. Miner brushed off the advice, saying, "Don, you tell me this house is only going to cost fifty thousand dollars, but it's going to cost a lot more, because if I buy that house, my wife will divorce me."[14])

As for the rest of Ellison's habits, there wasn't much that Lucas could do. He believed that Ellison eventually became "more sophisticated" in the things he said, though he acknowledged that "Larry likes to spark controversy." Ellison's personal life became, if anything, less stable and got even shakier from there. As for Ellison's being late, Lucas never made a dent. Ellison almost never arrived on time for anything. Once, when the board of directors met at Lucas's ranch, Ellison did not show up at all.

Those quirks aside, Don Lucas always believed that Larry Ellison had many of the attributes necessary for business success. According to Lucas, Ellison had "leadership, drive, imagination, determination. The attitude that we can win. That we will win."

To Ellison, life was a never-ending contest, every day a new opportunity to prove himself. He would compete with anyone, anytime, over anything. One day Miner and Oates were in the computer room at 3000 Sand Hill Road building something out of lumber. Ellison burst into the room. "Larry comes in and says, 'Bob, I can drive nails faster than that,'" Ed Oates said. "Then he picked up a hammer. Whack-whack-whack. They started to compete over how many hits it would take to get the nail flush." Miner, whom one early employee described as "absolutely a hundred percent as competitive as Larry,"[15] was game for the challenge. "I assume that they both believed they won the contest,"

Oates said. When asked about the hammering contest years later, Ellison laughed and said, "Mea culpa."

Sometimes people got into contests with Ellison without knowing it. Once, when the company was just getting started, Stuart Feigin joined Larry and his wife Nancy for a weekend bike ride. Eventually the three came to a steep incline. When Stuart and Nancy reached the top, Larry was far behind, laboring to catch up. The next time Feigin asked Ellison to go riding with him, Larry begged off.

"Larry didn't ride with me for a while," Feigin said. "And the next time we rode, he left me in the dust. He was practicing. He was working. He knew there was a problem, and he fixed it." Even during a casual outing with friends, Feigin concluded, "he has to be in front."

The stories of Ellison's competitiveness became legend at RSI. At 3000 Sand Hill Road Ellison often had chin-up contests with Bob Brandt, one of the early software developers. Brandt, the superior athlete, usually won, but Ellison never stopped trying to beat him. Ellison's tennis matches and basketball games were equally intense.

People who played games with Ellison said he was a tough competitor but not a dirty one. The same could not always be said about the way he competed in business. At least once Ellison publicly paraphrased Genghis Khan: "It is not enough to succeed; everyone else must fail."[16] He later denied having said those words, but there can be little doubt that he believed them. Ellison bragged endlessly about his software and derided that of his competitors, even when his products were mediocre; he talked openly about wanting to "kill" his competition—not just defeat it but kill it—and he rejoiced in the hardships of other software companies. It was not enough for his opponents to lose; he had to humiliate them. Every businessperson wants to do well, but for Ellison, to do well was never enough. To win. That was the company's reason for being, and Ellison's too. He would win cleanly if he could. But he would definitely win.

From the beginning Ellison seemed willing to do or say almost anything to get business. One day some potential customers paid a visit to 3000 Sand Hill Road. They were considering buying Oracle but wanted to have a look at the company first to see if it seemed legitimate. Ellison showed the visitors around, then invited them to lunch. The rest of the guys in the company decided to go along. As the group was leaving the building, somebody locked the office door.

"Why are you locking the door?" one of the visitors asked.

"We're all going to lunch," Stuart Feigin answered.

"I thought there were fifteen employees," the visitor said. In fact there were half that many.

Recalls Feigin: "I just looked at Larry, and Larry said, 'They're all working from home today.' "

Bruce Scott told a similar story. He and Bob Miner were meeting with some potential customers when Miner left the room to get coffee. When the customers asked about the size of the programming staff, Scott innocently told them the truth. "There were basically four or five programmers in the whole company. And Larry was telling them we had about fifteen or sixteen. They were just flabbergasted. Larry got really, really mad. He told Bob I would never meet with a customer again because of that."

When I asked Ellison about those stories, he denied that he had ever lied about the number of programmers. He reasoned that exaggerating the number of programmers by eight or nine would not have impressed anyone. "I don't think that would do the job," he said.

Maybe not, but it seemed unlikely that both Feigin and Scott would have pulled those stories out of thin air. And Scott had a witness: He told his friend Gary Kennedy, who later went to work for Oracle, about the incident when it happened. Years later Kennedy still remembered it. "I heard that story from Bruce before I joined the company," he said, "and I heard Larry say that because of what Bruce did, he would never let him talk to another customer."

Some of the people who worked for Ellison had a theory about why he said these things. They believed that Ellison lived in the future. When he exaggerated the number of his employees, he was not really lying; he was just getting ahead of himself. "He had a problem with tenses," Oates said, echoing a joke that Ellison sometimes made about himself. "It was like, we will have fifty employees, so we might as well say we have them now." Oates, for one, admired Ellison's audacity. It was Ellison's chutzpah, after all, that made him the company's president and majority stockholder. Jenny Overstreet, Ellison's longtime assistant, summed things up this way: "There's so much wishing he could make it so. . . . He doesn't live in today, because there are problems today and there are solutions tomorrow."

Maybe Ellison did live in the future. But some of his dissembling was purposeful. Oates also said the company exaggerated the number of customers it had. These fibs were told "to establish the organization as a credible place to buy a product from." If customers had known how small and shaky Ellison's company really was, they never would have taken a chance on the product, many early employees believed. If customers were told the truth, they might well have bought a database from one of Oracle's competitors, which, Feigin noted, "were also lying." In the software industry exaggerations and outright lies were commonplace.

The idea that somebody else might take away Oracle's business was poison to Ellison. He understood the importance of locking up a large share of the market early. "How much does it cost Pepsi to get one half of one percent of the market from Coke once the market has been established?" he once asked rhetorically. "It's very expensive. This market is being established. If we don't run as hard as we can as fast as we can, and then do it again twice as fast, it'll be cost-prohibitive for us to increase market share."[17]

Exaggerating was part of running fast and hard. It was something Ellison did—something he felt he had to do—to get business. If he did not do it, he might fail. And there was no way Larry Ellison was going to fail. He made that point to Bruce Scott several times. "He would say, 'You know, all the competitors are doing the same thing, and if I don't do it, somebody else will,' " Scott said. "He used to tell me, 'I cannot run this business and tell the truth to the customers. It's not possible.' " Years later Ellison had no recollection of saying any such thing.

* * *

In ways big and small Ellison tried get the most from his employees. Every time RSI hired a new person, the company had to buy two new computer terminals: one for the office and one for the new employee's home. The home terminals were connected to an RSI computer by a phone connection. Ellison referred to the home machines as "a company benefit." What he should have said was that they were *a benefit to the company:* Putting a computer terminal in employees' homes ensured that they never really left the office. "Everybody worked all the time," said Roy Bukstein, the company's first accountant. But most didn't mind, he said. For most people, working for RSI "was a passion."

Oddly Bob Miner was not one of those people. He worked extremely hard in the early days but was never comfortable with the sometimes crushing demands Ellison laid on his employees. "Bob didn't like to see people work too hard because he thought the company was taking advantage of them," Ellison said. "Even in the early days he thought it was wrong for people to work late hours and for us to ask them to work late hours. He thought people should be home with their families, and if they didn't have families, they should be out surfing or something."

Miner's engineers were profoundly loyal to him, and it was no wonder. When Ellison demanded to know what certain people were working on, Miner always had a ready answer. "Bob would say, 'I don't know what they're doing, exactly, but I'm sure they're working their asses off. All those guys are really working their asses off.' And he'd say that without knowing what they were doing," Ellison said. "Bob would always—sometimes armed with information, sometimes not—defend his people." Ellison believed that Miner was "loyal to the people before the company."

That might have been the starkest difference between them. Ellison's entire being was committed to the success of RSI, and his deepest loyalty was always to the company and, in a sense, to himself. Ellison was the company, and the company was Ellison. His commitment to RSI was appropriate; he was the founder and president, and people depended on him to run the business well. But some employees believed he cared about them only because they were doing something for him. When they were no longer working to fulfill his vision for the company, they meant nothing to him, or so they believed.

By contrast, Miner was seen as a nice guy, more interested in people than in what they produced. He took the company only half-seriously at the outset, and the more successful it became, the less seriously he took it. "He was definitely unassuming," Oates said.

Ellison thought that people liked Miner partly because Miner did not do any of the company's dirty work. "Larry was very jealous of the difference in their jobs, which required Larry to do things that were unlikeable, like firing people. Bob never had to do that. Larry = Bad Guy. Bob = Good Guy," Ellison's longtime assistant Jenny Overstreet told me in an electronic mail note. "One of the sorest points in Larry's memory was his lost friendships with the cofounders of Oracle Europe"—Brian Cassidy and Bo Ryden, whom both Ellison and Miner considered friends. According to Overstreet, "The business ultimately required that Larry ask them to leave the company. End of relationship for Larry. Years after, whenever I would mention that we were all off to dinner with these guys from Europe, who happened to be in town, Larry would feel sorry for himself and jealous and wistful and pissed that it was he who had to be the bad guy and lose their friendship," while Miner had not lost anything.

Yes, it was lonely at the top. But if Ellison sometimes felt sorry for himself, he had only himself to blame. After all, he had insisted on being the boss (he had founded the company for that purpose), something Miner never desired. Ellison wanted it both ways: He wanted the authority to hire and fire at will and wanted everyone to adore him anyway. (That was another way in which he was like the fictional Kane: He wanted love on his own terms.) Ellison may have resented Miner's popularity, but he had reason to be grateful for it. There were some people who simply would have quit if Miner had not been around.

"People who had problems or who had had just about enough of Larry and the way Larry did things would come into Bob, and Bob would calm them down," said Roy Bukstein, one of the first few employees of RSI.[18] "A lot of people stayed, I think, longer than they would have because of Bob."

That is not to say that no one at RSI liked Ellison. Most people did like him. He pushed people hard but rewarded them well. More than that, they were amazed by him, energized by his unstinting energy, cheered by his optimism, enraptured by his predictions that someday they all would become millionaires. (Virtually all the early employees did.) Ellison was a high-tech pied piper, a mesmerizing figure people followed for reasons they could not always explain. Bukstein liked to say that if he had only thirty minutes to address the President of the United States on a matter of paramount importance to the planet, he would ask Larry Ellison to go in and do the talking for him.

"There's no one like him. There's no one that combines intellect, energy, sense of humor, and timing like Larry," Bukstein said. "I don't care who you are, you will stretch to go for the brass ring with Larry."

Six

ONE OF ORACLE'S DISTINGUISHING FEATURES IN ITS EARLY DAYS WAS that it didn't work—or didn't work consistently. The first version, written on the DEC PDP-11 computer at Precision Instrument Omex, was "the roach motel of databases," Stuart Feigin said. "The data went in. It didn't come out." It didn't matter; the company wasn't going to sell it to anybody. Besides, as Bob Miner said, "When you're a start-up and you've got a brand-new technology and there's nothing else out there, you can afford to build something less than an industrial-strength piece of software." In the history of high technology there had probably never been a computer program that worked perfectly right away.

The software that Oracle sold to the CIA and the Navy was not much better than Version 1. It took the company a couple of years to deliver something to the CIA, but what it delivered "was really not usable as a database," the CIA's Dave Roberts said. And the Navy people quickly realized that they were debugging the product for RSI and paying for the privilege of doing so. "We were teasing them about being their testing and evaluation arm," said John Schill, who oversaw the contract. "We had what we called the tape du jour. We would [load it onto the computer] in the morning and then call up Ed Oates in the afternoon and tell him what our problems were. The guys at RSI would fix it in the evening, and then they'd get it down on the next flight in the morning. We almost did this on a daily basis." For RSI, this was a nifty, cost-saving arrangement. Having customers debug the product obviated the need for a quality assurance department. The company's attitude was summed up by programmer Kirk Bradley: "We don't have to do QA [quality assurance]. They'll just tell us [what's wrong with the product] and we'll fix it."

In the early 1980s RSI's potential customers were all buying VAXes as fast as Digital Equipment Corporation could build them, but RSI still did not have a product that worked well on the VAX. Version 3 of Oracle was going to be it. That version was created under grueling conditions. The company was running short of money—there were times when money manager Roy Bukstein did not think RSI would make payroll—and desperately needed something to deliver to customers. For months Ellison had been selling a relational database system that didn't exist, and wouldn't until Miner and Scott finished building it. The word in the software industry for what Ellison was peddling was vaporware. "He would make the sales based on the features that we were currently working on," Scott said. Ellison reluctantly conceded that it was true. "In the early days we did not scrupulously make clear what was in the design spec and the language spec and what was actually in the current version of the product," he said. "Someone once jokingly said that in 1992 we actually finished the implementation of the 1977 brochure."

While working on Version 3, Miner and Scott often logged sixty- and seventy-hour weeks—a long time to spend staring into a numerical sea of computer code. "I just got burned out. It was destroying my life," Scott said. He had never been entirely comfortable at the company anyway. Scott was a Latter-day Saint—a Mormon—and working alongside the company's decidedly nonreligious founders always made him feel like a stranger in a strange land. "Ed Oates was very much a humanist. He knew that Latter-day Saints didn't believe in artificially limiting the growth of a family. He'd say, 'The earth is overcrowded, blah-blah-blah.' He'd just go on and on." The pressure to rewrite Oracle was so intense that even the normally affable Bob Miner got a little crazy. Whenever he found a bug in Scott's code, "He'd call me at home and yell at me for a while," Scott said.

Late in 1982 Scott got fed up and quit—but not before he lined up another job. The timing could not have been worse for RSI. Version 3 was still not done, and Scott's departure meant that Miner had to finish it by himself—a tricky job, to say the least. Working with Scott's code was going to be like trying to write the last few pages of *Ulysses* for James Joyce. The code was that dense, that arcane. Miner was furious that he would have to go it alone. To make matters worse, Scott had submitted his resignation to Miner by electronic mail, an act that he was still embarrassed about fifteen years later. ("I was very immature," he said.) Scott later sent Miner a letter apologizing for the way he had left. Miner, still fuming, responded with a letter pointing out bugs in Scott's code.

Soon after he resigned, Scott brazenly showed up at the company's annual meeting. "I think that ticked them off a lot," he said. Ellison and Miner didn't want to see him anymore, so they arranged to have Don Lucas buy back his stock for five dollars a share. Scott realized more than four hundred thousand dollars from the sale—hundreds of millions less than he would have made if he had held on to his 4 percent stake. But he had no regrets. "I'm not bitter about it at all. That was my decision," he said. He used the money to buy a house and start a new company.

Ellison and Miner did not speak to Scott for years after that. As bad as their parting was, that always perplexed Scott. "For me, I was just changing jobs. For them, I was betraying them and leaving the family," he said. After Scott left, he heard rumors that the company was blaming him for problems in the software. Once, he said, an RSI employee asked him if it was true that he had stolen some of the code when he left. Of course not, he said.

Miner and the remaining engineers managed to get Version 3 done early in 1983. Years later Bob Miner admitted that the software "wasn't very reliable. . . . In the database world there are a couple of things you can never do," he said. "One is you can't lose data, and the other is you can't return wrong answers. In the early days we had problems losing data and returning the wrong answers."

Shortly after the release of Version 3, Miner, Ellison, and some Oracle salespeople met with users in Europe. The users were "fairly hostile" because the software didn't work the way they wanted it to, according to Brian Cassidy, then part of Oracle's European sales team. Someone in the audience asked why. "Larry attempted an answer at this and wasn't getting very far because it was technical," Cassidy said. Miner grabbed the microphone and tried to explain further. When he was done, he whispered to Cassidy, "I've just talked a whole lot of balls there."

Afterward an angry Oracle user cornered Miner and demanded to know why a certain function did not work as expected. Brian Cassidy overheard the exchange. "Bob said, 'It's a bug.' The guy persisted. Bob said, 'You're not listening to me. It's a bleeping bug. Version 3 is full of them. What do you expect, blood for your money?' "[1]

Years later, in a speech to the Commonwealth Club of California, Larry Ellison summed up his company's early approach to customers this way: "Here's our software. Use it. I dare you."[2]

Only a start-up with everything to gain and nothing to lose could get away with such a cavalier attitude. IBM, by contrast, could not and would not say that to its customers. This was another reason IBM did not get to market with a relational database until 1982—four years after Oracle did. Yes, IBM was hopelessly bureaucratic, and politically complex. But it was also deeply committed to its customers and therefore was not about to ship a product that someone might later refer to as a roach motel. Thomas J. Watson, Jr., the legendary IBM leader who got the company into the computer business, once reduced his business philosophy to three brief statements:

> Give full consideration to the individual employee.
> Spend a lot of time making customers happy.
> Go the last mile to do a thing right.[3]

IBM went the last mile when it labored to turn System R into a commercial product. According to IBM's Mike Blasgen, System R was tested at only a few carefully chosen customer sites, under close supervision by IBM employees. If anything went wrong, the IBM engineers were there to make it right. At the time Blasgen argued that IBM should go ahead and start selling the database or risk losing precious market share. But the company culture would not allow it.

Said Blasgen: "People were concerned about putting out a product that actually worked."

Larry Ellison was too, but he was just as concerned about winning market share. Because of this, some early customers received software that worked fairly well, or not too well, or not at all.

Sometimes the customers did not even receive software. When RSI had a deadline to meet, when it absolutely had to ship something in order to fulfill a contractual obligation, the company was not above shipping a blank or unreadable tape just so it could not be accused of violating the agreement. "When the customer called, we'd say, 'Oh, there wasn't anything on there? Sorry,' " Kirk Bradley said. It was a way of buying a couple of extra days to work on the real product, he said. That apparently did not happen often. But RSI's willingness to ship an unreadable tape said something important about the company culture. At RSI the idea was to ship something, get market share, *win.*

In Miner's opinion, Oracle really wasn't "industrial-strength" until Version 4 came out about a year later—five and a half years after the founding of the company. (Others believed that Version 5, introduced in 1986, was the first version that could be trusted.) According to Miner, a lot of Version 4 customers "actually went live with some pretty scary things."[4] NASA once used Oracle to keep track of the quality testing on a modular spacesuit that was going to be used on the space shuttle.

"I got a call once from Hamilton Standard, who was the contractor for the spacesuit," Miner said. "And they said, 'We did this query on an Oracle database, and it returned the wrong results. The astronaut is supposed to be using the new spacesuit in the cargo hold of the shuttle in three weeks. We have to tell him whether or not he can do that, and we can't do that unless you fix this problem. And if you don't fix this problem, you'll be in a whole lot of trouble with the U.S. government." Miner spent hours debugging the database in his basement in San Francisco. Finally he discovered the problem and fixed it. A few weeks later Miner was listening to his car radio when he heard that the space walk had been aborted. "I thought, oh,

God, it's us," he said. But it turned out that the walk was canceled for other reasons.

There were plenty of times when the problem *was* Oracle. Jeff Walker, an executive who joined the company in the mid-1980s, once asked Ellison about his early customers. "I said, 'Larry, back in the early days of Oracle, before my time, I know you shipped stuff that didn't work. I know that customers got really unhappy. Didn't anybody ever call you and threaten you and try to get their money back?' And Larry said, 'No. I don't think so. But I do remember that people used to call up and say, "Could we please have our data back?"'"

The funny thing was, the early customers did not really mind losing money or data. Even though RSI never really delivered what the CIA wanted, the agency was not upset. Engineer Kirk Bradley remembered a CIA employee telling him, "We bought this thing knowing damn well it wasn't going to work. We're buying an idea." Ed Oates's tape du jour did not satisfy the Navy either, but it did not matter. "We were buying a Wright Flyer with the expectation that we would probably have an airline in a few years," the Navy's John Schill said. The Navy understood that Oracle might crash and burn at any time. Yet it was "exciting" to try to make it fly, Schill said, with a hint of irony.

Larry Ellison, an avid pilot, also used an aviation metaphor when he talked about the early product. "It had lots of nifty, whizzy features, a little bit like the fastest airplanes, but occasionally the wings fall off. So make sure you have a parachute," he said. Twenty years later a lot of customers expected their database systems to run perfectly and continuously every day of the year. "You reach a point where the demands get almost impossible to achieve. But back then the requirements were pretty relaxed. If the software did all these really cool things, and if the plane flew most of the time, that was OK. And if the wings fell off every couple of weeks, no problem. We'll just put the wings back on."

Geoffrey Moore, the author of the high-tech marketing book *Crossing the Chasm,* has a name for people who used such unreliable products: "innovators." He describes them as people "who appreciate the technology for its own sake." According to Moore, such people are in many ways ideal customers. "They are the ones who will spend hours trying to get products to work that, in all conscience, never should have been shipped in the first place. They will forgive ghastly documentation, horrendously slow performance, ludicrous omissions in functionality, and bizarrely obtuse methods of invoking some needed function—all in the name of moving technology forward."

That was what Oracle's early customers wanted most of all: to move technology forward. As the CIA man told Kirk Bradley, they were buying an idea, not a product. When the tape du jour arrived from 3000 Sand Hill Road, customers knew they would have to spend long hours fiddling with it. "They know it's not going to work first time every time. There's going to be problems," said Katherine Daugherty, who worked in customer support for RSI. "That's why they have that saying. 'The pioneers are the ones with the arrows in their backs.' But the people buying the software know this. We know this. This is the deal. If you want to do it first and you want to do it fast, and if you want to get out there on the bleeding edge of technology, this is what you do."

Still, even innovators have standards. According to Moore, technology enthusiasts demand only a few things when testing a new product. First, he writes, they want "the truth, and without any tricks." Many people continued to deal with Larry Ellison because relational technology held so much promise. Innovators also want to get the new technology before anybody else does, and they want it at a low price, says Moore. No problem there. Everybody who bought Oracle in the early years was getting it first; for a while there was a new version almost every day. And for government agencies buying technology during Ronald Reagan's defense buildup, price was not an issue.

Finally, Moore writes, when innovators have a technical problem, "they want access to the most technically knowledgeable person to answer it." Customers certainly got that from RSI. For the first three years Ellison, Miner, and the other guys answered the phones themselves. Even after that it was easy to get a technical wizard on the line. But the advice these wizards gave did not always please the callers. Often, RSI tackled technical glitches by telling customers to reload—in other words, empty the database, put all the data back in, and start over. This was like telling someone to fix a leak in the roof by tearing down the house and building a new one. But the innovators were pleased because "they were talking to the guys who wrote the code," Kirk Bradley said. They did not mind rebuilding their houses if the architects were willing to help.

Oracle in the early days was by no means a worthless product; when people got it to do something, they were astounded. Lots of government agencies got it working well enough to perform some Cold War applications. "Oracle was tracking everything. Everything. If there was anything up above, on the water, or underneath, Oracle was tracking it," Katherine Daugherty said. Once an oil and gas company used Oracle to bid on the right to explore off Alaska. The areas that could be explored were divided into plots. Every time a competitor placed a bid on a plot, the company entered the information into the database. That way the company could keep track of what its competitors were doing as they were doing it. The information helped the firm get drilling rights at bargain prices. "It probably saved that company millions of dollars in doing bids," Daugherty said.

Still, most people in high technology were extremely skeptical—not only of Oracle but of relational technology in general. It had a lingering reputation for being horribly slow; there was nothing worse than sitting around waiting for a computer to do something. And even though corporate information managers weren't entirely satisfied with the database packages marketed by IBM and Cullinet, they were pretty sure those products wouldn't eat their data. If you managed information for a big company, losing data usually meant losing your job.

Ellison's job, then, was not just to market Oracle but also to sell relational technology. (Henry Ford had a similar problem at the beginning of the twentieth century. He could not get people to buy Fords until he persuaded them to buy cars.) Ellison met the challenge cleverly and with resounding success. In the early 1980s he regularly gave speeches about relational technology at computer trade shows. In the audience were corporate database administrators, systems managers, programming managers, and sometimes a few senior programmers—in short, Oracle's potential customers.

Ellison's standard speech was called "What's Wrong with Relational?" This appealed to the technical people, who all came to these shows thinking there was something wrong with it. Ellison would give a long and entertaining talk, full of witty asides, about the intractable problems of relational technology. Then he would explain how his company had solved every one of them. People who came expecting to hear a discussion of the demerits of relational technology got something more than that. "What it really was was a sales pitch," said George Schussel, the promoter who organized many of the shows.

As always, there were some differences between what Oracle actually did and what Ellison said it could do, but that was beside the point. The point was that a lot of people finally put aside some of their doubts. Ellison somehow knew that he could sell relational technology by talking about it in a *negative* way, which was an example of his uncanny sense of communication and marketing.

Ellison had a second, equally effective trade show shtick. While other speakers merely talked about what their products could do, he actually gave a demonstration. He set up a personal computer and projected the on-screen image onto a wall. Then he typed a simple relational query—for example, he would ask the computer to list all the employees in a certain department who earned more than twenty thousand dollars—and waited for the answer. People always oohed and aahed when the computer returned the right information. After all, this kind of information could take systems managers months or years to get. Of course executing a simple query was one thing, and managing the information in a huge corporate database was another thing entirely. Whether Oracle could do the latter was questionable at best. But again Ellison's audiences were impressed, and that was what counted.

People knew Ellison was giving them a sales pitch. But what

they didn't realize was he was also training them to become relational database users. From his demonstrations, people learned how to write a simple query using the SQL language. Ellison recognized that people would not buy his software until they were comfortable using it. So he made them feel comfortable. He said, in effect, "See? It's easy."

In our interviews Ellison made it clear that he thought of himself as a superb technologist. At one point he described himself as "a *really* good programmer," but that wasn't surprising; just about all the programmers I interviewed thought they were really good. Ellison also made a point of saying that he ran Oracle's technology division and that it was the job he enjoyed most. Certainly he understood technology; nobody disputed that. But Ellison's success was not principally about technological innovation. He didn't invent the idea for a relational database; it was borrowed from IBM. And though Oracle eventually released some excellent products, Ellison himself did not build them. He provided direction, and others built them.

Ellison succeeded not as a technologist but as a marketeer. He did not have any special convictions about technology; Silicon Valley was where he ended up. If he had wound up somewhere else, he might well have succeeded in peddling other kinds of ideas—political ideas, or literary ideas, or whatever. Even so, he didn't seem comfortable thinking about himself as a marketeer, and that was understandable. Technologists were solid, upstanding citizens; hell, they were scientists. Ellison had always wanted to be a scientist. Marketeers, on the other hand, were a little like con men, and who wanted to be a con man? Still, science did not have much to do with Ellison's success. People bought Oracle not so much for what it could do, as for what Larry Ellison said it could do.

"Oracle was in the right place at the right time, with a person, Larry, who put marketing first and everything else second," said Schussel, the trade show promoter. "Average technology and good marketing beat good technology and average marketing every day."

* * *

In 1980, after three years of running an exclusively male office, Larry Ellison decided to hire a woman. Actually he decided to hire a receptionist. There was a feeling among the guys that "a receptionist should be a woman," salesman Bob Preger said, so Ellison started interviewing women.

He found two promising candidates. One was Marcia (mar-SEE-a, not Mar-sha) Wells, then thirty years old. She was, in her own words, "prim and proper" but also bright and dependable, the sort of person who could impose order on an unruly workplace. Wells had been doing temporary work for Ellison's first wife, Adda Quinn, who recommended her highly. Ellison interviewed her over lunch at the Sundeck, a restaurant at 3000 Sand Hill Road. (In those days every prospect got lunch.) She found him enchanting: "Everything about the way he talked to me was charming and confidence-building."

The other candidate was Barbara Boothe, twenty-five, a Stanford graduate who had held a couple of responsible jobs but whose real ambition was to get married and have a family. Boothe, who grew up in Portland, Oregon, said her parents sent her to Stanford so she would get a good education and marry a young doctor- or lawyer-to-be. "I was brought up not to work," she said. She was married right after college, but the marriage did not last, and now she needed a job. Ellison "was real expansive, and he had a vision of where he wanted to go, even then," she said. She liked him and felt that he liked her—precisely the problem.

"[The other guys] didn't want to hire me because I was just what Larry liked: tall and blond," she said. They wanted a receptionist, not a companion for Ellison. Even Ellison had reservations about the striking young woman. When he and Boothe spoke, he told her that she might be too young and inexperienced to manage the office.

Finally Adda Quinn persuaded Ellison that Marcia Wells was the right person for the job. Yet he had concerns about Wells too. Was she too prim and proper? Would she be offended by the guys' rough language and raunchy jokes? (They weren't about to tone things down for her.) Wells soon got wind of Ellison's doubts. "I really wanted the job, so I wrote them a letter telling them why they should hire me," she said. "I don't remember what I said, but I ended it with some dumb thing about how I could fart and belch with the best of them. It was a total lie."

Ellison hired her. And though he did so on the sexist assumption that a receptionist should be a woman, Wells quickly learned that Ellison did not believe the opposite was true: He did not think a woman had to be the receptionist. (There was some question, as the years went by, whether Ellison believed women could be senior executives or directors of the company.) In the early days at RSI, Wells said, she thought she could do whatever she had the talent and energy to. She began by doing the things that the men in the office expected of a woman: answered the phones, scheduled Ellison's appointments, and brought cake to the office on special occasions. At the company Christmas party she insisted that everyone sing carols. "They were a bunch of bachelors who weren't used to doing anything cozy, and I made them do cozy things," she said.

Soon Wells assumed more important responsibilities. She arranged for the company's insurance, bought new office furniture, and—a vitally important job—negotiated contracts with customers. When she oversaw the contracts, she often sparred with salesmen who, in her view, gave away the store just to close deals. When she complained to Ellison about the problem, he said supportive things but never actually told the salesmen to cut it out. "It seemed like he was counting on me to say it but not telling anybody that he agreed with me," Wells said. When Ellison disappointed her, she did what other people did: poured her heart out to Bob Miner.

One of her unofficial titles was staff apologist. Ellison often erupted when someone did something he did not like or said something he considered stupid. He often used that word, *stupid.* (Steve Jobs and Bill Gates, the men who became respectively Ellison's best friend and his greatest rival, were also known for describing certain ideas that way.) "There was a lot of intimidation, a lot of uncomfortable intimidation. I didn't like that," Wells said. But she didn't believe that Ellison meant to hurt anyone. He only wanted "to create an atmosphere of the very highest expectations," a place "where the limits of what you can do are your own."

She believed that Ellison succeeded in creating such an atmosphere. "It was my best job in the whole world. The most perfect job I could have asked for," she said.

Yet because Wells was female, the job was far from perfect. Her experience at the company probably said as much about the corporate world at large as it did about RSI. "What I learned was that if you don't know how to manipulate men, you're not going to get ahead in the business world," she said. So she acquired what she considered the essential skill for a woman in the corporate world: Allow men to believe they have all the power even when they don't. As the person in charge of contracts Wells had the authority to send one back if she didn't like it. But simply pulling rank never worked, she said; it often led to bloody battles between her and the male salespeople. Eventually she learned what to do. When she rejected a contract, she tried to make the salesman think she was merely suggesting changes.

Wells would have appreciated it if Ellison had stood by her during those battles. But if he didn't do so, it wasn't because Wells was a woman. It was because Ellison never wanted to do anything that would get in the way of closing a deal.

Within a few months Wells took on so many other duties that Ellison had to hire another receptionist. This time he hired Barbara Boothe, who had called the company asking about openings every week since her interview. RSI's tenth employee, she started at twelve hundred dollars a month.

Boothe said she was "not even remotely" attracted to Ellison at first. He was ten years older, and besides, she was seeing someone else. "At that time I read him as just being a really nice guy," she said. Others could see that he was very much interested in her, though they questioned his reasons. "I just thought he was interested in what she represented, what she looked like. I didn't think that they were ever kindred spirits that would make a good match," Marcia Wells said. She spent much of her time "trying to keep Larry and Barb apart, for Barb's sake. Then, when I relaxed, thinking, *Oh, this isn't going to happen,* it happened."

It happened, oddly enough, when Ellison and Boothe were three thousand miles apart: He was promoting Oracle at a trade show in Boston, and she was back in California. One evening after the show Ellison signed on to his computer and saw that Boothe was also logged on. The RSI computer system allowed people to conduct on-line conversations, with each new message appearing above the previous one. Ellison's on-line name was PELLISON, and Boothe's was PBOOTHE; the *P* stood for "production." (The development people had a *D* in their screen names. Ed Oates's cryptic user name was DED.)

Ellison and Boothe began exchanging messages. They had talked on-line before, but now "the tone changed," Boothe said. Work talk changed to flirtation. Sitting a continent apart, typing instead of talking, Ellison and Boothe found that they could say things they would not have said face-to-face. They went on that way for an hour and a half. By the time Ellison got home, the two were involved—at least electronically.

About the same time Boothe broke up with her boyfriend, leaving her without a place to live. In June 1981, before they had even had their first date, Ellison invited her to move into his guesthouse in Woodside. "We kind of did everything backwards," Boothe said. "My parents weren't pleased." She did not stay in the guesthouse for long. Early in their relationship Ellison drove her to Orinda to see his house there. "I don't think I said more than three words the whole way up," she said. But Ellison talked "constantly. All the time. He was really sweet. He kind of picked up that I was real shy and uncomfortable and commented on it and said, 'It's OK. We'll get to know each other.' "

From the beginning Ellison set the agenda for the relationship, as he had with Adda Quinn. He was an irresistible force, deciding where the couple went, what they did, whom they saw. Boothe liked most of his decisions. He took her to nearly every movie that came out (they saw *Superman* on their first date), introduced her to Japanese culture, and bought her a takeout Chinese dinner every night from Su Hong in Menlo Park. (Fifteen years later the staff still remembered Boothe and always seated her immediately.) Being with Larry was like riding a roller coaster; all she had to do was strap herself in. All along she kept working for the company, part of the time as his assistant.

She learned quickly that Ellison's work came first in his life. He was building a company, and nothing could get in the way of that. Once Ellison scheduled a two-week trip to Japan. He planned to work for a week, then have Boothe meet him for a week's vacation. "I got over there, and he said, 'Sorry. Business.' " She ended up taking tours every day, meeting him only for dinner.

Ellison would make any sacrifice for the company, and Boothe was expected to do the same. When Canadian Kirk Bradley moved to California to work for RSI, Ellison let him live in the now-vacant guesthouse. Ellison and Bradley began spending long hours together writing documentation, sharing ideas, and running to McDonald's for Chicken McNuggets. "I felt like I had two boys," Boothe said. She liked Bradley but missed the intimacy she and Ellison had shared. She missed having Ellison all to herself.

One day Adda Quinn visited the RSI offices, as she often did. According to Boothe, "She said, 'Be careful. He won't stay faithful to you.' And here I am, twenty-five or twenty-six years old, thinking, *Yeah, yeah. You're ten years older than me. I'm young*. She tried to warn me, but I didn't listen."

Boothe believed the relationship was important to Ellison. "He wanted children with me," she said. "From the minute we started going out, he wanted kids with me. And I was ready to have kids."

A few months after Boothe moved in with Ellison, she became pregnant. The couple bought engagement rings and made plans to be married on New Year's Eve 1981. But three days before the wedding, and fourteen weeks into the pregnancy, Boothe had a miscarriage. "He was so kind. We went to Stanford [Hospital], and he was by my side the whole time, cracking all these jokes." The jokes were corny, but they took her mind off the pain. After they got home,

Ellison went out and picked up food from Boothe's favorite Japanese restaurant.

The couple did not go through with the wedding. According to Boothe, Ellison did not want to marry her until he was sure she could have a baby.

In April 1982 Boothe became pregnant again. Though the pregnancy went well this time, she always knew where Ellison's priorities lay. He cared for her and seemed genuinely excited about becoming a father, but she knew his true loyalty was to RSI. "He once turned to me—I was like five or six months pregnant—and said, 'If this [RSI] thing doesn't work out, don't expect me to stick around.' " She knew what he meant. If the company collapsed, he was not going to wallow in the failure. "He was going to leave and start over again someplace else. I mean, just get out of here, clear out. And clearing out meant leaving everything, including me."

Early in the morning of January 9, 1983, Boothe woke up Ellison and told him it was time. They went again to Stanford Hospital, where they filled their assigned roles: He tried to be supportive, and she screamed at him because of the pain. A mirror was positioned so that Boothe could watch the baby being born, but she was not interested. "Larry said, 'Barb, look! This is so cool!' I said, 'Get out of here! I'm busy!' " It was a boy. The parents named him David, after Ellison's brother-in-law, David Linn. Though it was only 5:00 A.M., Ellison called everybody he knew.

Boothe, who gave up her job to be a full-time mother, wanted more than ever to be married, but Ellison was unwilling to set a date. "I told Larry he had until David was a year old, and if he didn't marry me by then, I was leaving. I had to get on with my life. [I told him that] I had already talked to my mom and dad about coming home to Portland and trying to find a job and get myself on my feet and raise David. I couldn't go on living like that."

That October the three moved into a big stately colonial house with pillars in front, a home befitting a successful young entrepreneur and his family. (The place was just down the hill from a much larger house owned by Steve Jobs, the young multimillionaire founder of Apple Computer.) As he did whenever he moved into a new place, Ellison immediately started doing renovations. "I liked this piece of property because it could have horses on it," Boothe said. "The first thing he did was put the tennis court up right where I would have put horses. He said, 'No, we're not having horses.' " Ellison said horses would attract flies, and he did not want flies.

With the year running out and Boothe's deadline drawing near, Ellison finally agreed to get married. The wedding was scheduled for 2:00 P.M. on December 4, 1983, at the new house. It was to be a small affair, just family and a few friends. Simple, elegant, memorable.

A few hours before the wedding, according to Boothe, Ellison handed her a typewritten prenuptial agreement and told her she had to sign it. He had mentioned a prenuptial agreement several weeks before but hadn't brought it up since. Now here was an eleven-page document, sprinkled with *whatsoevers*, *wheresoevers*, and *with respect tos*. Boothe had to sign it now or leave her child without a father.

Why did Ellison spring the prenuptial agreement on his bride-to-be? He declined to discuss it. "I don't want to get into disputes between me and Barbara which will affect the kids. I think it's completely inappropriate for me to attack the mother of my children in any way. If that means I'm just not going to answer or defend this at all, then I'll have to live with that," he said.[5]

The prenuptial agreement listed all the assets Ellison was bringing to the marriage. According to the document, he had $1 million in real property, $600,000 in cash, $250,000 in miscellaneous things, and $350,000 in stock. But his real money was in his company, the document said. According to the prenuptial agreement, Ellison owned 2.2 million shares of Oracle stock. He estimated that it would fetch $11 million to $16.5 million in a private sale and would be worth $33 million to $44 million, or $15 to $20 a share, if and when the company went public. Ellison was probably overestimating the value of the stock at that point. In fiscal 1983 the company had rung up only $5 million in sales, arguably not enough to justify a price of $15 a share. A few years later Barbara Boothe Ellison's lawyer made precisely that point, arguing strenuously that Ellison's stock gained most of its value during the marriage, not before.

To Ellison, marriage was at least partly a business proposition, one that held the possibility of serious losses. In a paragraph titled "Larry's Efforts Prior to Marriage," the document said, "During the years prior to his marriage to BARBARA, LARRY's estate has grown significantly as a result of hard work and diligent attention to his business and investments, particularly in connection with his stock in Oracle Corporation. LARRY and BARBARA therefore agree that any increase in his estate that may occur during his marriage will not be a result of his efforts during the marriage relationship, but rather the result of his own hard work [before the marriage]."

Having established that Ellison's assets belonged to him and only to him, the document specified what his financial responsibilities would be. "LARRY will be solely responsible for support of the community household and will pay all living expenses,' " it said. The money he made from investments or from the sale of stock would remain exclusively his; he would share only his salary and bonuses with his wife. The document listed Ellison's salary at a modest $130,000. If he died while married to Boothe, 10 percent of his estate would be held in trust for her. The document said Ellison would use twenty-five thousand shares of Oracle stock to establish a trust for David.

What if the couple got divorced? In that "unlikely and unhappy event," Boothe would receive $100,000 for each year the couple was married, but in no case would she receive more than $1 million. Ellison would also buy Boothe a $250,000 house in her hometown of Portland and give her $50,000 to furnish it. According to the document, "Reasonable support payments, if any, shall be agreed to and determined by the parties or by court order" at the time of the divorce, "and shall be based upon BARBARA's needs rather than upon LARRY's ability to pay." Perhaps Ellison could see that his ability to pay would soon have no limit.

According to the document, its purpose was to "promote marital harmony and understanding." It did not have that effect when Barbara Boothe read it. She felt cornered, and that, she said, was how Ellison wanted her to feel. "At the time it seemed pretty calculated. He put me in a position where he knew I wanted to get married, and there was no way I was going to back out. Nowadays if that happened to me, I would say 'screw you' and 'good-bye.' I would walk out. Back then I still wanted to be a good girl and go by the rules."

Boothe needed a lawyer. At 11:00 A.M. she got one: her father, who had arrived early to help with the preparations. He read the agreement, then suggested a few changes. There was no time to retype the document, so he scribbled in the margins. The Boothe family gave the document back to Ellison, who was being represented by his brother-in-law, David Linn, the Chicago lawyer and judge. Boothe was horrified; what was supposed to be the happiest day of her life was deteriorating into a cold business deal, with wedding guests as negotiators. The whole scene was like something out of . . . what? Kafka? A Woody Allen film? Ionesco? *Dallas*? Meantime florists and caterers, unaware that the whole affair had a good chance of being canceled, began to arrive with bouquets and trays of food. Then the guests started showing up. What would Boothe say if the wedding fell through? *Sorry, the negotiations have reached an impasse, you'll have to go home.* No, that could not be allowed to happen. The humiliation would be unbearable. Something had to be worked out.

The paper shuffling continued. The Boothes won a few concessions. Ellison agreed to establish trust funds for any future children as well as for David. He also crossed out a paragraph that said he would not support Boothe after a divorce if she was capable of supporting herself. Boothe added a sentence saying she would be under no "obligation to be employed while raising children." Having ironed out the details, Ellison and Boothe initialed each page of the prenuptial agreement and fixed their signatures to page 9.

Then they stood before their family and friends and pledged their eternal troth.

Barbara Boothe Ellison arguably was never the most important woman in Larry Ellison's life, nor was she even the one who knew him best. Those distinctions belonged to a woman with whom he had a more complex relationship: his executive assistant, Jenny Overstreet. Ellison's lasting success came in his professional life, not his personal life, so it was perhaps not surprising that the woman who meant the most to him, and did the most for him, could be reached at the office, not at home.

Overstreet burst out of Mills College in Oakland in 1982. She wanted to go someplace, and fast. She set out to find a job where "people were as excited about working hard as I was." That was about as specific as she could be about what she wanted to do. Like a lot of liberal arts graduates, Overstreet had a lot of ability in general and no skills in particular. One day she saw a newspaper ad for a job as an assistant to one of the senior executives (not Ellison) at Relational Software Inc. Soon after, she went to 3000 Sand Hill Road for an interview. "As I sat there waiting for Larry for a very long time, I knew without ever even talking to anybody that I really wanted to work there because people were running from place to place. And they looked like whatever they were doing, they were having a good time doing it," she said.

Finally Ellison invited her into his office. She was struck by how fast he talked (a thousand words a second, or so it seemed) and how many things he talked about: Thomas Jefferson, the Louisiana Purchase, how nervous Overstreet seemed, relational technology (she had no idea what it was), and on and on. The discussion lasted two hours, but Overstreet knew in minutes that she wanted to work at RSI. She wasn't going to, at least not right away. Ellison called her later and told her he couldn't hire her yet because the executive who needed an assistant didn't like her.

A couple of months went by. During that time RSI changed its name to Oracle Systems Corporation, which later became just plain Oracle Corporation. The reasons for the change were simple. Naming companies after their main products was in vogue in high technology, and scrapping the RSI initials helped people distinguish between Ellison's company and Relational Technology Inc., the maker of Ingres.

In February 1983 Overstreet finally got a job at Oracle. (She was employee number thirty-five.) She accepted the position without knowing the salary, her title, or her responsibilities. Her first day on the job was Presidents' Day, and her boss, Umang Gupta, wasn't around. Ed Oates—who was back at the company, this time as an employee—let her in the door, handed her an Oracle manual, and said, "Here. Read this." She did. Then she edited it. It was the first work she did at Oracle.

A year later Ellison made Overstreet his executive assistant. She was everything he wasn't: punctual, detail-oriented, thorough, discreet in her communications. Ellison never had an organized day in his life (in that way he was a typical entrepreneur), but Overstreet managed to make him seem organized. Well, almost.

In time people in high technology came to identify Oracle almost as closely with Overstreet as they did with Ellison. In 1993 a former Oracle executive named George Koch published a technical manual called *Oracle: The Complete Reference*. In the acknowledgments he wrote that Overstreet "quietly runs Oracle and most of the western world. She solves problems before anybody else has even realized there are problems." Koch wasn't exaggerating by much about Overstreet's influence. At Oracle Overstreet was a power unto herself (but as one person pointed out, "It never went to her head").[6] Part of her job was to take calls from Oracle employees who wanted to speak with Ellison. If she liked the people and what they had to say, she granted an audience. If not, not. There were plenty of times when she kept people away from Ellison without first asking if he wanted to see them; he trusted her to do that. "She's a very good guard," Stuart Feigin said.

Best of all for Ellison, Overstreet admired him. Ellison said and did a lot of things that could be interpreted in more than one way. When he released product brochures saying Oracle had certain features that it didn't yet have, some people thought he was lying, while others thought he was just being optimistic or just getting his tenses mixed up. ("Most of the things Larry says are innocuous enough that it doesn't matter if it's a lie or not," Oracle engineer Kirk Bradley said.) When Ellison withheld stock options from people about to leave the company, as he did repeatedly, some people thought he was a prudent businessman, while others thought he was petty and vindictive. Overstreet usually gave Ellison the benefit of the doubt—and more. It was not that he could do no wrong in her eyes—he often ticked her off—but her loyalty to him never wavered.

She often got the best Ellison had to give. "He is the most *interested* person that I have ever known," she said. "He is as mad about the poems of Poe and A. E. Housman as he is talking about what's going to happen with this new chip, as he is talking about the doll collection at Windsor Castle, as he is talking about how much graphite is in the new mast for the world's fastest sailboat. He's interested in *everything*." From time to time Ellison called Overstreet late at night to read a stanza from his favorite Poe poem, "Annabel Lee," in which he was constantly discovering nuances. Ellison couldn't contain his enthusiasm, and Overstreet didn't want him to.

She enjoyed the childlike manner that Ellison sometimes had, his sense of wonder. One time, after Oracle had moved into larger offices at 2710 Sand Hill Road, the company's entire California staff gathered in a conference room to celebrate someone's birthday. When Ellison walked into the room and saw fifty people there, he was wide-eyed; for the first time, he realized his little company was becoming big, his baby was growing up. "He was really shocked," Overstreet said. "He said, 'You know, Bob and I just never figured we would ever need any more than fifty people to do everything we ever wanted to do.' "

Over the years Overstreet did just about everything for Ellison. When he needed place settings for a dinner party at his home, she chose them. When he insisted on finding the best interior decorator or boat designer or auto mechanic or whatever, she found them. When he got seriously hurt in separate accidents in the 1990s (once body surfing and once bicycling), she took care of him.

One thing Overstreet could not do was get Ellison to show up for things on time or, sometimes, at all. If, on his way to a meeting, he found something better to do (and someone better to do it with), he might disappear for hours or even days. Overstreet said Ellison had a good reason for missing the board meeting at Don Lucas's ranch: His doctor saw something unusual on his EKG and told him not to go. But she acknowledged that he sometimes no-showed "just because he was being a jerk." Once I asked her in an E-mail message what it was like to work for such a quirky person. She wrote back, "The cardinal rule of being an assistant is that your boss gets to be as quirky as he or she damn well pleases." Jenny Overstreet was not a fool.

Ellison made no apologies for his quirks. If anything, he seemed amused at Overstreet's concerns about time management. "Jenny and I approach things very differently," he said. "Jenny feels that she has to be exactly on time all the time. Jenny feels if there are a hundred things you have to know, you have to know all hundred of them. If there are a thousand things you're doing, you have to do all thousand of them. My view is different. My view is that there are only a handful of things that are really important, and you devote all your time to those and forget everything else. If you try to do all thousand things, answer all thousand phone calls, you will dilute your efforts in those areas that are really essential."

Almost any relationship between a boss and an employee functions mostly on the boss's terms. The relationship between Ellison and Overstreet was that way, only more so. She was at his beck and call. "He will call her at times when he doesn't call anybody else," said Kirk Bradley, one of the early employees of Oracle and Overstreet's longtime companion. "At two in the morning he'll call. Not two in the morning his time, of course. It'll be eight o'clock wherever he is. He won't think twice about what time it is for *her,* because he needs to talk." Bradley said Overstreet didn't mind taking the calls because "she believes that's her job."

Overstreet once described herself and Ellison as "the world's oldest married couple." What she meant was that they knew each other as well as any married couple did and sometimes attacked each other as fiercely. "They're like two pussycats, you know?" Bradley said. "They can sleep side by side, and you can see that they're nice and cute together. They can be friendly, and then they can hit each other in the eye."

Being "married" to Ellison meant that Overstreet had to take the good with the bad. Yes, she felt privileged to share her boss's joy in Poe and to converse with him about ballet and Japanese art. But if she made a mistake—if he wanted something done and she didn't do it—"he'd call and rail," Bradley said.

"He can be unbelievably cruel," Bradley went on. Overstreet was one person with whom Ellison was usually emotionally honest, not always a nice thing for her. "If you only have one outlet for anything, you know, it all gets directed [at that person] . . . Probably because he wants to be nice to everybody, he kind of builds up this cloud, and then he rains on her." Bradley summed up by saying, "There's happy Larry, there's crushing Larry, and there's Jenny, who knows Larry better than he does."

Overstreet never talked to me about the downside of being the most important woman in Larry Ellison's life. She preferred to talk about Ellison's charms, which were considerable and which, to her way of thinking, more than made up for the pain.

Seven

LARRY ELLISON NEEDED A NEW SALES MANAGER. BOB PREGER, THE company's first salesman, was not going to be able to carry out Ellison's plans, and both men knew it. Preger had done a good job; in fiscal 1982 he and his small staff had doubled the company's sales, to $2.4 million. But he wasn't interested in managing a big sales force, and besides, he did not have the sort of personal drive Ellison was looking for. Ellison wanted Oracle to be larger than Cullinet, a $100-million-a-year company that was then the premier software company in the world.

"To my way of thinking, for us to get bigger than Cullinet, there were a lot of mountains we had to climb," Preger said. "But Larry was thinking of that from day one. So maybe that's part of his genius."

Ellison soon found someone who could help him climb those mountains: P. Michael Seashols, who had several years' experience in sales and marketing at software companies. From the beginning Ellison knew Mike Seashols was his man. Seashols was young, trim, good-looking, aggressive, and full of himself—an alpha male, just like Ellison. Their first conversation quickly turned into blood sport, a competition to see who was brighter, quicker, wittier, more knowledgeable. There should have been a referee.

"We met for about an hour and a half," Seashols said. They talked about technology, sports, family, and especially religion. Seashols was a member of the aggressively evangelical Covenant Church; Ellison didn't hold with any religion and didn't mind explaining why. "We were sort of a couple of roosters walking around a table, trying to figure out who was better than the other guy," Seashols said. "In those days I was driving a Porsche and he was driving a Mazda. So he really didn't have a claim to fame."

Ellison hired him. Almost immediately Seashols got a lesson in Ellison's unpredictable ways. Ellison asked the new man to come to the Oracle Christmas party in December 1982 so Ellison could introduce him to Preger, who would be working for Seashols, and to Kathryn Gould, who would be Seashols's vice-president of marketing. Seashols went to the party late because he wanted to be sure that Ellison was there when he arrived. The strategy did not work: When Seashols walked in, he didn't see Ellison anywhere. It was a small party; Seashols soon found himself in a conversation with Preger and Gould.

According to Seashols, Gould said, "Who are you?"

"Vice-president of sales and marketing," Seashols said.

"Oh, really? Well, what does that mean?" Gould said. It was clear that Ellison had not told her about her new boss.

"Well, I run sales and marketing," Seashols said.

"Do I report to you, or do you report to me?" Gould said.

"I have no idea," Seashols said. He was fibbing, of course; he knew but did not want to be the one to tell her.

The whole episode felt strange to Seashols. But the more time he spent with Ellison, the more he understood that overlooking such details was just Ellison's way. His vision was that Seashols would run sales and marketing. Communicating that vision to people who were affected by it was, to Ellison, just a niggling little detail, and he did not concern himself with such things. Besides, telling Gould that she had a new boss might not please her, and Ellison very much wanted to please people. He still had the need to be the boss and the conflicting wish to be loved.

"Larry didn't want to confront them," Seashols said. "He left me to do that."

After Seashols made it clear who reported to whom, he began building a sales organization. In the early days Ellison "had just kind of gone on charisma and positioning," Seashols said; he did not know the basics of sales management.

"Mike really brought a kind of big-company discipline to Oracle," said Tom Siebel, a former Oracle salesman who once worked under Seashols. One of the first things Seashols did was create a written compensation plan, in which salespeople received small base salaries, with large bonuses for reaching or exceeding their sales quotas. For Oracle, having such a plan was "like some Stone Age guy discovering fire," Siebel said. "It doesn't seem like a big deal if you have a gas range at home. But believe me, it was a big deal."

Seashols also instituted the practice of selling high—that is, pitching Oracle to people at the highest levels of an organization. Oracle software may have been intended for use in small corporate departments, but in most cases it was not the department heads who had the authority to buy it. If Oracle was going to grow as quickly as Ellison wanted it to, it had to get the attention of top corporate decision makers: chief information officers, chief financial officers, and chief executive officers. Seashols was the first Oracle salesman to enter the mahogany-filled rooms of the executive suites. Ellison almost couldn't believe the size of the deals Seashols tried for. Once, after Seashols briefed him on a deal he was working on, an excited Ellison called Jenny Overstreet and said, "Mike asked for seven hundred thousand dollars!"

Seashols eventually asked for much more than that. According to Siebel, "Mike could walk into a large multibillion-dollar company, sit in the boardroom, and lay out the context of a long-term strategic business partnership. Mike was pretty amazing at that." Among the companies Seashols sold successfully was IBM, which installed Oracle database software in its (ultimately unsuccessful) RT PC project. In high technology most companies were partners as well as competitors.

Seashols often called Ellison in to help him close those big deals. CEOs like to talk to other CEOs, after all. But Seashols quickly learned that Ellison was not the right guy for every sales situation. "If we were talking to a marketing guy, he'd be a positive. If we were talking to a businessman, he'd be a negative, because he was flippant. He'd say anything," Seashols said. "Finance guys don't like that. Marketing guys might think that's cute. Technical guys think that's a scam." Seashols devised a strategy: Allow Ellison to meet with the top-level people, then "keep him in a bottle."

Ellison and Seashols got to know each other well. They had breakfast almost every morning at a Denny's restaurant near Oracle and worked together all day. On weekends they had dinner, went to the movies, visited San Francisco, rode bikes, and played tennis. Sometimes the two men and their wives went on double dates. "We were very close," Seashols said, so close that Ellison shared some of his feelings about his childhood and about his father, something he rarely did with others. Seashols came to believe that Ellison was still trying to impress Louis Ellison, long since dead. "He wasn't loved or comforted or mentored," Seashols said. "And that motivated him to be successful, you know, to prove something."

Seashols continued. "I found Larry to be a real intriguing, interesting, bright, sensitive person. He made a comment to me that really had an impact on me early on. He said, 'Mike, we're all going to become millionaires. I know that, and you know that. The only thing is, I don't want to become a millionaire by myself. It's too lonely.' That gives you a lot of insight. It's kind of a sweet moment that tells you where he was. It was a very serious comment. There is a sensitivity to Larry that is very genuine."

Ellison was convinced they all were going to become millionaires because he had developed a marketing strategy that could not miss. When Seashols heard about the strategy, he passed it along to his salespeople and told them not to deviate from it when talking to prospective customers. "Everybody said the exact same thing, so you had a very tight marketing message," Seashols said. Ellison's strategy was to sell Oracle software on the strength of three characteristics: portability, connectability, and compatibility. These qualities became known at the company as the abilities. The sales pitch was ingenious, effective, and, like a lot of the things Ellison said, mostly accurate.

Portability—the idea that Oracle software worked on many different kinds of machines—was not new. Ellison had been talking about the "promiscuity" of Oracle ever since Version 3. About the time Oracle began selling Version 4, Ellison claimed that the software functioned on a dozen different operating systems.[1] Some people doubted that Oracle was as portable as Ellison said it was. One analyst said he didn't think it was possible for Oracle to work really well on so many different machines,[2] and the experience of some Oracle customers—banks, government agencies, the military, big corporations—said he was right. A New Jersey company found that some of its Oracle applications worked beautifully on Hewlett-Packard hardware, but poorly on certain DEC machines.[3] Even so, Oracle's software was far more portable than any of its competitors' database programs. Stuart Feigin liked to say that portable software is like a dancing pig: The amazing thing is not how well the pig dances but the fact that it dances at all.

Oracle was connectable too—more or less. Connectability meant that users could link Oracle databases running on a couple of different machines. This was useful when a computer operator wanted to get information from his or her own computer and from another database somewhere in the network. But Oracle was not nearly as connectable as it was portable. With Version 4, for example, Oracle could make connections only between certain IBM microcomputers and the VAX. If your company happened to use IBM microcomputers and, say, Data General minicomputers, Oracle wasn't connectable, at least at first.

When Ellison talked about Oracle's compatibility—the last of the three abilities—he meant that it was compatible with IBM's relational database software. Oracle made this claim in an oblique way as early as November 1984, in an industry trade magazine. The company placed a tiny front-page advertisement with this headline:

IBM SQL/DS & DB2
DBMS NOW ON PC

It was an eye-popping headline. A couple of years earlier IBM had released its first relational product, SQL/DS, which ran on a few small mainframes. It was about to begin selling DB2, its long-awaited relational system for big iron. The idea that these programs were available for use on a personal computer was tremendously exciting to anyone who made a living managing information.

According to the ad copy, "The ORACLE relational DBMS [database management system] is 100% compatible with IBM's SQL/DS and DB2. SQL/DS and DB2 run only on mainframes. ORACLE runs on IBM mainframes, DEC, DG, HP and most other minis. And all of ORACLE—not a subset—runs on the IBM PC XT and AT." The ad gave Oracle's address and phone number.

The headline was absurd: SQL/DS and DB2 were not "on the PC." Oracle advertising man Rick Bennett, who wrote the ad with Ellison, believed it was justifiable because Oracle's version of the SQL user language was based on an IBM specification. Ellison and his partners had copied what IBM had done, so why not imply that Oracle, SQL/DS, and DB2 were the same? At least one part of the ad was true: An IBM-*like* product was indeed available on the personal computer. Oracle was offering a PC version of its software. The early PC version was, in Bennett's estimation, "a piece of shit," too small and slow to be of any real use. But then the purpose of the ad was not to sell the PC version of Oracle but to bring attention to the company, and claiming "100%" compatibility with IBM software was an effective way to do it. The ad "had the phones ringing off the hook," Bennett said. "It was real simple, and I knew we could get away with it once."

Oracle talked about compatibility more than once; in fact it did so continually. That ad, or a similar one, ran many times after that. And before Oracle went public in 1986, it was required to file a prospectus with the Securities and Exchange Commission. The claim that Oracle was compatible with IBM's products appeared several times in the document, including a reference in the summary on page 3: "ORACLE has the same user interface, the SQL command language, as IBM's latest generation of relational database management system products for its mainframe computers, SQL/DS and DB2." The company used similar language later that year, in its 10-K annual report filed with the SEC.

But Oracle's real compatibility with SQL/DS and DB2 was, as

always, a little fuzzy. Oracle's Kirk Bradley thought the claim was misleading because the user interfaces in SQL/DS and DB2 weren't even fully compatible with each other, much less with Oracle. Once, at a company meeting, an Oracle engineer asked Ellison why he suggested that they were. According to Bradley, Ellison "exploded. He said, 'All companies do this. It's standard stuff. You don't know anything about business.' " Eventually, Oracle backed off the compatibility claim, at least a little bit. Beginning in 1987, its Securities and Exchange Commission filings no longer claimed that Oracle had "the same user interface" as the IBM software. Instead Oracle now said only that it was the first product to use SQL, "later also commercially marketed by IBM."

To Ellison, the whole discussion about compatibility was just semantics. "What does compatibility mean?" he said. "Does it mean queries will run? Updates will run? Programs will run? Most programs will run? Every last program has to run? Since it's open to interpretation, I took a liberal interpretation."

He had a point. In the end customers didn't care whether Oracle was really portable, connectable, and compatible, and fortunately for Ellison, they weren't going to analyze every last bit of code to see if it was. Ellison met his customers' needs; that was what mattered. "Whatever they needed was being satisfied. And what they needed was not necessarily the truth," Chicago consultant Richard Finkelstein, who specialized in relational technology, said.

One of the things customers wanted was software that conformed more or less to IBM's standards. That was vitally important. Early on corporate information managers chose Oracle largely because it was (or claimed to be) like SQL/DS and DB2. There was a saying in the computer industry: "Nobody ever got fired for buying IBM." If Oracle's stuff was anything like IBM's, people would remain employed. Much later, when Oracle became the market leader, people bought the software partly because other people were buying it. Either way Ellison met their needs by providing something safe, something they felt comfortable with. For people making complex purchase decisions, "what a relief that is," Finkelstein said. "Now you don't have to worry anymore."

People also wanted software that worked reasonably well. This was of course a problem at first. In Seashols's own estimation, Oracle's early products were "very unstable." His salespeople did not emphasize that in their sales pitches; instability was not one of the abilities. "Sure," the salespeople said, "there are bugs, but we're fixing them." According to Seashols, Oracle told its customers, "By the time you're ready to deploy, [the software] will work. It's going to take you a year to build your applications, so don't worry about a few bugs today."

Seashols had to think that way. The only way to get people to accept relational technology was to talk about what it could do for them. "What Mike was selling was the idea, the future," said Bob Ney, who worked closely with Seashols and Ellison. "SQL and relational hadn't happened yet. I'd call him an evangelist. That's what I would call him. He knew everybody needed relational, but they didn't know they needed it yet."

Ellison was evangelizing too; when he talked about the abilities, he was promising businesspeople a sort of technological heaven. But like most evangelists, he did not dwell on the unpleasant fact that people would have to die to get there. Ellison, the classic entrepreneur, was trying to change the world or at least a small piece of it. He couldn't be responsible for every little bug—or even every big one. So he told people what they wanted to hear and what he believed would be true—someday.

"If there are this many billions of bugs in today's product, those can be fixed. That's uninteresting [to him]," Jenny Overstreet said. "What's interesting is all the features we're going to be able to cram into the next release. [He thinks,] 'I know the guys are going to be able to solve these bugs. Don't bother me with that.' "

Oracle's customers did sometimes get less than they expected. All that talk about the abilities pretty much ensured it. Even so, most of them got something useful, something that made a difference in their careers. It was wrong to say, as some of Ellison's adversaries did, that he merely exaggerated, merely made promises. If Ellison had sold only vaporware, he never would have made the Forbes 400.

Still, he was aware of his reputation. Once a reporter came to the Oracle offices to interview Ellison and Jeff Walker, then a top executive. According to Walker, the journalist said, "Larry, we hear a lot of comments from your competitors that Oracle is selling shelfware and Oracle is selling vaporware."

Ellison leaned forward and spoke softly, as if telling a secret. "You know, it's true," he said. "We don't actually have any software. This year we sold a billion dollars' worth of stuff and we never delivered anything. It's a really great business."

Walker turned to the reporter and said, just to be safe, "He's kidding."

Oracle's chief competitor during most of the 1980s was Relational Technology Inc., which most people referred to by the name of its product, Ingres. In the first half of the decade Ingres was the hotter of the two companies. In 1984 it roughly tripled its sales, to $9 million, while Oracle's sales doubled, to $12.7 million. Ingres's revenues doubled in 1985; Oracle's fell just short of doubling. Oracle consistently had higher sales numbers, but Ingres boasted faster growth. If the trend continued, Ingres would soon surpass its archrival.

That never happened. Again IBM was a large part of the reason.

In the early days of relational technology there was great debate in the marketplace over which user language would prevail. Would it be SQL, developed by IBM, or QUEL, invented up the road in Berkeley? Oracle was of course rooting for SQL, because it had based its system on that language. And SQL got a boost when IBM released SQL/DS in 1982. But because that product ran on only a few machines, the market was not sure SQL would be widely accepted by the industry. QUEL still had a chance.

About the same time the committee that set standards for database technology in the United States began looking into relational technology. The committee, made up of bureaucrats and computer industry people, wanted to establish a standard specification for relational databases, a standard that any hardware or software company could use when developing new software products.

IBM had to make a decision. Should it offer SQL as the standard, thus giving its competitors the benefit of its innovation? Or should it keep SQL to itself and take the risk that another language—most likely QUEL—would be adopted as the standard? It turned out to be an easy decision. Because IBM had published an early specification for SQL in the System R papers several years earlier (that was why Oracle existed), there was no point in withholding the specification now. "The cat was already out of the bag," said Phil Shaw, IBM's representative on the database standards committee. With the approval of IBM's top management—the decision was reviewed in IBM offices around the world—Shaw presented the SQL specification for use as a standard. Then IBM waited to see whether Ingres would present QUEL. Shaw, for one, hoped it would not. The Ingres group had spent years refining QUEL, and by 1982 Shaw was not eager to put SQL up against it. "The fact is that QUEL is a somewhat better language than SQL," the IBM man said.

But Ingres did not show up at the committee meetings because founder Mike Stonebraker detested the idea of having technology standards. Stonebraker was vocal about it. He thought they inhibited innovation and artificially restricted what got to the marketplace. Maybe so, but his hard-line position probably did not help his company. Don Deutsch, who served as chairman of the database committee, summed things up this way: "I tell you, QUEL was a much nicer language than SQL. No rational person would have chosen SQL instead of QUEL. . . . Ingres was stupid."

So it was that the database committee began evaluating SQL. The committee spent two years "filling in the holes and leveling out the bumps" in the language, Deutsch said. After all the fixes were made, the committee gave SQL a new plain brown wrapper of a name—Relational Database Language—and prepared to anoint it as the industry standard.

Then something funny happened. People on the committee—people from such hardware companies as Digital Equipment Corporation and Data General and Unisys—began telling Deutsch that the new, improved SQL was not what the industry needed after all. What the industry needed, they said, was a version of SQL that was exactly like the one IBM was going to use in its soon-to-be-announced relational database product, DB2. If ever there was a clear expression of IBM's power in the computer industry, this was it. What the hardware and software people were saying was that they wanted to do what IBM was doing—even if what IBM was doing was not as good as what they had just finished doing themselves.

So the database committee threw away two years' worth of work and started working on a version of SQL that was just like IBM's. "We pushed the reset button, and we literally put everything back to the way it was when we started," Deutsch said.

Where was Oracle during this time? The company had a representative on the standards committee, but "I don't recall Oracle being a conspicuous participant," IBM's Shaw said. The committee's most conspicuous participant—its one-thousand-pound gorilla—was Shaw. That was fine with Oracle, which was riding on IBM's coattails anyway. According to committee chairman Deutsch, it was "absolute serendipity for Oracle" that IBM decided to push SQL as a standard. Had Ingres done the same and had it prevailed, he said, "Oracle would have been a big loser."

Such is fate. In 1985 IBM announced the release of DB2, a relational database management system based on the SQL user language. DB2, the product of years of testing and development, was intended for the kind of big mainframe computers used by some of IBM's most important customers. Its release made SQL the de facto standard in the industry. It was clear now that IBM was sprinkling holy water on the SQL user language.

Ellison saw this as a marketing opportunity, and as usual he made the most of it, using every opportunity to connect Oracle and IBM in people's minds. "The Company believes that IBM's introduction of DB2 indicates a continuing trend toward relational database management systems and toward SQL as the standard language. The Company further believes that compatibility with IBM's SQL gives

ORACLE a competitive advantage over DBMS products that do not support SQL," Oracle said in a company document.

Oracle was right. After 1985 fewer and fewer customers were willing to buy a relational database product that was not based on SQL. IBM's choice of SQL for its database products, and its push to make SQL the industry standard, created what Mike Seashols called a herding effect. Everybody in the industry wanted to do what IBM was doing. Oracle engineer Roger Bamford, a former IBM man, summed it all up: "If IBM said it was real, it was real."

The truth of this was apparent in Ingres's revenue figures for the next few years. Ingres, which had at least doubled its revenues every year until 1985, saw sales increase 63 percent in 1986, 63 percent in 1987, 89 percent in 1988, and 49 percent in 1990. Nothing wrong with those numbers—except that Oracle's sales jumped 139 percent, 137 percent, 115 percent, and 102 percent in the same period. After the introduction of DB2, Oracle steadily stole market share from Ingres, and Ingres never got it back. Ingres eventually released a SQL-based product, but by that time "the game was half lost," Stonebraker said.

In October 1986 the database committee made SQL the official standard for relational query languages. Larry Ellison had been right all along: SQL was indeed the future.

Years later Mike Stonebraker could not believe Larry Ellison's luck. It was, he said, a "happy accident" that Ellison, Miner, and Oates had stumbled upon SQL instead of QUEL. After all, the specification for QUEL was published about the same time the System R papers were. Oracle's early adoption of SQL "had nothing to do with Larry Ellison being smart. It was a lucky accident," Stonebraker said. IBM's push to make SQL a standard was another bit of good fortune, he said. Maybe so, but you don't get to be a billionaire without being at least a little bit lucky.

IBM's endorsement of SQL was crucial to Oracle's success, but it did not end the database wars. Other companies, such as Sybase and

Informix, were starting to enter the fray, and Oracle still had to fight to keep taking market share from Ingres. Like politicians trying to wear down their opponents with personal attacks, Ellison and Seashols discovered Ingres's weaknesses and exploited them.

One was that Ingres was slow to develop a version of its product for the personal computer. If there was one thing that Larry Ellison and Bill Gates had in common, it was that they both understood how the PC was changing the workplace. Early on Oracle introduced an inexpensive version of its database software for the IBM PC, the machine with Microsoft's MS-DOS as its operating system. While Oracle liked to say that all versions of its software were "functionally identical," the company also pointed out that "size of database and complexity of applications may be limited by the capacity of the computer on which ORACLE is running."[4] That was an understatement. Trying to run Oracle on the IBM PC of yesteryear was like trying to dance *Swan Lake* on a tabletop. The PC just wasn't big enough for the job. The early PC version of Oracle "never really did run. It never did," Mike Seashols said.

Yet it served its purpose. For one thing, it gave Ellison and Seashols a way to differentiate Oracle from Ingres: Oracle ran on the PC, and Ingres did not. So what if Oracle's PC version worked poorly? The PC version was mostly a marketing tool at first.

And what a tool it was. The PC version of Oracle allowed the average corporate employee to become familiar with the SQL language and with relational database technology. People tinkered with it at work, and some even loaded it onto their PCs at home. They learned how to put information in and how to get it out. After a while these people began to consider themselves Oracle experts. That made them feel good about themselves—and about Oracle. Marketing a product this way was "enormously visionary," said consultant Richard Finkelstein. The PC version couldn't do much, Finkelstein said, "but that was not the point. The point was that Ellison put the product in front of everybody at almost no cost." As a result, the folks around the water cooler knew all about Oracle, but a lot of them had never heard of Oracle's competitors.

Of course Oracle had an Achilles' heel of its own. User languages aside, Ingres was generally perceived as the faster and more reliable product. Ingres, Seashols said, was "much better. It was faster, and it had better features." But Relational Technology Inc. lacked any real sense of marketing. "They would try to throw it around and say, 'Let's run a benchmark [a measurement of the software's speed]. Or let's do a feature comparison.' I'd say, 'Wait a minute. If they can't run on a PC, who cares about speed?' Whoever made the sandbox won. And it was our sandbox."

Many customers chose Ingres anyway. Seashols had a strategy for dealing with that too: He forbade his people to lose a deal on price. If a customer was about to buy Ingres because it was cheaper than Oracle, Seashols's salespeople were under orders to lower the price until they made the sale. "Whatever it is, go below them," Seashols told them. He did not care if Oracle did not make much money on some of the deals. Like Ellison, Seashols knew the importance of winning market share, of being seen as the industry leader. (Oracle continued to do this after Seashols left the company. In the late 1980s Oracle offered 85 percent off the list price to some government agencies, according to John Schill, the early Navy customer. "The Sybase guy came in and said, 'I'm the new guy on the block.' [And the customers said,] 'Can you match eighty-five percent off list?' ")

The mid-1980s were heady days for the company; Oracle was really starting to gain an advantage on Ingres. About that time Ellison and Miner called the company's senior managers to a conference room at 2710 Sand Hill Road. Oracle would soon become a public company, and everyone in the room stood to profit enormously as a result. Seashols recalled the meeting vividly. "Larry said, 'I want you to look around the room. Everybody in this room Oracle has made a millionaire. Do you understand that?' " Seashols said. "And he just sort of let that time freeze in everybody's memory."

But for Seashols that sweet moment passed all too quickly. One day in early 1986, only days before the company went public, Larry Ellison took Mike Seashols to breakfast and fired him. Seashols remembered the conversation this way:

"You're not my guy anymore," Ellison said.

"OK," Seashols said.

"I want you to resign."

Seashols refused, saying, "I didn't do anything wrong." According to Seashols, Ellison "gave me a very short severance and said, 'Sue me.' " Ellison didn't remember the conversation.

Years later the two sides gave different reasons for the firing. According to Jenny Overstreet, Seashols was fired because "he stopped making his numbers. There has never been another reason than that when it comes to a salesperson." Seashols denied that. And it was hard to imagine what number he didn't make: During his time at Oracle domestic sales jumped from $3.8 million to $8.4 million to $15.6 million to $36 million. If Ellison expected more than 100 percent growth every year, Seashols said, he wasn't being realistic.

Seashols claimed his relationship with Ellison broke down over issues related to the company's initial public offering. Seashols thought that some of the things in Oracle's prospectus, or red herring, were misleading. "The red herring would come out and make all these statements," Seashols said. "And I said, 'Larry, verbally communicating [something] to our customer, or doing some marketing hype, is one thing. Writing something in a red herring . . . is very different. This is a legal document. That's a marketing position. Those two things are very different. And when you make these kinds of statements in a red herring, they're really inappropriate.' " What exactly was misleading? Seashols wouldn't say, and Ellison saw that as proof that his charge was nonsense.

On Seashols's last afternoon at Oracle, Bob Miner stopped by his office and gave him a hug. He and Seashols had always gotten along well. Miner had tears in his eyes, Seashols said. "He said, 'I just can't believe you guys can't get along. This is just a shame.' And he walked out. He was sick. He was just *sick*." Maybe so, but there was not much Miner could do about Seashols's firing. He may have been a founder, but he left those decisions to Ellison.

Seashols landed on his feet: He went to work for Oracle's archenemy Ingres. Relational Technology had made him a good offer.

Besides, working for Ingres would be a good way to get back at Ellison—or so Seashols thought.

As soon as Seashols started work at Ingres, he realized he was too late; the war would soon be over, and Ingres was going to lose. Even if IBM had not chosen SQL as the standard user language, Ingres would not have stood a chance; Seashols could tell that just by walking through the door. Ingres was not Oracle, that was for sure. At Oracle people had expensive haircuts and designer clothes and designs on their bosses' jobs. You didn't dare turn your back at Oracle for fear that somebody would stab it. One person who worked at Oracle in those days said there were people who thought you were a lightweight if you asked them how their weekend was.[5] This was in the days when Larry Ellison was still practicing what he later regretfully called "management by ridicule."

"Because I personalize a lot of Oracle and personalize a lot of the things we do, I was not terribly forgiving of mediocrity," Ellison said. "I was completely intolerant of a lack of effort. And I was fairly brutal in the way I expressed myself."

Ingres was a much kinder, friendlier place, and that was why Seashols thought it smelled like death. The company—which, after all, had its beginnings as a college computer science project—still had the feel of a laid-back university campus. People wore casual clothes and Birkenstocks, and nobody ever called anybody else stupid. The people at Ingres had an academic mentality: They wanted to fine-tune their product, get it just right. But from Seashols's point of view, they were not overly concerned about selling it. This was exactly the reverse of Larry Ellison's approach. "It was not a great environment. The intensity wasn't there; the drive wasn't there. And that just percolated," Seashols said. Industry consultant Jeffrey Tash agreed. "The technologists all loved Ingres," he said, but "their management was always highly suspect."

Ingres had board meetings once a month, and to Seashols, it

seemed that half the company came. These meetings had a communal, touchy-feely aspect, as if the top managers were afraid they might offend people by excluding them. It was not an environment in which to make tough decisions. "You'd sit there and you'd talk to them about marketing things, positioning things, concept things, and perception things," Seashols said. "And they'd say, 'Well, wait a second. If we build it, they will come.' That was the mentality. 'We have the best product. We have the best performance. We have the best features. We think it's a sales problem.' I'd say, 'No, it's a perception problem. . . . I'd talk to them about perception issues, and it was sort of just like I was talking Swahili."

When I asked Mike Stonebraker about competing with Oracle, he said that basically *it wasn't fair.* To hear him tell it, Ingres was always the loveliest and most talented contestant in the beauty contest, but Oracle won anyway. "Throughout the eighties Ingres constantly had a superior technical product. If you ask anybody, any watcher of the scene, they will all say exactly the same thing. It was not only better, it was way better," Stonebraker said. "Larry Ellison was always able to make up for his deficiencies with superb marketing."

Stonebraker gave the story of the query optimizer as an example. Relational queries were often highly complex. Let's say you wanted your database to give you the name, salary, and job title of everyone in your Chicago office who did the same kind of work as an employee named Allen. (This example happens to come from Oracle's 1981 user guide.) This would require the database to find information in the employee table and the department table, then sort the data. How quickly the database management system did this depended on how cleverly the system was constructed. "If you do it smart, you get the answer a lot quicker than if you do it stupid," Stonebraker said.

He continued. "Oracle had a really stupid optimizer. They did the query in the order that you happened to type in the clauses. Basically, they blindly did it from left to right. The Ingres program looked at everything there and tried to figure out the best way to do it." But Ellison found a way to neutralize this advantage, Stonebraker said. "Oracle was really shrewd. They said they had a syntactic optimizer, whereas the other guys had a semantic optimizer. The truth was, they had no optimizer and the other guys had an optimizer. It was very, very, very creative marketing. . . . They were very good at confusing the market."

"What he's using is semantics himself," Ellison said. Just because Oracle did things differently, "Stonebraker decided we didn't have an optimizer. [He seemed to think] the only kind of optimizer was his optimizer, and our approach to optimization wasn't really optimization at all. That's an interesting notion, but I'm not sure I buy that."

Sometimes Oracle did not address its shortcomings at all, Stonebraker said. Suppose you instructed the database to give a raise to everyone in the shoe department. Then suppose that you added two new employees to the department. If the database worked properly, either both new employees would get the raise, or neither would. Otherwise you could have employees getting raises they weren't supposed to get. "Oracle blissfully ignored this problem. And so their system couldn't handle this at all. Every other database system did. Oracle was unique in being sloppy," Stonebraker said. "Oracle just said, 'To heck with it.' "

But if that was so—and Ellison strenuously denied it—why did customers choose Oracle? "Oracle routinely—I wouldn't say lied to their customers, but they certainly misled their customers," Stonebraker said. In the 1980s customers began demanding that their database management systems have something called referential integrity. Suppose you owned a lumber company and one of your customers placed an order for some roof trusses to be delivered to a construction site next month. Right away you would instruct your database to place a hold on the inventory. Now imagine that the customer went bankrupt a week after placing the order. You would delete the customer from the database. If your database didn't have referential integrity, it would not automatically remove the hold on the roof trusses and you would wind up holding inventory for a customer that no longer existed.

Whether Oracle had referential integrity was something of a mystery for a while, which, Stonebraker believed, was how Oracle wanted it. According to Stonebraker, the company gave people the impression that Oracle had that feature. But when the user manual came out, there was a note saying, "This feature is not yet implemented."

Oracle was infamous in the industry for saying that it had, or would soon have, a version of its software that ran on the MVS operating system for IBM mainframes. "It was six months away for at least three years," Stonebraker said. Actually it was even longer than that. "We had a data sheet"—a marketing document—"on the MVS version of Oracle in probably 1981 or 1982," longtime employee Bill Friend said. But the product wasn't available for at least four years after that. Oracle's red herring, the document that described the company when it went public in early 1986, did not list MVS as one of the operating systems on which the software worked. When Oracle filed its annual report with the Securities and Exchange Commission that August, MVS was finally included.

"Someone once said that in our industry we don't miss schedules by weeks or months; we miss them by years. In those days Oracle missed schedules by years," Ellison said. "By the way, we're in great company. Our big friends to the north do the same thing." He was referring to Microsoft, which had a reputation for releasing products late. The most notable example was Windows 95, which was so late it almost ended up being Windows 96. Oracle no longer releases products years late, Ellison said. "Now we miss by a few months. And that's a bad slip for us. We're much more careful in scheduling."

As late as it was, the MVS version of Oracle probably should not even have come out when it did. In 1986 a New Jersey company called Timeplex went shopping for a relational database management system. Timeplex, which manufactured modems and other telecommunications devices, planned to use the system to run virtually its entire business. The Timeplex people entertained proposals from Ingres, IBM, and other companies but settled on Oracle because they believed Oracle would run on all kinds of hardware, from the MVS-based mainframe to the personal computer. (Chalk it up as another victory for portability.) According to Danny Turano, the Oracle salesman who made the deal, Timeplex agreed to pay $325,000 for the system, which included licenses for the PC, Hewlett-Packard, and Digital minicomputers and the IBM mainframe.

Oracle's gain turned out to be Timeplex's loss. For starters, the PC version of Oracle was not what Timeplex thought it would be. "They neglected to tell us that it had a memory requirement that far exceeded anything we had on the desktops in the company. I guess they weren't totally forthcoming," said Al Guibord, then Timeplex's technical guru. Eventually Timeplex upgraded its personal computers, with Oracle paying for part of the upgrade, and the PC version started working.

The same could not be said for the MVS software. For a year Guibord labored to make Oracle work on Timeplex's big iron. He never succeeded. Oracle sent a succession of people to New Jersey to listen to his complaints, but the software never worked. Guibord, who had recommended Oracle to his superiors, was under intense pressure; his job was on the line. Finally the situation got so dire that Ellison flew in for a meeting.

"It wasn't a real pleasant session," Guibord said. "I basically unloaded on him. He was very noncommittal. 'Jeez, I can't understand why you're having these problems.' Larry can talk around issues but never talk right at them."

Salesman Danny Turano was at that meeting. After they left Guibord's office, "Larry said something to the effect that this guy is an airhead or this guy is a loser," Turano said.

Guibord, feeling the heat, left Timeplex without ever getting satisfaction from Oracle. "I knew that my future there was probably not going to be too bright," he said. Nor did Timeplex ever get a working version of MVS. Instead Oracle settled things by giving the company credits for nine hundred thousand dollars in minicomputer software—software that presumably did work. According to Turano, Timeplex never used the credits.

Of course, for every Al Guibord, Oracle had many, many customers who were immensely satisfied (though none, in those days, who ran Oracle software on MVS). The company could not have survived if it had made every customer as miserable as it made Timeplex. But the deals that went wrong created a lingering odor, one that Oracle could not easily wash away. Corporate information managers were a tight-knit group; Guibord told his peers about his experience with Oracle, and no doubt they repeated the story to other people. After a while Ellison got a reputation for promising the moon and delivering—well, cheese. That wasn't always what happened—some customers actually got the moon—but that was the reputation. Virtually every major article published about Oracle in the 1980s made reference to it. The company's motto might as well have been "Oracle: Not exactly what you thought it was."

To Mike Stonebraker, the founder of Ingres, it was maddening to compete with such a company and lose anyway. "If I had been an Oracle customer, once I got lied to once or twice, I would just refuse to use the product. But they got away with it, and I don't quite know how," he said. "He absolutely dodged the bullet. And we were sort of sitting on the other side of the table just wondering how he continued to dodge the bullet."

In January 1985 Oracle treasurer Roy Bukstein wrote a memo to his boss, Robert Oster, then Oracle's vice-president and chief operating officer. Bukstein, who was among the first dozen or so employees of the company, was concerned about the way Oracle was doing business. The company was spending an inordinate amount of time trying to collect its receivables. Bukstein's memo explained why and suggested how to bring the money in faster.

The biggest problem, Bukstein wrote, was that Oracle was forever sending bills to customers without making reference to a purchase order number. That, he thought, was inexcusable. When accounts payable clerks received such bills, Bukstein pointed out, they threw them in the garbage. After a while someone at Oracle would have to call the customer to find out why the invoice had not been paid. It was all a waste of time and money.

What Bukstein called special deals were another problem. In their efforts to close deals, Oracle salespeople frequently offered special payment terms to customers without telling the collections department and sometimes without telling the sales managers. When Oracle's invoice wasn't paid on time, the collections people would start calling for payment only to have the annoyed customer explain that the salesman had offered special terms.

Sometimes, Bukstein said, Oracle did not get paid simply because it was slow to send bills. Other times the collections staff would look up a sales record in the customer database only to find that some of the information was incomplete or inaccurate. "In some cases, the entire customer history is just plain wrong," he wrote. It went without saying that a database company should be embarrassed if it couldn't maintain customer records. (In the memo Bukstein referred to Oracle's order-processing system by its acronym: OOPS.)

Bukstein suggested some obvious solutions. First, make sure that every customer invoice has a purchase order number. Oracle might even consider withholding commissions until salespeople communicated purchase order numbers to headquarters, he said. Also, Seashols's salespeople should be required to tell Oracle headquarters about special deals. "Consideration should be given to instituting some sort of penalty for salespersons who consistently ignore this rule," he wrote. Finally, Bukstein said, the sales force needed to be more diligent about keeping customer records up-to-date.

"As Oracle grows, its accounts receivable will grow," he wrote in conclusion. "Our objective is to make sure we have a working system which results in the collection of these obligations on a timely basis. The system should be constructed so it is in everyone's best interest (salesman to developer) to collect our receivables at the earliest possible date." Bukstein said requiring purchase orders "will reduce the average time it takes to collect our receivables by 50 percent. Such a reduction would have a positive effect on our cash flow and our ability to continue to finance Oracle's growth."

Roy Bukstein's memo was probably just one of many written that day and one of hundreds generated at Oracle that year. It is unlikely that anyone at the company regarded the document as anything other than routine. The treasurer was just doing his job. Yet the memo was extremely important, even historic. The business practices that Bukstein mentioned—the sloppy billing, the careless record keeping, the hush-hush deal making—wouldn't change anytime soon. Indeed they only got worse. People at Oracle did not think much about cash flow or about record keeping. They thought about victory: about conquering Cullinet and killing Ingres. They believed they were at war. And if salespeople had to make special deals to win the war, then they would. The idea that a salesperson would be *penalized* for being too aggressive was laughable.

Making the product and selling it: Those were the things that mattered. Let those dullards in finance worry about receivables and purchase order numbers and all that other boring, pedestrian stuff. Certainly that was Larry Ellison's attitude. He held the title of chief executive officer, but he wasn't really a chief executive, he was an entrepreneur, a supersalesman, a samurai. If Bukstein's warnings had any effect on him, it didn't show in the way he did business. Ellison's goal—a wildly ambitious one—was to double the size of his company every year.

With Ellison's leadership, Oracle achieved that incredible growth for years to come. Then the company learned, in the most dramatic way possible, that Roy Bukstein had been right all along.

I interviewed former Oracle sales chief Mike Seashols twice. We met both times at a breakfast place just off Highway 101, the long crooked spine of Silicon Valley. From Seashols I learned about Ellison's peculiar ways, his competitive drive, and the ease and comfort with which he manipulated reality. But the insights into Ellison were not Seashols's only contribution. More than anyone I talked to, he got me interested in a man named Gary Kennedy, who was, next to Ellison, the person most responsible for Oracle's dominance in its part of the industry. Kennedy, Oracle's sales chief after Seashols left the company, was, like Ellison, a tangle of contradictions. He was generally so mild-mannered that he was almost bland, yet he scared some of his employees half to death with his intensity; he set a high moral standard for himself yet worked each day in an industry that was often amoral at best; he was sometimes repulsed by Larry Ellison, yet he and Ellison had at least one thing in common: Like Ellison, Gary Kennedy was absolutely not going to lose.

The second time I had breakfast with Mike Seashols, he told me a story to illustrate that point. One day Seashols took Kennedy out to dinner. Kennedy, who was only about thirty years old, had just been put in charge of half the United States, and Seashols, his boss, was meeting with him to discuss his compensation plan. Seashols had worked with Kennedy long enough to know that he was ferocious, but it was not until that night at dinner that he learned just *how* ferocious.

"He said, 'Mike, I'm going to trust you, but I only trust people one time. And if you screw me, you're dead meat.' Now, he's talking to his boss. *He's talking to his boss.* And he says, 'I'm going to trust you and I'm going to trust you one time. And if I don't have a compensation plan that's fair, you're dead meat.' What a way to build a relationship. He was serious. He was really serious . . . I saw in Gary's eyes and his demeanor a maliciousness that I think was really inside this guy."

Later I called Kennedy at his home in Salt Lake City and read him Seashols's quotation. I was expecting him to deny having used such a threatening tone with his boss, but he didn't. Far from it. "Look," he said, "the presumption with me is trust. But if you don't take care of me, that presumption goes out the window."

Still, Kennedy insisted he never used the phrase *dead meat.* That just wouldn't have been gentlemanly.

Eight

JANE KENNEDY, A MEMBER OF THE CHURCH OF JESUS CHRIST OF LATTER-day Saints, liked to tell the story of how the Mormons built their first temple. She was reminded of the story whenever she thought about her experiences with Oracle.

In the early 1830s Joseph Smith and his fellow believers moved the Mormon church headquarters from New York to Kirtland, Ohio, where they set out to build a temple. The church was new and small, and resources were scarce. The men put in a day's work on the building whenever possible, and the women supported the men in any way they could. In a stirring testimony of their commitment the women eventually smashed all their china so the men could use it to make mortar for the building. The shards of porcelain gave the building a lasting shine.

A century and a half later, in 1982, Jane Kennedy's husband, Gary, became the fifteenth employee of what soon became Oracle Corporation. He stayed eight years, during which time Jane often felt that she was as much a part of the company as he was. At first that was actually true: she briefly worked as his secretary, taking messages for him while he was out on sales calls. As Gary got more and more responsibility in the Oracle sales organization, she supported him in other ways. When he traveled on business, she stayed home with their small children. When the company transferred him from California to Illinois to Washington, D.C., and back to California, she followed him. When one of the Kennedys' children was born near the end of a fiscal quarter, a hectic time for Gary, he made it to the hospital for the birth, then hurried back to the office at Jane's insistence; she could get by on her own. On the evenings when Gary was home, he talked business with her so often that she began to think of herself as an uncompensated company executive. Yet she never grew tired of her husband's work, never wished he would move on. "Oracle was a dream come true; you could be whatever you wanted to be," she said. How could she deny her husband his dream? For the Kennedys, building Oracle was not a job but a calling, as it was for Larry Ellison.

When the eight years were over, Jane Kennedy often could not stand to look at the Oracle campus as she drove past it on Highway 101. But sometimes she would look at those shiny blue towers, those gleaming cathedrals, and think, *I gave my china for those buildings.*

Gary Kennedy was born in 1953 in Randolph, Utah, a tiny ranching town known as one of the coldest places in the United States. He was reared in a Mormon family, though not a particularly observant one. His father commuted to work in the coal mines in southern Wyoming, forty miles away, and his mother was a housewife. The Kennedys were poor. As a boy Gary Kennedy spent his summers driving a tractor to earn money; later, when he went on a church mission to Brazil, his mother got a part-time job collecting garbage to support him. Kennedy always felt out of place in Randolph. "It was a hard place to grow up. If you got a good grade on a test, people would make fun of you. Mediocrity was expected," he said. He considered the townspeople unambitious and narrow-minded and ill spoken. If Randolph had had a theme song, he said, it would have been "The Way We Was." What Kennedy dreamed about during those long summers on the tractor was getting out of town.

Nothing was going to stop him. From his youngest days Kennedy was fiercely and sometimes frighteningly competitive. "I forget if I was all-state basketball and all-region football or all-state football and all-region basketball," he said. In his senior year in high school his fellow students elected him class president. There were only twenty-five students in the class, but then, Kennedy was the kind of person who probably would have been elected even if there had been twenty-five hundred.

Once, Kennedy said, a man who had just been released from prison came to Randolph and started terrorizing people. "I was sixteen or so, and he was twenty-one or twenty-two, and everybody just assumed that he'd rather shoot you than look at you," Kennedy said. "I confronted him. I told him I wasn't going to put up with it. He could push everybody else around, but he wasn't going to push me around. And I just beat the hell out of him. He didn't throw a punch on me; I just about killed him." According to Kennedy, the man later sneaked up behind him and broke a two-by-four over his head and then ran away. "He left town that night," Kennedy said. The story was impossible to confirm, of course. But even if it was pure fiction, it said a lot about Kennedy's self-image. In his days at Oracle people knew that if they were going to take him on, they had better bring two pieces of lumber.

Randolph was a dinky town; Kennedy wanted to prove himself in a more competitive environment. He got his chance at age eighteen when he went to São Paulo, Brazil, as a missionary for his church. The purpose of the mission was to establish the church in southern Brazil and spread the Gospel of Jesus. Kennedy did those things. He also used the opportunity to school himself in the art of persuasion. He proselytized people, got them to join the church, and then looked after their needs until a local leader could be trained and installed. He also contended with the local pastors, who were skeptical at best and derisive at worst, and pleaded with the press for positive coverage. "We spent a lot of time convincing people that we didn't have horns and weren't polygamists," he said. "It was a positive but hard experience for a nineteen-year-old kid from a small town." Most of the time he got people to do what he wanted them to do, a feeling he enjoyed immensely.

Kennedy spent a couple of years in Brazil. On the plane home he made a list of the things he wanted to accomplish in his life. For one, he wanted his face to be recognized by the majority of the people on the streets of Salt Lake City. (Years later Kennedy was embarrassed that he had wanted that; how, he wondered, could he have been so vain?) Another goal was someday to become a Mormon mission president; he eventually did. He also wanted "to marry somebody whom I considered to be a significant step up." Within a couple of years of his return he married Jane Adams, the daughter of a prominent Utah rancher. Finally he wanted to run a billion-dollar-a-year company by the time he was forty years old. In Gary Kennedy's last year at Oracle Corporation, during which he ran the entire U.S. sales organization, the company's revenues reached $916 million. He was thirty-seven.

After he arrived home from the church mission, Kennedy breezed through the University of Utah, then took a job as the Omaha sales representative for Procter & Gamble. He kept the position for eighteen months, just long enough to sharpen the skills that later made him one of Oracle's best salesmen. After a brief stint in Idaho running a family business, he enrolled at Northwestern University, outside Chicago, to get his master's in business administration. At first Kennedy felt like a rube at Northwestern, and he probably was one. Once his class was discussing the case of a clothing company that wanted to establish itself as a high-end clothier. "I made a comment consistent with someone who had never bought a two-thousand-dollar suit, and a couple of people ripped me to shreds," he said. Kennedy didn't get mad; he got even. By the time he got his master's, he said, "I felt able to go head to head with anyone there." To Kennedy, nothing mattered more than being able to compete.

He passed up some enticing consulting offers to take a job as a salesman at "a flaky little company in California called Intel." The flaky little semiconductor company soon teamed with Microsoft and IBM to become a dominant force in the fledgling personal computer business, but Kennedy did not stay around long enough to see that happen. When he began to think that he wasn't being paid fairly—Kennedy always insisted on getting what was coming to him—he started looking for another job. That was when his friend Bruce Scott told him about Relational Software Inc., the company that became Oracle.

Kennedy called Larry Ellison, who told him that he believed a company ought to be like a baseball team, wherein the people who performed the best got paid the most. That was just what Kennedy wanted to hear. Still, when Kennedy asked around about Ellison, he heard such adjectives as "charismatic," "blatant," and "amoral." "I think you'd be hard pressed to find anybody who has a totally positive or totally negative view about Larry," Kennedy said. Even before he joined the company, he understood Ellison's ambiguous nature.

Ellison immediately understood Kennedy too, understood that he would let nothing stand in the way of his success. In their first interview Ellison asked Kennedy what he hoped to achieve at Oracle. "I want *your* job," Kennedy said. He wasn't smiling when he said it. Yet the cocky Ellison was not threatened; he had little to fear from this twenty-eight-year-old kid.

Given Kennedy's strong religious background, one might have expected him to find a company with a squeaky-clean reputation, but he didn't. "I have to bet on someone, I'll bet on the smart guy every time," he said, speaking of Ellison. "I saw myself as a person who could serve as a catalyst for culture changes in the company," he said. "I think nobody really wants to be amoral. . . . I think people are basically good and want to do what's right. And if someone could show them that you can be successful being honest, why wouldn't they do it?" When I spoke to Kennedy, it occurred to me that there were a lot of people who got rich failing to change the culture of Oracle.

Kennedy joined the company in 1982 as vice-president of marketing. Soon after he started, he realized that somebody else already had that job. He did not believe that Ellison had deceived him. There had been an innocent mix-up; somebody who was planning to leave the company changed his mind at the last minute, and as a result, the marketing job wasn't open. So instead of becoming vice-president, Kennedy moved his family to Chicago and became an Oracle sales representative.

"For the first year I was just a disaster. I was the worst salesman in the company," he said. Kennedy was not given to making excuses, but this time he had a good one: The software he was selling didn't work. At the time the company was selling Version 2 of Oracle, which was far from stable. Kennedy believed he wouldn't be able to sleep if he sold the software without any sort of guarantee, so he promised to give the customers' money back if Oracle did not work.

It didn't. Once, when Kennedy called home, his wife told him he had received two discouraging phone messages from customers. One customer said, "Tell Gary my database is corrupted." The other said, "I have four words for Gary: dead in the water."

According to Kennedy, the customers were all "early adopters, people who knew they needed a relational database. And they were willing to put up with a high level of aggravation. But they drew the line when the software ate their data. That's when they got really upset."

Kennedy was upset too. He had closed a few deals but did not keep any commissions because he had been obliged to refund the customers' money. (Roy Bukstein, who handled Oracle's finances at the time, recalled refunding money to at least one of Kennedy's customers.) This was no way for Kennedy to make a living. After a year at Oracle he lined up a job in the venture capital business, then flew to California to resign from Oracle. But when he got there, he said, Ellison talked him into staying. Version 3 is coming out, Ellison told him, and when it does, it will work. Then you'll be able to make some money.

"That turned out to be one of the few things that Larry said that was precisely accurate," Kennedy said.

Version 3 wasn't great either—even Bob Miner said it "wasn't very reliable"—but a dedicated technophile could make it work. Finally Kennedy had a product he could sell in reasonably good conscience, and that was what he did. "I went from the worst salesman in the company to the best salesman in the company," he said. Soon he became a sales manager. Among his first hires were Sohaib Abbasi and Tom Siebel, technical wizards who went along on sales calls to explain how Oracle worked; both went on to become top managers in the company. Years later, Kennedy gave Abbasi and Siebel much of the credit for his early success as a salesman. But his own determination was mostly responsible. "Larry saw a tenacity and an aggressiveness and a focus on results in Gary that he loved," said Mike Seashols, Kennedy's boss at the time. "All you had to do with Gary was aim him. That's all you did. And whatever was in the way moved." Jane Kennedy said her husband "picked out his dream" every night before he went to sleep. His dream in those days was to sell Oracle software.

Oracle was a dubious environment for someone like Kennedy, who thought of himself as a moralist. Ellison had a way of talking about Oracle's products that left people with the impression that it could do certain things that it couldn't really do. Everyone at the company accepted that.

"The trouble with the culture that had grown up around Oracle was it was a little bit of wink, wink, nod, nod, so everybody could sort of say, 'Oh, yeah, we have that, sure, sure.' And you could make a case that we had it. That just wasn't exactly the way it was presented," Kennedy said. "To me, honesty means that you convey a true impression. And if that is the standard you use, there was a lot of dishonesty at Oracle. There were a lot of defensibly accurate things said which conveyed an absolutely false impression."

Yet Kennedy did not think Ellison actually meant to lie when he created some of those misleading impressions. "I think he really believes that these things can be done. When he says it's going to be done, he just says it with too much conviction for me to believe that he's misrepresenting. And if Larry were doing the work, it probably would be done then. But he's working with mortals, and mortals don't do things as fast as Larry does."

About 1984 Ellison rewarded Kennedy for his good work by making him the Washington, D.C.–based manager in charge of the eastern half of the United States. It wasn't a huge job—the entire company employed only 150 people at the time, the sales division far fewer—but it soon became one. The Kennedys "had a feeling" that they should not move the family to Washington, according to Jane, so for seven months Gary commuted halfway across the country for work. He was away from home for days at a time.

Soon Kennedy received what he interpreted as a sign that his decision to stay in Chicago was right. When he was about thirty-one years old, Mormon leaders in Chicago asked him to become a bishop, an honor normally reserved for much older people. As bishop Kennedy would be responsible for the well-being of the entire Chicago-area congregation, thousands of individuals. He would counsel them on marriage and family issues, look after their spiritual needs, and even come up with money for those in need. He was inclined to accept the appointment but wanted to clear it with Ellison first. Becoming a bishop was bound to take time away from Oracle.

One might have expected Ellison, an avowed rationalist, to chuckle at the fundamental Mormon belief that God and Jesus Christ appeared before fourteen-year-old Joseph Smith in the woods of rural New York in 1820. But if Ellison thought the story was implausible, he didn't say so to Gary Kennedy. Instead, he encouraged Kennedy to become bishop and continue doing the best he could for Oracle. "Larry professed a tremendous respect for the church," Kennedy said.

Then again, Ellison probably would have respected anyone who contributed to Oracle's bottom line as much as Kennedy did. "Larry used to say, 'I love Kennedy. He brings in all his Mormons, and I can trust them. I love that Mormon ethic. You can trust them, and they work hard,' " Seashols recalled. According to Jenny Overstreet, "It was never an action item to seek these people out. But there was this sense that wow, these people work incredibly hard, they do what they say they're going to do, and a commitment—man, oh, man—is a commitment." Ellison himself said he couldn't make generalizations about the Mormon work ethic.

Certainly that stereotype applied to Kennedy, for whom the rewards kept piling up. In the fiscal year ending May 31, 1985, he earned $182,837 in cash. The next month Ellison made him the Chicago-based national sales manager, still reporting to Mike Seashols. Just before Oracle went public in 1986, Seashols was fired, and Kennedy, still only thirty-two; became vice-president of U.S. sales and service. His salary that year was $231,654. And the salary was gravy: By then Kennedy had amassed 260,000 stock options, exercisable at a measly $2.66 each. Kennedy eventually sold many of those shares, netting millions. Had he held on to his shares for the next decade, as Larry Ellison did, he would have been worth hundreds of millions.

A lot of people at Oracle retired after the company went public, but not Kennedy. Like Ellison, he wasn't motivated by money but by the desire to win. The similarities in the two men were most apparent on the basketball court. In the 1980s the frequent pickup basketball games among Oracle executives became the predominant metaphor for life at the company. Most of the top managers, every one of whom was male, were between thirty-five and forty-five—a little creaky, but still lively enough to shove each other around. In the basketball games, as in business, Oracle people competed at full speed, with elbows flying. And nobody was more competitive than the founder and his hard-charging young sales executive.

Once, when Mike Seashols was still at the company, he and salesman Craig Conway took the court against Kennedy and Ellison, who were much taller and more skilled than the challengers. "For some reason that day Mike and I were just playing good basketball, and we won. So of course we had to play two out of three," Conway said. "Then we won the second game. . . . Larry and Gary Kennedy were getting more and more angry because these short guys were beating them in basketball." After the second game Seashols and Conway asked the others to join them for a water break. "They wouldn't go with us," Conway said. When Seashols and Conway offered to bring some water back, Kennedy and Ellison refused. They were still smoldering about their loss.

Another time, at a sales meeting in Hawaii, a group of Oracle people played in a pickup game that was so rough that it was almost violent. Conway remembered thinking that people seemed to be playing for their very lives. Ellison and Kennedy, their elbows swinging, did as much as anyone to create that feeling.

They were intense in business too—and were effective as a result. Early in 1985 a man named Ken Marshall flew to Oracle headquarters in Belmont to interview for a job as the company's Boston-area sales manager. (Oracle had opened its first branch office several years earlier, in Washington; by now it had several.) At the time Cullinet, based near Boston, was the largest independent software company in the world. If Marshall went to work for Oracle, he would be selling business software right in Cullinet's backyard. He asked Ellison, "How can I expect to be successful?"

Ellison looked Marshall in the eye and said, "Cullinet is dead."

"What?"

"They're not an issue. They're a dinosaur. They're dead."

Dead? Cullinet's sales the previous year were $184 million. Oracle's were $12 million. How could Ellison say Cullinet was dead?

Later that day, Kennedy explained everything. Cullinet, which made network database software for mainframe computers, was missing the trend, he said. Businesses everywhere were buying smaller computers and equipping them with relational databases. Relational technology was the future, and Cullinet didn't see it. Ellison and Kennedy were so sure of themselves that Marshall eagerly took the job. His starting salary was $36,000, with a total compensation package of $125,000 if he made his number. He did.

The Oracle people weren't just confident about their chances against Cullinet; they were right. In 1986 mainframe software orders slowed considerably, and Cullinet's net income plummeted from $24.7 million to $15.3 million. About that time *Business Week* published an article which began, "Cullinet Software Inc. . . . is dead in the water."[1] That news was followed by a string of quarterly losses, layoffs, and desperation management changes. Cullinet finally introduced its software for VAX minicomputers in 1988, way too late to catch up with the competition. The company was eventually sold for scrap.

Barbara Boothe Ellison had the feeling that she was not what her husband wanted. The prenuptial agreement that he presented as a wedding gift was the first clue. But even after that crisis passed, she could not seem to make him happy. "I tried everything. I tried to be everything he wanted. I used to try to read the books he wanted me to read, and watch the shows he wanted me to watch, and dress the way he wanted me to dress," she said. But she couldn't shake the feeling that he didn't love her. She came to the conclusion that there was something seriously wrong—with her. If only she could be perfect, things would be different.

"I was still young enough to think, I can change him," Barb Ellison said. "He just needs to feel secure. He just needs to know that

I love him, that I'll always be there for him. He just needs someone to stand by him, believe in him, you know, be there. I used to feel that everything was my fault. You know, God, if I hadn't done that, this wouldn't have happened, he wouldn't have reacted this way. But that was being young and stupid, I think."

The Ellisons had some good days too. Their second child, a girl, was born on January 31, 1986, at Stanford Hospital in Palo Alto. The labor was only an hour and a half, just enough time for the Ellisons to get to the birthing room. Sarah Crawford, an Oracle lawyer, happened to be in the hospital that day to see a friend who had given birth. "Larry came up and kissed me. He was truly happy that his wife had borne this daughter and everything was OK," Crawford said. The couple named the girl Margaret Elizabeth but right away began calling her Megan.

As happy as they were about the baby, the Ellisons still were not happy together. Ellison, who was busy preparing for Oracle's initial public stock offering, was rarely home. Barbara spent "a lot of time in tears. A lot of time wishing things could be different." Also, "There were other women, and that was hard for me." Still, Barb Ellison wanted to work things out. "We went to see a psychologist, trying to work out our marriage," she said. "And Larry got me so mad. He sat there in the psychologist's office, agreeing with her. And when we got home he said, 'Forget it, I'm not going to work on that stuff.' I was so crushed, because I so wanted it to work. I had the two kids, and it just wasn't going to work."

On April 21, 1986, a month after Oracle went public, Larry Ellison filed for divorce on the ground of irreconcilable differences (the only other choice under California law was "incurable insanity"). He asked that he and Barb share legal custody of the children but that she have sole physical custody. He said he wanted "reasonable" visitation rights and was prepared to pay child support.

He was not prepared, however, to share his wealth with his ex-wife. The petition listed only four things that Ellison considered community property: part of the couple's joint investment in Pacific Monolithic, Inc., a high-tech company in nearby Sunnyvale; the family Volvo; a Thoroughbred horse that was kept in a stable near their home; and a joint bank account. Everything else—including and especially the 4,506,658 shares of stock in Oracle, then worth about $67 million—belonged exclusively to Ellison, on the basis of the terms of the prenuptial agreement. That was his claim anyway.

Two months after Ellison filed his papers, Barb Ellison filed her own suit for divorce, also on the ground of irreconcilable differences. That was about the only thing they agreed on. While Larry's suit said the couple had separated a year earlier, Barb's said they were "Not Yet Separated"; it was not clear how two people could disagree on that point. While Larry asked for joint custody, Barb wanted sole legal and physical custody. She also wanted him to pay spousal support, something he had not volunteered to do. Among her monthly expenses were a $5,378 mortgage payment, $800 for baby-sitting, $500 for clothes, and $2,300 for entertainment, "including vacations/horse."

The big dispute was over who owned what. Under the heading of "Community property," Barb Ellison listed the following: Oracle stock; the Woodside house; other stocks and securities; and cars, personal effects, and bank accounts. From her point of view, everything was up for grabs.

She made this clear in a separate lawsuit that accompanied the divorce petition. When Ellison claimed, in the prenuptial agreement, that his Oracle stock was worth between $35.2 million and $46.2 million, he knew it wasn't true, according to the suit. (Nobody would have paid that kind of money for it on December 4, 1983, that was for sure.) "The true facts were that [Larry] would be required to perform substantial labor following his marriage to [Barbara] in order to prepare his company, Oracle, for public offering," the lawsuit said. Sure, Ellison had owned the stock in Oracle before he married Barbara. But the stock became truly valuable because of the work he did after he married her, and therefore she was entitled to half of it. That was the idea.

Barb Ellison asked the court to throw out the prenuptial agreement, give her half the property Ellison had accumulated in the five years they had lived together, and award actual and punitive damages. Seventy million dollars would be fine.

Along with the lawsuits, Barb Ellison submitted a declaration in which she gave all the gruesome details of her wedding day. It was all in there: the groom's threats about the prenuptial agreement ("[he] told me that if I did not sign it, he would not go through with the wedding"), the bride's tears, the impromptu negotiations between Barb's father and Larry's brother-in-law, the amendments scribbled in the margins, and the arrival of the florist, who was expecting a wedding but nearly got a wake. "At the time [Barbara] signed the agreement, she was extremely distraught," her lawyer deadpanned in the lawsuit.

The case never went to trial. On December 30, 1986, Larry and Barbara Ellison reached a settlement. Not all of the details were revealed, but it was clear that Barbara got more than she was entitled to under the terms of the prenuptial agreement. Ellison agreed to establish a trust (the value was not disclosed) for her benefit. He also agreed to pay her eight thousand dollars a month in spousal support for five years, but he had the right to end the payments when the trust was funded with two million dollars. As for the children, Barb Ellison would keep them, and Larry would see them or call them at least once a week. He would pay two thousand dollars a month in child support until the children were eighteen, which was a long time off. The divorce became final on January 30, 1987, the day before Megan Ellison's first birthday.

I interviewed Barb Ellison (she kept his name) at the house where they lived together in Woodside. She had used part of the settlement money to buy the place from him and another substantial part to have it renovated so that any memory of their lives together would be paneled or Sheetrocked or painted over. When I arrived, she had just finished exercising one of her many horses, which she coddled like children. At forty-one, she was tall, trim, bow-leggedly athletic, and blond—in short, everything Ellison liked in a woman, except more mature. It was easy to see why everybody at the office had tried to keep him away from her fifteen years earlier.

At times she talked about being married to Ellison in the way that a Navy officer might reminisce about his plebe year at Annapolis: It was something she had to endure, yet it built character. She said she once told him, "Being married to you, as hard as it was, and as detrimental as it was at the time, made things in the rest of my life seem so easy." When her friends complained about their spouses, she always told them, "You guys, all you need to do is be married to Larry for a while. This is nothing."

Like Adda Quinn, she was happy being the former Mrs. Larry Ellison. After the divorce they rarely argued anymore; Ellison became just an old friend who lived across town. Sometimes she and the kids called him on the spur of the moment and asked him to join them for a movie, and often he went. He was generous with the kids; he took David flying all the time and bought Megan a horse for Christmas in 1996. He was good to Barbara too; according to her, he once promised her that he would always take care of her, even after the divorce settlement expired.

Barb Ellison has never remarried; she told me she didn't even date seriously. Once you've gone vertical in a fighter jet, who wants to chug along in a biplane?

"I've never met anybody like him, ever," she said. "Not even close. And that is hard when you're trying to go out with other people. They pale in comparison. Or I go out with extreme opposites of Larry, but then they're not interesting enough. It's really tough."

Larry Ellison and Gary Kennedy may have shared a competitive nature and confidence in Oracle's future, but those were apparently the only personality traits they shared. Ellison's family life was a train wreck; Kennedy and his family were living in Ozzie and Harriet bliss. Ellison was brash, charismatic, and sometimes wonderfully funny; Kennedy was stolid, inscrutable, and deadly serious. He stood over six feet tall and had nondescript brown hair, an expressionless face, and an inflectionless voice. If the Oracle headquarters had had gargoyles on it, they would have looked like Gary Kennedy. "I found him humorless," Ellison said. Jenny Overstreet agreed: "Gary wasn't fun. Gary didn't have a fun bone in his body." Jane Kennedy did not see her husband that way, but she admitted she knew what they meant. If you don't know him, "he is *so* intimidating," she said. One of her roles as Gary's wife was reassuring people that yes, he did indeed like them. It just *seemed* as if he didn't.

"Gary Kennedy was a tough guy—tough, impatient, highly demanding, not particularly warm. Just a bottom-line results kind of guy," said Craig Conway, who worked for him for several years.

When Kennedy did something warm, he usually kept it to himself. Once Jenny Overstreet was in line at her bank when she saw a woman and a bank employee engrossed in conversation. The woman, obviously destitute, was begging that her account be kept open even though she had no money in it. Overstreet told the story later that day in a meeting at which Kennedy was present. A couple of hours later Kennedy called Overstreet's office and said, "Which bank was it?" Overstreet told him. She later learned that Kennedy had tracked down the woman and given her a job in his home.

Still, Kennedy had a way of communicating in business that wasn't exactly what people expected from a leader of his church. The things he said tended to be, well, unsweetened; he just said what he thought. In 1987 a woman named Kitty Cullen applied for a job selling Oracle to the federal government. Kennedy interviewed her at Oracle's Bethesda, Maryland, office. He started by closing the blinds, saying he had a terrible headache. With the room suitably darkened, he began to grill Cullen. "Ten minutes into the interview he took my résumé, flipped it across the table, and said, 'Well, if this is just a puff of smoke, we'll find out,'" Cullen said. *Welcome to the team, Kitty!*

Cullen did not see Kennedy for several months after that. The next time she saw him, he greeted her warmly and asked about her kids—by name.

Kennedy could be kind and thoughtful, and certainly he had a good memory for names, but he scared his employees out of their wits. At the beginning of each year, he met with his sales managers to establish quotas for them and their employees. He always went into those meetings looking ravenous, as if he wanted to eat someone. Meeting his expectations was hard. A quarter was about sixty-one business days long, after you subtracted weekends and holidays. To make a sale, an Oracle sales rep had to get a meeting with the prospect (not always easy), pitch the software, fend off the competition (Ingres, Informix, and Sybase, mainly), and close the deal. If a customer took a sick day or, God forbid, a week's vacation, tough luck. Sixty-one days. That was all the time any salesperson had.

Danny Turano, at the time one of Oracle's top sales managers, said those quota-setting discussions usually went something like this: "I think we can do about two million this quarter," Turano would say. He considered two million dollars in sales an ambitious but attainable figure and expected Kennedy to be impressed with it.

"That's not enough," Kennedy would say gravely. That he did not raise his voice made him all the more terrifying. Then he would sit there, stone-faced, his hands folded on the table, and allow the silence to swallow Danny Turano.

After a while Turano would try to explain. "But two million—"

"Unless you want to fire half your group, two million isn't enough," Kennedy would say.

"Fire half my group?"

"I need three-point-two million from your group this quarter. Otherwise we have to shut down half your organization."

Turano would be reeling. He had tossed out what he thought was an ambitious number, and now half his people would be facing pink slips. He would look at his list of sales prospects and say he could probably find a way to do $2.5 million or so.

"Unacceptable." Kennedy seemed to consider a $2.5 million quarterly sales goal to be a personal insult to him.

After a few minutes of negotiation Turano would sigh that he would try to do $3.2 million in sales that quarter.

"Don't try," Kennedy would shoot back. "Do it. You're making a *commitment.*"

Kennedy established what he called "a commitment culture" at Oracle. If you said, even under duress, that you were going to sell a certain amount of software, by God, you had to follow through. When Kennedy checked up on his people at the end of each quarter, he was "almost religious in his zeal," former Oracle executive Ken Cohen said. "The worst thing that could happen to you was having to sit down across the table from Gary Kennedy and tell him you missed your number by five percent. You knew Gary would say, 'You gave me your word. You made a promise. You promised me you were going to bring in twenty million bucks in sales. You made a promise. How could you go back on your promise?' "

Kennedy was uncomfortable with these stories years later; he said he was much less threatening, and much more encouraging, than Turano and Cohen remembered. Probably he was. After all, in Turano's interview with me he was reconstructing a long-ago conversation from memory. Still, whether Kennedy threatened people or not, they certainly *felt* threatened. Kennedy was a scary guy, and his employees were always afraid of what would happen if they let him down.

And that was ironic, because in most cases nothing happened. "Gary, for as much as he intimidated people, fired almost nobody. But the *feeling* was that you had to deliver or you were out of there," former salesman Tom Siebel said.

At one point Kennedy came up with a slogan describing what Oracle hoped to do to its competitors: "Cut Off the Oxygen." If that seemed to be a violent way for a former Mormon bishop to talk about business dealings, well, Kennedy did not care. "I think that competition is not a place for the faint of heart," he said.

Kennedy liked to say that working for Oracle reminded him of the opening scene of *Raiders of the Lost Ark,* the 1981 film by Steven Spielberg. At the beginning of the movie Indiana Jones (played by Harrison Ford) scrambles to get out of a cave before he is flattened by a giant rolling boulder. "When you're doubling, the ball's rolling real fast," Kennedy said. "Now, if you've got real fast legs, you can stay ahead of the ball. In other words, you're real good. Or if you've got a long head start, you can stay ahead of the ball. In other words, you're overqualified to do what you're doing. Otherwise you get crushed." Salesman Ken Marshall said Kennedy believed so completely in the metaphor that he actually went out of his way to hire athletes. "He wanted a lot of jocks, quite frankly, because he knew they were competitive and high-energy," Marshall said.

Oracle's adventure movie pace helped it become the market

leader in the relational software industry. But the relentless drive for more revenue also caused problems—such as negative cash flow, the booking of questionable deals, and a go-for-broke corporate culture—that very soon threatened to bring down the company. Doubling every year was impressive, but in hindsight it might have been more impressive to grow at "only" 80 to 90 percent.

Once I asked Larry Ellison if he thought it had been a mistake to keep pushing for more and more revenue.

"I didn't push for more and more revenue. That's a big myth," he said.

"You didn't?"

"No. I never gave anyone quotas. Huge myth."

"You never—"

"Never. In fact quite the contrary, I almost always tried to get people—if I participated in that at all, I tried to get them to lower their numbers. Not often did they have to lower their numbers, but I never, ever assigned those numbers. Never."

That was not how his lieutenants remembered it. According to them, Ellison not only pushed for more revenue but pushed hard, categorically rejecting any suggestion that it might be prudent to grow more slowly.

Tom Siebel: "Larry set the revenue objectives every year. I was there. Larry walked in with the number, said, 'This is it,' and told everyone to go meet it."

Sales and marketing executive Craig Conway: "I have been in sales meetings where managers would say, 'Let's lower expectations.' Larry wouldn't stand for it."[2]

Gary Kennedy: "I can't imagine how many times [international vice-president] John [Luongo] or I must have said to Larry, 'Larry, Wall Street doesn't care if we grow a hundred percent a year or sixty percent a year. It just doesn't matter.' In fact we might have had more credibility growing sixty percent a year than a hundred percent a year. . . . We would come in with plans that showed sixty or seventy percent growth, which is not bad. And we would leave with plans that showed a hundred percent growth."

Stephen V. Imbler, vice-president of corporate finance: "It's almost like [Ellison] is saying that Gary Kennedy was Frankenstein's monster, that he went out and did all these things [on his own]. Well, Larry was Dr. Frankenstein. He really created Gary. He put the pressure on him. I was in meetings where Gary would say, 'I can only do thirty-five million in revenue this quarter,' and Larry would say, 'You're going to have to do forty.' "

Ellison's statement to me that he "tried to get [people] to lower their numbers" was also contradicted by something he said in a 1993 interview with the magazine *Forbes ASAP*. He told the magazine that people urged him to slow the pace of growth but said he rejected their advice.

ELLISON: We were an adolescent company. Our goal was growth, much like an adolescent's. Every year we were a year older and twice as large. . . . And every year, people said, "Larry, you should stop trying to grow so fast." I mean, how insane.

ASAP: Who was insane—you or the critics?

ELLISON: The critics! By far the larger risk was in growing too slow. Look. I watched Ingres implement a much more conservative strategy. People forget that one year Ingres was $8 million and Oracle was $13 million. Now Oracle is ten times larger. So you tell me, which strategy had the higher risk? We aggressively pursued the market and growth because we thought we had to get up to a certain scale to survive the shake-up in the computer industry. *We pursued growth relentlessly* [author's emphasis.] Growth was the center of our culture for a long time. I knew eventually this would hit us—at some time we would invest for very rapid growth and we would get nailed. And when we hit the wall, we hit it very, very hard.

Clearly Ellison was driving Oracle's growth. If he had wanted people to lower their numbers, he could have ordered them to do so; he was the boss after all. So why did he deny it when I asked him about it? Probably because we were talking about Gary Kennedy at the time. In the years after Oracle's near collapse, Ellison often graciously accepted responsibility for what had happened. "In the end I could only blame myself," he told me. But there were also many times when he assigned the blame to someone else—namely, Kennedy. The discussion we had about revenue was an example. Ellison's point was that it was Kennedy who had pushed the company to the breaking point. But the people who were there thought that Kennedy was merely doing what Ellison wanted him to do. As Mike Seashols said, "With Gary, all you had to do was aim him." Well, it was Ellison who aimed him.

"In the final analysis, what did Gary know?" former marketing man Ken Cohen said. "It wasn't like he spent months every year at American Management Association retreats learning sales management. He spent seven days a week selling Oracle."

Actually he spent seven days a week pushing Danny Turano to sell Oracle.

When a computer industry headhunter asked Turano if he would like to interview for a sales job with Oracle Corporation, Turano didn't know what to say. Oracle? Turano had never heard of it. Still, he told the headhunter yes. At the time he was working for Applied Data Research, selling software to the IBM mainframe marketplace. Business was lousy. Instead of buying mainframes, a lot of companies were now encouraging their department managers to buy minicomputers—Data General machines, Digital Equipment Corporation VAXes, and so on. (This was what was killing Cullinet.) Turano had no intention of hanging around to administer the last rites to the mainframe business. Besides, the headhunter said Oracle was hot.

Turano was told to meet Gary Kennedy for breakfast at 7:00 A.M. at the Helmsley Palace Hotel in New York, where Kennedy stayed when he was in the city. Turano was at the restaurant promptly at 7:00, but Kennedy was not. By 7:15 Turano was thinking he might be in the wrong hotel. He called and asked for Kennedy's room. The phone rang several times before Kennedy picked up and croaked hello. Turano had woken him up. "I'll be right down," Kennedy said. So far Turano was not impressed.

A few minutes later the elevator doors opened and Gary Kennedy burst out. "He's like a freight train," Turano said. Kennedy introduced himself, hustled Turano into the restaurant, and started pounding him with questions. *What makes you think you're so good? How do I know you're of the caliber we want? What makes you think you can meet our standards?* This wasn't an interview; it was an interrogation. After thirty minutes Turano wiped his mouth, stood up, and said, "I gotta go. This is not working out. You and I will never work together because I don't think I can deal with your personality."

Kennedy seemed to like that. Immediately his demeanor changed. "No, don't go. Let's try this again," he said. Suddenly, instead of grilling Turano, he was trying to sell him on Oracle. "This guy was Sybil," Turano said. Kennedy was persuasive. Turano soon became convinced that he could sell a lot of Oracle software.

The reason was portability. When Turano learned that companies could—at least theoretically—use Oracle on whatever hardware they had, he thought, *Holy smokes, do these guys really know what they have?* Turano's contacts in the business world were going crazy trying to rewrite their applications for various machines. With Oracle they wouldn't have to do that. "I was thinking, even if this company's not for real, I could make a fortune on commissions by selling this story." Turano joined the company as vice-president of the New York region. Six months later he was overseeing sales throughout the East.

Danny Turano became a legendary figure at Oracle. A driven salesman who was always working his cell phone or banging out E-mail, he quickly started doing deals worth $150,000, $200,000, even $500,000. His efforts showed up on the bottom line—Oracle's and his. In his four years there Oracle's sales jumped from $23 million to $583 million, and Turano sometimes socked away more than $1 million a year in commissions and stock options. He was not the only excellent salesman at Oracle, but he was among the best known, and the most admired. "[His attitude was] 'We're defining the market, we're leading the market,' " former Oracle sales chief Mike Seashols said. "You know, it was just a mind-set. Just a friggin' mind-set."

Turano's experience reveals just what a pressure-packed, manic, and hyperambitious place Oracle was in the late 1980s. It also shows what can happen when a hard-charging CEO drives his people to win, win, win but does not really tell them how. Though Danny Turano and Larry Ellison worked thousands of miles apart, Turano felt Ellison's powerful presence every day.

Turano's job was to help Gary Kennedy fulfill Ellison's vision for the company. Turano joined Oracle in the year that Ellison was trying to double revenues from $23 million to $50 million. It exceeded the goal by $5 million. The next year Ellison set a goal of $110 million. No way, Turano thought. Then he and the other salesmen went out and did $131 million. Ellison told them to double that the next year. *It's never going to happen. That's crazy,* Turano thought. The sales force exceeded the goal by $20 million. When Ellison set the 1989 revenue goal at half a billion dollars, and the salespeople did $583 million, Turano became a believer. "At that point I stopped raising the flag," he said.

Still, he was worried. "I gotta be honest with you, Turano said. "We were running and gunning and growing so fast that I don't think anybody expected the company to have staying power. It was like, get what you can get today, because I don't know if there's going to be a tomorrow."

Turano knew how to keep the money flowing in. He was a dynamo, careening from one meeting to the next with one hand on the wheel and the other on the buttons of his cellular phone. "It was just a rush to be with him," Oracle support executive Bob Ney said. "Just getting to the meeting was a rush. The guy goes a thousand miles an hour." While riding the elevator to meet with a prospective customer, Turano would call the customer's secretary and pump her for information. Was the boss a fisherman? A golfer? Who were his friends in the company? If he had time, he would call the friends too. That way he could walk in and say, "Hey, I was just talking to your friend so and-so; he says hello." Turano was always searching for an opening line, a way to ingratiate himself with the prospect. It was important, he knew, to show an interest in people. "In many cases I was being genuine," he said.

After Turano convinced his customers that Oracle was portable, he gave them an incentive to buy lots of it. Early on he dreamed up something called the Corporate Credit Purchase Plan. Under this scheme a company would pay Oracle, say, $250,000 up front. For that sum it would receive credits for $335,000 worth of software. The company could use the credits whenever it wanted; it could spend them right away or use them whenever it needed more software. The deal made both sides extremely happy. Turano got credit for a $250,000 deal, and the company got a big discount on the software. A lot of corporate information managers liked the deal because they never knew when their companies might cut their budgets. These people wanted to spend the money they had lest they lose it. Danny Turano had no problem with that. For a while, he said, the scheme brought on a buying frenzy among corporate information managers; the deal was that good. "People were buying our product just because they didn't want to be the only one without it. They had no idea what to do with it. They had nowhere to run it," Turano said.

The Corporate Credit Purchase Plan further complicated an already complicated business. Now, instead of selling licenses to use software, Turano was selling the right to acquire licenses to use software sometime in the future. In other words, he was selling software futures. This raised some interesting questions for Oracle's accounting people. If Oracle sold software to a customer but did not deliver it, could it recognize the revenue from the deal? Under the accounting standards in place at the time, it could. So it did. Oracle did not have to book the revenue up front—it could have waited until the customers received something for their money—but Oracle was not a cautious company. It couldn't double its sales every year without being aggressive about booking deals.

Selling futures was just one of Turano's tactics. He also liked to offer extended payment terms. Instead of asking customers to pay their bills in thirty or sixty days, he would sometimes let them pay for their futures over a period of a year or more. When he did deals that way, the customer didn't get any software, and Oracle didn't get any money. Nobody got anything. Nobody except Danny Turano, who got a commission.

Turano's employees were making deals the same way. "Revenue per head was going through the ceiling," he said. Salespeople were earning two hundred thousand, five hundred thousand, even one million dollars a year.

Even as Turano and his people were making the deals, he knew there was at least a chance that some of the deals would go bad. After all, the longer you give a customer to pay, the greater the chance that you will not get paid. Ellison also understood this. For all his talk about doubling revenues, he was at least briefly concerned about the way Turano was doing things. Once he called Turano and said, "Aren't we mortgaging the future?" Well, yes. That was exactly what Turano was doing. In order to double sales every year, Turano said, "you somehow had to reach forward into the future and pull the revenue into the present." Selling futures and offering extended payment terms were ways of accomplishing that. Turano said that satisfied Ellison.

Nobody at Oracle ever asked Danny Turano to make deals less aggressively or go for a lower number, that was for sure. "When you closed a deal and beat a competitor, they made you feel that you were golden," he said. Sometimes Turano knew his competitors all too well: They were his colleagues. In addition to the field sales organization, for which Turano worked, Oracle had a group that sold to certain industries, a headquarters sales unit, a telesales group, and so on. "In any given deal there could be five competitors. And three of them could be Oracle," he said. "You wouldn't go to sleep at night for fear that one of the other guys would end up taking your deal." That was fine with Ellison and Kennedy. They didn't want salespeople who slept. They wanted salespeople who made deals.

While Danny Turano was selling Oracle in the field, Rick Bennett was selling it through advertising. Bennett, who had helped write the ad equating Oracle to SQL/DS and DB2, possessed all the qualities Ellison was looking for in an adman: He was bright, he was creative, and he was by no means a slave to the literal truth. Before

Bennett arrived, Oracle rarely advertised. But during his years at the company, advertising expenses boomed. In the fiscal year ending May 31, 1985, Oracle spent only $330,000 on ads (including the small charge for the IBM ad, the first one Bennett placed for Oracle). It spent $1.33 million the next year and $4.8 million the year after that.

Bennett's mentor was Tony Schwartz, who created the famous daisy ad that portrayed Barry Goldwater as the presidential candidate most likely to push the nuclear button. In 1980 Bennett and Schwartz were hired by high-tech companies in Massachusetts to promote a tax limitation initiative. The initiative passed. "Tony and I became friends, and he taught me how to kill for a living," Bennett once wrote. Bennett eventually formed a one-man advertising agency that catered to Silicon Valley companies. He worked in exchange for a small creative fee—and a ton of stock options. Bennett—who after all made his living on hyperbole—once described his work this way: "My ads attack like a pack of speed-crazed wolverines and have the same general effect on your competition as a full moon does on a werewolf. I'm looking for clients with killer technology . . . and a taste for blood." Bennett, a Mormon, came up with a catchy slogan for his ad agency: "Because God Hates Cowards." Yes, this was just the man Larry Ellison needed to publicize his company. The impressions Rick Bennett created lingered for years, and not all of them were positive. Years after he left the company, Jenny Overstreet would hear Bennett's name and sigh, "He brought out the worst in Larry."

With their ads, Ellison and Bennett established an aggressive and extremely arrogant public image for the company. Instead of promoting Oracle on its merits, they created the high-tech equivalent of political attack ads: They tried to raise Oracle's stature by smearing the competition. They did this cleverly, obnoxiously, and to great effect.

"Larry played a zero-sum game: You have to destroy your competition. It's quite Darwinian," Bennett said. "The concept of firing a token shot is alien to Larry's mentality. If he's going to take a shot at you, he's going to go for the heart, or the head, or the dick. He's not going to try to wound you."

In the mid- and late 1980s Ellison and Bennett fired some shots that the industry never forgot. They were in Ellison's office one day, trying to think of a new campaign, something macho, something memorable. A science fiction film called *The Last Starfighter* was playing in theaters at the time. That gave Bennett an idea for a theme. He mentioned the movie, then said, "Oracle is the last DBMS [database management system]."

Shouted Ellison: "Yes!"

The two men immediately put together an ad. It is unlikely that readers knew that the headline THE LAST DBMS was a reference to a movie, but it did not matter; the advertisement on which it appeared was utterly clear. The top half of the ad carried a picture of a pilot giving the thumbs-up sign from the cockpit of a fighter jet. Painted on the gleaming flank of the jet were the names of Oracle's competitors—Cullinet, Information Builders, and so on. The names of all the competitors were crossed out. Oracle had shot them down; that was the idea. In the decidedly unsexy world of information management, that was a pretty exciting message. As if the picture were not enough, Bennett threw in this "disclaimer" at the end of the ad copy: "ORACLE is a registered trademark of Oracle Corporation. Our use of Cullinet, Information Builders, etc. trademarks in this advertisement is the least of their problems." Bennett said one of the companies wrote a letter complaining about the fake disclaimer. Said Ellison: "Hey, I'm paying for a litigation department. Let 'em litigate!"

The "last DBMS" ad was merely the first in what Bennett would later call "sorties" on the part of the "Oracle Air Force." In the mid-1980s the software company Ashton-Tate virtually owned the market for database software that ran on personal computers. Over the years the company had sold hundreds of thousands of copies of its dBase program; by the end of 1988 Oracle had sold fewer than twenty thousand copies of its PC program.[3] At the time Oracle was too busy ringing up sales to VAX minicomputer customers to think much about the PC market. Then Oracle received the unsettling news that Ashton-Tate was developing a version of its software for—yes—the VAX. "Now the VAX was Oracle's bread and butter. Oracle was not making substantial profits on the PC. So the strategy was simple: Destroy Ashton-Tate before they became a threat on the VAX," Bennett said.

How did Oracle accomplish this? First, Ellison and Bennett came up with a "last DBMS" advertisement that showed the Oracle fighter jet shooting down a bright red triplane labeled "Ashton-Tate." (In another ad Ashton-Tate was represented by a biplane.) Black smoke issued from the fractured tail section of the Ashton-Tate aircraft. The ads appeared in all the trade magazines. Then the Oracle sales force began pointing out technical flaws in the Ashton-Tate product. Finally Oracle sold its PC software at cost, as it could easily afford to do because it was raking in profits on the VAX. The idea was "to make the price point so low that Ashton-Tate didn't exist," Bennett said. Within a few years Ashton-Tare was acquired by another company. Oracle's "air attack" was one reason; another was that Ashton-Tate, a poorly managed company, self-destructed.

"We girded our loins, we put on our armor, and we got ready to go out into the field of battle and do battle in the PC market against this big, bad mother of a juggernaut called Ashton-Tate," former Oracle executive Tom Siebel said. "And by the time we got out there to do business, Ashton-Tate went into a tailspin on us." Siebel's aviation pun was unintended.

The fighter jet campaign was effective, but like the IBM ad of a couple of years earlier, it upset some people in the company. Don Lucas, the longtime board chairman, later said, "The fighter plane ad would have been a great ad when we were a dinky little four-, six-, eight-million-dollar company. When we were a hundred- or two-hundred-million-dollar company"—as Oracle was at the time—"it was totally unacceptable. It was a very unsophisticated message. And the truth—the *truth*—was not that they had a biplane and we had a jet." But that *was* the truth according to Ellison. When I asked him about the Ashton-Tate ads, he referred to Oracle as "modern technology" and dBase as "ancient, dangerous technology," adding, "dBase can be dangerous to your health."

The purpose of the ads was to help software buyers make decisions, Ellison said. Each one had extensive copy explaining why Oracle was better than other products. "It's side-by-side comparison. And we do those same ads today. [We say,] 'This is what we have, this is what they have. Take your pick.'" The fighter jet pictures were just a way of getting attention, he said.

The "last DBMS" ads were not just attention-getting; they were outrageous. Oracle may have had $131 million in sales in 1987, but it was still a minor player, at least compared with Cullinet and Ashton-Tate. For Oracle to suggest that it could overtake such legendary companies was almost laughable; it was as if a mom-and-pop burger joint had said it was going to outsell McDonald's. "Larry fundamentally believed even at that point that his company was going to be more important than IBM, that his company was going to be more important than Ashton-Tate, that his company was going to be more important than Cullinet. You can't imagine how farfetched those ideas sounded," Tom Siebel said. Sometimes Ellison's predictions actually damaged Oracle's credibility. "You had to be careful. If you took him with you to see a customer, he'd say this sort of stuff in front of a customer. He would say he was here to become the largest software company in the world." People were taken aback.

Even as Ellison was trying to bring down the mightiest names in software, he was still warring with RTI, the maker of the Ingres relational database management system. About 1987 word got out that the Ingres database would soon have a sexy new function: It would be able to do distributed queries. This meant that the database could find information even if bits of it were stored on many different computers tied into a network. RTI announced that the feature was coming soon; everywhere the Oracle salespeople went, customers kept asking when Oracle would have it too. The sales force had no answer.

Back at the office, Ellison told Bennett to prepare an advertisement announcing Oracle's distributed capability. Then he assigned an engineer to whip up a distributed feature so the company would actually have something to sell when the ad appeared. Ten days later Bennett's advertisement hit the trade press. "Oracle Announces

SQL*Star," it said. "The First Distributed Relational DBMS." For a company that sometimes missed deadlines by months, getting the product ready was impressive. Sort of.

"The fact of the matter was Oracle didn't have anything," said George Schussel, the trade show promotor who had followed Oracle from the beginning. "But that was the way they worked. Everything was marketing, everything was image. You simply announced the product and then figured out later how to deal with it from a technological point of view."

Ellison did that a lot: imagined the way things could be, then tried to make reality conform to his vision. When he succeeded, he looked like a visionary. When he failed, he looked like a liar. Ellison reminded Rick Bennett of the character Marlon Brando portrays in the 1990 film *The Freshman,* with Matthew Broderick. Brando plays a Mafia don, much like the one he portrays in *The Godfather*. At one point a dumbstruck Broderick asks Brando if he can believe what Brando says. "Because I have said it, it is true," Brando replies. Bennett could imagine Ellison saying the same thing.

"At least once a week he and I would go out to lunch. One day he told a couple of us, 'You guys, mark my words, within ten years this is going to be a ten-billion-dollar company.' That was 1988," Bennett said. "He made it happen. He willed it to happen. Anyone else would have been a liar and a flimflam man. And they would have been out of business by now."

Maybe SQL*Star wasn't much good at first, but again that wasn't the point. To longtime Oracle employee Ron Wohl, who worked closely with Ellison for many years, the SQL*Star episode was just another example of Ellison's marketing genius. "That was an example where Larry was not about to allow a major differentiator for the competition. His attitude was, and is—it hasn't changed at all—we must have a lot of differentiators against our competition, and if we're doing our jobs well, we don't allow them to differentiate against us."

Ellison and Bennett attacked RTI mercilessly—and gleefully. When it was becoming clear that Ingres was losing ground to Oracle, Bennett came up with an idea to humiliate the competitor. One day he and his son drove to the RTI offices. While his son waited in the car, Bennett jumped out, ran to the Relational Technology sign in front of the building, and placed a small sign there. It said, FOR SALE BY OWNER: CALL 1-800-4INGRES. Then he took a snapshot of the sign and hurried the film to a developing place. A couple of hours later he gave Ellison the photograph. Soon after, Ellison showed the picture to his sales force at a rally in Hawaii. "He showed the sign, and he brought down the house. It was glorious," Bennett said. According to Bennett, Ellison told the group, "We're going to run them out of business and buy that building, which we're going to bulldoze. After that, we'll salt the earth. Then we'll go after their families." (Ellison did not recall these events, but said it was possible that Bennett was right.) It was just talk, just something Ellison said to whip up the sales force. But there was an element of truth to what Ellison said: He really did want to crush the enemy.

During his last four years at Oracle Bennett received options to buy Oracle stock, with the options vesting at 25 percent a year. For the first three years everything went fine: Bennett exercised his options and made handsome profits. He planned to work the final year, exercise his final stock options, then move on. At the beginning of the fourth year he notified Ellison that he would be leaving after his options vested.

But with sixty days remaining in that final year, Ellison suddenly fired him, effectively denying him the last 25 percent of the stock options. This was going to cost Bennett a bundle. If the options had vested, he would have been able to buy Oracle stock at forty-seven cents a share, at a time when shares were trading at twenty-seven dollars. Ellison told me that he canceled the options because he thought that Bennett was "no longer contributing" and was "just hanging around to vest."

Bennett said that wasn't so. "I said to myself, 'I could get pissed off and hire a lawyer,' " he said. Instead he relied on his Mormon beliefs for a response. He thought of the Bible story in which Joseph is sold into slavery by his brothers. Bennett decided to follow Joseph's example by being the best slave he could be. During his final two weeks at the company he worked eighteen hours a day finishing ads, he said. At one point he received an E-mail message from a woman in marketing who had heard of his dismissal. "Larry is one bloodthirsty troglodyte," the woman wrote. "Yes," Bennett wrote back, "but he's *our* bloodthirsty troglodyte." Having nothing left to lose, he sent copies of both messages to Ellison. Later Ellison changed his mind and allowed the options to vest.

Bennett was a salesman and a self-promoter, but he was also a Mormon, and he took his faith seriously. Years after he left Oracle, I asked him how he thought God would have looked upon his work there. He never lied for Ellison, he said. He may have embellished or misdirected—that is what admen *do*—but believed he had not lied.

Nor did he believe Ellison had done anything to deserve condemnation. "He basically uses people and spits them out. If that's a moral flaw, then so be it," he said.

Bennett was forgiving of Ellison partly because he saw him as an important figure in the life and mission of the Church of Jesus Christ of Latter-day Saints. The reason was fascinating. The three principal goals of the church were to perfect the living, proclaim the Gospel, and redeem the dead. Redeeming the dead meant posthumously baptizing the legions who died without accepting the truth as the Latter-day Saints saw it. To keep track of all the people who had lived and died beginning with Adam, the church used various kinds of database software, including some made by Oracle, which of course was founded by Latter-day Capitalist Larry Ellison. You might say he *begat* the relational database industry. That was why Rick Bennett believed that Ellison was "nontrivially involved in a full one third of the mission of Mormonism." From Bennett's point of view, Ellison was not merely a clever marketer or a great motivator or an astute technologist. He was—this was simply Bennett's opinion, not the official statement of the church—"*in some way an instrument of God.*"

By the mid-1980s Bob Miner was exhausted. Though he occasionally managed to make time for chess and backgammon and always made it home for dinner at six-thirty, he also spent many long nights in his home office, writing and debugging versions of the Oracle software. That grueling work was always done under pressure. He and his developers were constantly scrambling to solve customers' problems and to make good on Larry Ellison's grand promises. By the time the company began to work on Version 5, Miner was ready for a break. "He was really burned out, and he pulled back and did more managing and not so much developing," his daughter Nicola said. "He told me he couldn't do it anymore. He just couldn't put the energy in. He was so tired."

Soon Ellison found someone to take over the day-to-day responsibility for developing products. That man was Wayne Harvey, an engineer in his early thirties who had worked at Ross Perot's Electronic Data Systems Corporation and at Tandem Computers Incorporated, a hardware maker. Harvey's management style could not have been more different from his predecessor's. Miner let people do things at their own speed, in their own time, understanding that the job would get done eventually. Harvey, by contrast, was just as interested in when work got done as in how it got done. "He made everybody sit in rows and work according to deadlines," said Sarah Crawford, then an Oracle lawyer. "He brought order to the chaos that was development."

Harvey demanded a lot from his people, but he could also be warm and thoughtful. On several occasions, when lawyer Crawford needed help understanding some aspect of Oracle's technology, Harvey took time from his busy day and answered her questions. When programmer Bill Friend left Oracle to start his own company, Harvey gave him a going-away present, a book about Silicon Valley success stories. Other people in the company said Harvey would often bring his son, Alexander, then just a toddler, to the office for a visit. "He was one of the nicest people I've ever worked for," said Habeeb Qadri, who ran Oracle's data center. "Oracle was one of the few companies that could absorb Wayne's energy and intellect."

Maybe so, but sometimes his energy was too much even for Oracle. Harvey was known for working two or three days without a break, taking a day off to sleep, and then doing it all over again.

That worried people. "Wayne didn't sleep. He didn't sleep," Jenny Overstreet said. Said Qadri: "Sometimes I'd kid him and say, 'Wayne, if you keep this up, you're going to explode, and I'm going to come to work and there's just going to be a pile of ashes.' " Harvey always assured Qadri that he would be OK.

Wayne Harvey was not OK. He had a long history of mental trouble and had been seeing professionals for years. He and his wife, Sandy, had recently separated and were selling the house they owned together. According to Qadri, Harvey was afraid that his wife would move his two children out of state and he would not see them anymore.

This may or may not have been a rational fear, but it was real to Harvey, and the strain on him showed. He was prone to sudden, explosive outbursts of temper at work; one minute he would be all right, and the next minute he would just detonate. Roy Bukstein, then Oracle's treasurer, remembered Harvey's getting upset about something, stomping out to his car, and peeling away, his tires smoking. Qadri also saw Harvey erupt over what Qadri thought were little things.

Harvey was often irate at Larry Ellison, and it was no wonder. As head of development Harvey was there to build whatever products Ellison wanted built. That was no easy job. The key to "the abilities," Ellison's marketing strategy, was portability, the idea that Oracle software could run on any kind of hardware. When the company was preparing to go public in 1986, it listed no fewer than twenty-five computer operating systems on which Oracle was supposed to work. And just as he had done when Miner was running development, Ellison was constantly selling new products to customers and then demanding that his developers build them, and build them quickly. "He'd send E-mail down to development and say, 'We need a port to the Maytag washer,' " one employee said,[4] and the developers would be expected to get right on it. Another employee[5] said Harvey would become "violently upset" when Ellison made commitments to customers without first checking with him.

Eventually Harvey rebelled. "He had tremendous integrity. When [Larry] said, 'We need Oracle to run on Nintendo,' he said, 'I'm sorry, I don't think that's going to work,' " lawyer Sarah Crawford recalled. "I think he said that one too many times."

Larry Ellison agreed that Harvey had a difficult job. "Wayne's experience as an engineering manager at Oracle was fairly typical. He was under a lot of pressure. Dates were foisted upon him, and he had to deliver. I would argue that they were not terribly unrealistic dates, but it's true that he didn't make his own schedules. Schedules were given to him. And the resources necessary to make the schedules were given to him also." Harvey was constantly struggling to meet Ellison's demands. Could he build the products? Could he make the deadlines? What would happen if he couldn't? He had trouble dealing with uncertainty, Ellison said. "And in our business, with the development of any new technology, there's tremendous uncertainty."

Finally Harvey cracked. "He just sort of lost it in a meeting," Ellison said. Harvey got angry about something, "and in a monotonous but very malevolent tone, he just kind of attacked me." Jenny Overstreet, who was in the meeting, remembered saying to herself, "I have no idea what's going to happen in the next thirty seconds." The scary moment passed. After that Ellison briefly considered buying a gun because he feared that Harvey would "walk in the door and shoot me." He and some of the other managers came up with a secret nickname for Harvey: Mass Murderer.

Ellison could no longer put up with Harvey's recalcitrance and incendiary temper. First, he demoted him to Oracle's data center, a bitter humiliation but one that Harvey hoped to endure. Habeeb Qadri used a hockey metaphor to describe Harvey's predicament. "He was in the penalty box," Qadri said. "But he was too smart, too vigorous, too everything to be kept in the penalty box for long. He thought if he stayed in the penalty box and kept his nose clean, he could come out in a few months." Harvey was wrong. After a while Oracle fired him.

Harvey seemed to deteriorate after his firing. One night Qadri took him out and bought him dinner. Afterward Harvey told him, "Next time we go out, my wife and I would like to treat you to dinner," Qadri knew that Harvey and his wife were separated and that the dinner would never take place. "He was sort of out of touch with reality sometimes," Qadri said.

Harvey soon got a job with another software company, but he

never really regained his balance. One night when five-year-old Alexander was staying at Harvey's house, Harvey went to his bedroom with a .22-caliber handgun and killed him. Then he knelt by the bed and shot himself several times in the chest. Harvey's estranged wife found them both in the morning. Harvey left a note telling his wife that he loved her and asking to be buried next to Alexander.

Some Oracle employees read of Harvey's death in the newspaper that Monday morning, and others got the news when they arrived at work. People were in shock; many had liked Harvey, and almost everyone had seen little Alexander toddling around the office. How did Ellison feel that day? As was often the case, it was hard to know. "Larry tried to get the company together again on Monday. 'Move it along, let's ignore it,' " one employee said.[6] Whatever Ellison felt, he kept it to himself.

Habeeb Qadri, who knew Harvey well, believed that Harvey might never have died if Oracle had not let him go. "If you take a high-energy person like that and then put them out to pasture, they just self-destruct, which is exactly what he did," Qadri said.

But the truth was that Oracle was never the right place for Wayne Harvey. "There are people who fit into the Oracle mode real easily and real well. And there are people who may be equally bright but don't fit in at all," Bill Friend said. "And he definitely fit into the latter category." Harvey could not take the pressure: the high expectations, the nonstop quarreling, the endless workweeks, the ever-changing demands. Larry Ellison's company was no place for the weak, and Harvey was, at least near the end of his life, extremely weak.

"There was a meeting at Oracle where we started talking about Wayne," Ellison said. "Bob Miner got up and said, 'Well, Wayne was really a nice guy and a great guy,' and all of this. And I stood up and cut him off. I told Bob to sit down and shut up. I said that he could make up the world as he wanted to see it, but Wayne was *not* a nice guy."

Several people from Oracle attended Harvey's funeral. Ellison did not.

Nine

IN JUNE 1988 GARY KENNEDY WAS, AT AGE THIRTY-FIVE, A SENIOR vice-president of a $282 million-a-year company and a millionaire many times over. Still, he wanted something: power. And he wanted to get it by being recognized as the number two man at Oracle. The trouble was that another one of Larry Ellison's lieutenants wanted the same thing. His name was Jeff Walker.

Walker was forty-two years old when he was hired at Oracle. He had been the founder of Walker Interactive Products, which sold financial applications software to big businesses. These applications—a general ledger program, accounts receivable software, and so forth—let companies keep track of their finances by computer instead of by hand. In 1985 Walker was ousted by the Walker Interactive's directors, who decided he wasn't the kind of person who could lead the company into the future. (One engineer who worked under Walker[1] said Walker was strong on strategy but weak on implementation.) Ellison hired Walker to develop applications at Oracle, believing—correctly—that there was a booming market for applications software, which Ellison wanted to capitalize on. When Walker started, he was the only employee of the division.

Walker soon became Ellison's golden boy. A year or so after he started, Ellison promoted him to chief financial officer *and* head of applications. One man, two demanding jobs. Walker believed he was up to the challenge, but others thought the job of CFO alone was too much for him. He was not an accountant and had no experience as a chief financial officer, and it showed. Once computer industry analyst Stephen McClellan went to Walker's office for a meeting and found him reading a cash management textbook. "I said, 'Oh-oh.' That told me that he was going to figure out software finance as he went along," McClellan said.

Ellison liked Walker partly because he wasn't an accountant. "Larry thought finance people were idiots. Bean-counting idiots, pedestrian shitheads," said Joe Costello, chief executive officer of Cadence Design Systems, a Silicon Valley software company. Costello joined the Oracle board of directors in 1990. "[Larry's] reason to have Jeff Walker as CFO is he's an operating guy, understands the business, and he's smarter than hell, he's the smartest guy. Larry's view at the time was that brains conquer all. . . . Well, Jeff Walker's a brilliant guy. But a horseshit finance guy." To be fair, even an experienced CFO would have had a hard time developing a whole new suite of complex software products while also running corporate finance. "I don't know anybody who can do that in a company that's doubling in size. I don't know anybody," Oracle board chairman Don Lucas said. "I can be criticized for allowing it to happen, and Larry can be criticized for doing it. It was too much for one person."

With the exception of Ellison, Jeff Walker was probably the most controversial person in the company. When I asked Oracle people about him, they talked not only about how smart he was but also about much they disliked him. His intelligence was undeniable. A graduate of Brown University, Walker made himself into a successful programmer just by reading textbooks about it. As a manager he had the ability to make extremely complex problems seem simple. "What Jeff was able to do was pull out a sword and slice through the Gordian knot," said Stephen Imbler, who was vice-president of corporate finance under Walker. "He was able to penetrate right to the heart of the matter and leave aside the irrelevant stuff." Unfortunately Walker also had "one of the fatal flaws of extremely smart people," Imbler said. "He assumed everybody else was stupid."

Nobody (with the exception of Ellison) became associated with the haughty, arrogant, self-congratulating side of Oracle more than Walker did. He had several ways of dealing with people who disagreed with him or just didn't understand him. One was patiently and reasonably to explain his point of view. Another was to scream.

Still another was to . . . speak . . . very . . . slowly, as if "he was your third-grade teacher trying to tell you why spray-painting obscenities on the wall was not a good thing to do," said marketing man Ken Cohen. Walker could be especially pompous when talking about himself. When I interviewed him, he said such things as "I'm a fairly well known industry figure" and "I was the very first person at Oracle who was what you might think of as a real sophisticated product manager." He also recalled a time when Don Lucas told him, "Jeff, you're a hell of a man." It must have sounded more convincing when Lucas said it than when Walker did.

Walker may not have been popular, but he had friends where it counted. "Larry, Bob [Miner], and I always felt like we were the only people who liked Jeff in the company," Jenny Overstreet said. When he was around the top brass, he was clever, articulate, witty, insightful—altogether good company. Perhaps not surprisingly, the people in the executive suites were never "victimized by his very caustic behavior and comments," Overstreet said. "He could rip you apart if he thought you weren't getting it or if you were in some way threatening him, or questioning, or anything like that. You get really good at that when you're really smart." Walker often quarreled with Larry Ellison's top developers, who were trying to make Walker's applications work with Oracle's core product.

Ellison was happy to have Walker as an intellectual sparring partner, but he could see why Walker had problems with others. "Jeff is a very smart, very hardworking, very disciplined guy who is used to relying on himself and his own judgment. . . . He wouldn't easily accommodate to someone else's view and wasn't a great compromiser," Ellison said.

The person he was perhaps most uncompromising with was Gary Kennedy. "Jeff and I had a lot of problems at Oracle," Kennedy said. "Both of us wanted to be the second guy. If there were a push between his organization and mine, he wanted Larry to believe him, and I wanted Larry to believe me." (When I asked Walker about Kennedy, he said it was "always a pleasure" to work with him. Memory can be that way.) What Walker and Kennedy fought about most often were Walker's financial applications. Walker wanted Kennedy to do a better job selling them, and Kennedy wanted Walker to do a better job building them. The tension in Oracle's management committee meetings was sometimes pretty thick. "Both of them were extremely territorial," Jenny Overstreet said, "and both were extremely covetous of anything that looked like comparatively greater strength or power."

Then came the shift in power. For a long time Kennedy had complained that Oracle's lawyers were taking too long to review and approve sales contracts. Some salespeople even referred to Oracle's legal department as the "un-sales organization"[2] or even the "sales prevention organization." Kennedy believed that moving the contract lawyers into U.S. sales organization would make the process of approving contracts more efficient. The meticulous bean counters who worked for CFO Jeff Walker (Imbler among them) were also cramping Kennedy's style, so Kennedy suggested establishing a separate finance organization under his control.

Walker argued strenuously against these changes. Some believed he did so because he did not want to yield power to his rival. But certainly Walker was smart enough to see the danger in what Kennedy was suggesting. If the sales organization started reviewing contracts and booking revenue—in other words, if the people who made the deals played a role in deciding whether the deals were good—the company would lose some of its internal financial controls. Giving U.S. sales that kind of power would be like "putting the fox in charge of the chicken coop," said Stephen Imbler.

That was exactly what Ellison did. "Horrible, horrible mistake. Horrible mistake. Horrible, naive. . . . I didn't know what I was doing," he said. He said things might have turned out OK if he had given Jeff Walker the authority to audit Kennedy's organization. So why didn't he do that? "I didn't even know that there was such a thing called internal audit," he said. "I had never heard the term 'internal audit.' Where would I have heard this?" Actually Ellison might have heard it if he had poked his head into various offices in corporate finance, where an internal audit organization was being built at that very time. (Ellison might never have heard the term "internal audit," but Walker and Imbler had.) Oracle's internal audit group was chartered to oversee international finance; by the time it got around to auditing U.S. sales, a disaster had already happened.

Soon after decentralization it became harder and harder for Oracle to collect the money owed by its domestic customers. Everything a sales organization does right or wrong is expressed in its receivables. If it is making solid deals, customers will pay their bills quickly. If it is making weak deals—offering extended payment terms, selling to cash-poor customers, and so on—payments will come in much more slowly, if at all. Larry Ellison's customers were taking forever to pay. Oracle finished fiscal 1989 with $583 million in revenue—and $262 million in receivables. The company ended 1990 with $970 million in revenue and $468 million in uncollected bills.[3] (By contrast, in 1994, after Oracle got its act together, it did $2 billion in business and finished the year with only $455 million in receivables.) At one point in 1990 Oracle was taking an average of two hundred days to collect from customers in the United States. The industry average was about sixty days.

The consequences of all this were horrendous. Oracle was spending like crazy to hire people and open new offices but was not taking in money at the same pace. "We front-ended all of our expenses," Imbler said. Cash flow slowed to a trickle.

"I didn't know how serious a problem that was," Ellison said. "Had I been an experienced executive, I would have known several quarters before we had problems that problems were on the horizon. I would have been able to slow the train down long before we ever hit the wall. But I wasn't an experienced executive, and Jeff Walker wasn't an experienced finance executive. None of us had been in big companies before."

Ellison may not have been seasoned, but it was hard to believe that he didn't know how bad the problem was. The Oracle board's audit committee was concerned enough to hold a special meeting about this in the fall of 1989, and in the early months of 1990 financial analysts raised the issue repeatedly. "There was a real fear that receivables could swamp the company," said Scott Smith, who reported on Oracle for Donaldson, Lufkin & Jenrette, Inc. Inside Oracle, Steve Imbler "was constantly on a warpath" over receivables.

He was hardly the first to go on a warpath. A full five years earlier treasurer Roy Bukstein had written his memo expressing concern over Oracle's sloppy way of doing business. "Our objective is to make sure we have a working system which results in the collection of these obligations on a timely basis," Bukstein had written in January 1985. Bukstein was now long gone—he had left the company in 1988—but the problem he had identified still existed. Maybe Ellison didn't know how serious a problem it was. The more believable scenario was that he knew but didn't do anything because he didn't want to inhibit sales.

As even Ellison later acknowledged, Oracle's growth was outpacing people's ability to manage it. It was one thing to double revenues from $1.2 million to $2.5 million, as Oracle had done years earlier, and quite another to go from $282 million to more than $500 million, as it was trying to do now. The company needed to exert more control over its finances, not less. But that wasn't what happened. As always, Ellison was interested in growth, and it was hard to blame him. The companies that cared less than he did about competition would soon be dead or on life support. Still, the unchecked growth was bound to catch up with Oracle sooner or later. The decision to decentralize finance ensured that it would be sooner.

One day during Oracle's boom years Larry Ellison received an important visitor. There was no special reason for this meeting; no business deals were pending. Bill Gates just wanted to chat.

"Bill gave me call. He wanted to talk about stuff," Ellison said. "Bill is relentless not only in gathering intelligence but also processing intelligence."

Though Oracle and Microsoft were in different segments of the software business—Ellison's company built database software for minicomputers, while Gates's wrote operating systems and applications for personal computers—the two companies were closely linked in history. Bill Gates and Paul Allen had officially formed their partnership in 1977, the same year Oracle was founded. The revenues of the two companies had grown at roughly the same pace into the early 1980s.

What was more, Oracle and Microsoft had gone public together. Oracle had made its initial offering on March 12, 1986, at an opening price of $15. The stock closed at $20.75 that day, giving Ellison a paper fortune of $93 million. The next morning, March 13, Microsoft went public at $21 a share, closing at $28. Gates's holdings were worth more than $300 million.[4] The race was on, with Gates in the lead.

At the time that Gates visited Ellison, Microsoft did not dominate the computer industry as completely as it did in later years. But Gates was a major player even then, and Ellison treated him as such. On the day of the meeting Ellison sent Oracle cofounder Ed Oates to San Francisco International Airport to meet Gates's plane. Oates was an appropriate choice because he was vice-president of the personal computer products division. It was his job to make sure that the PC version of Oracle worked with Microsoft's DOS operating system and the OS/2 operating system then in development. As Oates waited for the plane, he thought jokingly about what he would do in the unfortunate event of a crash: Call his broker and go short on Microsoft.

Gates arrived without an entourage. Oates led him to a Porsche 911 Turbo cabriolet, a speedy (and very expensive) little convertible. Oates knew Gates would appreciate riding in a Porsche; at the time, according to press reports Oates had seen, Gates was awaiting delivery of his own customized Porsche, a 959. At the end of the twenty-minute drive to Ellison's home Oates delivered Gates to Ellison's front door, watched as the two men shook hands, and then left. Ellison had made it clear that this would be a private meeting.

Even the superwealthy Gates must have been impressed when he saw Ellison's house in Atherton, which Ellison had bought and spent several years and several million dollars improving. Modeled after the Katsura Villa in Kyoto, Japan, the finished house had white oak floors, shoji screens on the windows, a gallery where Ellison displayed his collection of antique Kabuto helmets, and a Japanese tea ceremony room. The guest bedroom had tatami mat flooring and a view of the garden. And what a garden it was. Ellison's landscape architect had created an authentic Japanese environment, using cherry blossoms, bonsai trees, and Japanese maples. In a magazine article about the house a writer breathlessly intoned, "The tranquil setting has the richness and delicacy of a Japanese brush painting."[5]

Years later Ellison would not say what he and Gates talked about, and Gates declined to be interviewed. One thing was clear: The exuberant, loquacious Ellison did most of the talking, and the sharp, cagey Gates did the listening.

"Larry liked him instantly," said Jenny Overstreet, who spoke to Ellison after the meeting. "God, this was a smart man. At that meeting and subsequently Larry knew that he really needed to be on his guard about revealing too much. Larry is such a stream of consciousness communicator that he was just going and going and going. Finally he just stopped himself. He realized Bill wasn't saying anything. He was just listening with his head tilted to the side. Larry said to himself, 'Shut up, you fool! Shut up!' This wasn't like all those other people that Larry spouts off to all the time . . . Bill wasn't talking back. He was just absorbing. Larry knew that there was a pretty awesome mind in there." Overstreet's conclusion: "It was a meeting of the minds. At least Larry met Bill's mind."

According to Ellison, Gates "was on an intelligence-gathering mission. He was curious. He was fully prepared to spend all of his time listening . . . I kind of broke off talking with Bill after a few of these things, for fear that in my own exuberance I might disclose something that might be useful to Microsoft. Let them figure things out for themselves. They've got so much talent and so many bright people and they work so hard that they don't need my help."

Even in that first meeting Ellison and Gates quietly competed. According to Overstreet, Gates "thought Larry's house was cool, then said, 'I'm building a bigger one.' " Ellison, showing off in his own way, drove Gates back to the airport in his Ferrari Testa Rossa.

When I asked Ellison about that first meeting, he seemed uncomfortable with the idea that Gates had got the best of him. He said Gates was "very focused, relentless, with incredible endurance," but would compliment him no further.

Then, without pausing, he said: "One story I remember. I was talking [by phone] to Bill Gates about an issue, and he and I disagreed about something, and he got off the phone. He called back two hours later and continued the conversation. He had thought about it for two hours solidly. Which is something I would never do. I can't imagine."

I asked what the issue was.

"I don't remember what the point was, but actually he conceded that I was right and then just moved on, which I found extremely scary: that he would consider something, consider his own position, consider this alternative view, weigh the merits of each argument, and then adopt the best view. And it was irrelevant whether the best view was his or someone else's." What Ellison wanted me to know was that he had been right, Gates had been wrong, and Gates had admitted it.

But Ellison was also making another point, one that had much less to do with his ego. His point—a valid one—was that Microsoft was known not for innovation but for brilliance in business. Microsoft did not create the DOS operating system but bought it from someone else. It was not the first to go to market with point and click PC software; Apple was. Microsoft did not pioneer the use of the World Wide Web browser; Netscape did. Microsoft did not conjure ideas, it took ideas and ran with them, much as Ellison had done years earlier. In his backhanded way Ellison was giving Gates credit for that.

"Bill does not suffer from the not-invented-here syndrome," he said. "That's what has made them such magnificent followers. *Embrace and extend.*"

Oracle released Version 6 of its database software in November 1988. Version 6 had the potential to be the company's best product yet. Ellison's top engineers had spent a couple of years rewriting the database kernel, the software's heart, and for the first time the software had a feature called row-level locking. In the past someone trying to make a change in a database couldn't do it if another person was working on a nearby column or row. (A hotel clerk might run into this problem when trying to book a reservation.) With row-level locking, a user would monopolize only one row in a database table, allowing other people to make changes to all the other rows within the same table. If Version 5 was the first Oracle release to work really well, Version 6 was the first that was supposed to be truly great.

It wasn't, at least not at first. "Everything was wrong. It wasn't working. Performance was bad, it wasn't reliable, and a lot of features were missing," said Richard Finkelstein, the Chicago consultant who specialized in relational technology. Right away Finkelstein told his clients not to upgrade to Version 6 until Oracle got the bugs out. A lot of customers who did try to use Version 6 were miserable. Large numbers of people would be sitting in front of their personal computers, working with the software, when the screen would freeze. No amount of key pressing or space bar banging or oath issuing would make something happen. "You can't do anything anymore. All you can do is sit there and stare at the screen or go out and get a cup of coffee," said Tony Ziemba, a longtime Oracle user from New York.

To Ziemba, the release of bug-infested Version 6 felt like a betrayal. Like a lot of other people, he had built his career on Oracle. A veteran of the financial services industry, he had been using Oracle software at work since the uncertain days of Version 4. In 1985 he managed to build a portfolio management application on the PC version of Oracle, making him one of the first Oracle customers to put that software to good use. By 1988 Ziemba knew people in Oracle management and was active in the New York Oracle Users' Group, an association of professionals who used Oracle at work. He was a loyalist. When Version 6 came out, Ziemba was insulted. How could Oracle do this to him?

He knew what had gone wrong. At the time Oracle was under constant attack from competitors. One of the ablest was a company called Sybase, whose database management system had some features that Oracle lacked. Sybase was still relatively small, not an immediate threat to Oracle, yet it was getting lots of attention. Oracle customers were jealous, and Oracle's sales force was screaming for something new to sell. So the company released Version 6 before it was ready. As always, "the salespeople were running the organization," Ziemba said. "They decided when things would be released because they wanted to keep that cycle of one hundred percent growth going." A lot of Oracle users were furious, Ziemba among them. He said he called people at Oracle to tell them that the product didn't work, "but they said, 'Oh, no. It's fine.' "

Even so, plenty of people at Oracle knew Version 6 wasn't fine. Andy Laursen, the release manager for Version 6, said he "got called every day with a broken database someplace. . . . I can't tell you how many databases I patched. It was unbelievable."

Soon Oracle started getting bad press—very bad press. "There was a stretch for about six months where trade publications were running stories that said, 'Users bash Oracle,' " Ziemba said. He was the person making sure those stories appeared. In the months after the release of Version 6 he made it his business to publicize Oracle's foul-up. He contacted all the reporters who covered Oracle and all the financial analysts who followed it. When the reporters published stories about the problems with Version 6, Ziemba clipped them out and showed them to other Oracle users as a way of spreading the news. Why did he do it? Because he wanted the company to care as much about its products as he did. He wanted Oracle's attention. "I figured the only way to get it was with a two-by-four to their head," he said.

It wasn't just the software that made Ziemba angry. As far as he was concerned, Oracle's technical support staff, which answered customers' questions by phone, wasn't good for much either. For Oracle, doubling in size every year inevitably meant that some of the support people were new on the job and therefore didn't know much about the products—or at least didn't know as much as the customers did. "It was always a crapshoot when you called," Ziemba said. "You'd end up on the phone for hours. You'd get bounced around, and six or eight hours later you still wouldn't have the system working." Equally maddening were the times when Ziemba called with a problem and was told that Oracle already knew how to solve it but had not passed the word to its customers. "I'd say, 'Oh, great. There was a known bug, and you didn't tell me.' You start getting real nasty with the support people after you have a couple of these experiences," Ziemba said. One oft-quoted industry observer liked to refer to Oracle's customer support as "a form of grief counseling": Your problem never actually got solved, but you somehow learned to live with it.[6]

The reason Oracle's technical help wasn't better was that the company wasn't doing much to make it better. It wasn't that it did not want to support its customers. It was simply thinking about other things—namely, building and selling products. "What I feel most badly about," Gary Kennedy said, "is that we made conscious decisions to grow and to abrogate other responsibilities. For example, if an outstanding job candidate came through the door and we had an opening in support and an opening for a presales person, we'd put them in presales. . . . There weren't enough people to go around. If we had to make that trade-off, in most cases we would push them toward presales. So we had insufficient resources in customer support." At the time Kennedy rationalized that Oracle software was allowing customers to do things they never could have done before; so what if the tech support wasn't the greatest? But years later he felt sorry that he had not done better by his customers. "There were people who bet their business on us—not just with the database but also with our application products," he said. "Some of them lost their jobs because we didn't perform to their expectations."

Oracle's top managers were working hard to soothe customers. A year after the initial release five Oracle executives, including Derry Kabcenell, a senior engineer, flew to St. Louis to meet with about fifty unhappy members of that city's Oracle Users' Group. The group was in an uproar over Version 6. A user from General American Life Insurance found that the release was actually slower than the previous version—hardly a strong selling point. "Oracle said our system was not tuned well, so we got an Oracle consultant to take a look at it. He did the best he could, and it made little difference," the user said.[7] An Oracle user from Monsanto was able to get Version 6 up and running, but only after "a long time and a lot of calls to Oracle." The database administrator for Florists' Mutual Insurance had so many problems that he decided to keep using Version 5 until Oracle made Version 6 work: "We couldn't afford to spend all day testing Version 6, and we didn't have enough room on our machines to have both versions up at the same time."

Kabcenell had to respond to those complaints, and more. There was little he could say. He admitted that even though the systems were advertised as being ready for production—ready to do the critical jobs that businesses needed done—they were still buggy. Oracle had not released the bad software on purpose, he said.

"If we had known about these bugs at the time we shipped them, we probably would not have called them production," Kabcenell told them.[8] "All software has bugs in it, and you know that. You guys write software. At some point you feel that it is reasonable to call the software production because virtually everybody can make effective use of it." Kabcenell said Oracle had made a misjudgment: "We sent out some releases that were labeled production which later turned out to be bugs."

When Kabcenell was asked which release of Version 6 was truly production-quality, he answered Version 6.0.27, meaning that the software had gone through twenty-seven fine-tunings since its initial release. The users were happy to hear that there was a production-ready version of the software at long last. But that would not be the end of Oracle's troubles, not by a long shot. While the product may have been fixed, the company itself still had a lot of bugs.

Oracle Corporation had come a long way since its precursor, Software Development Laboratories Inc., was founded in a small corner of the Precision Instrument building in 1977. It had moved from there to 3000 Sand Hill Road in Menlo Park, then down the street to 2710 Sand Hill Road, and then to an eighty-four-thousand-square-foot building at 20 Davis Street in Belmont, twenty miles south of San Francisco. As the company grew larger and larger, its headquarters continued to drift north, partly as a concession to Bob Miner, who still lived in a Victorian house in the city.

In 1989 it was time to move again. On May 31 Oracle employed 4,148 people, of whom 2,151 were in the United States and 1,997 were in twenty-four other countries. The company needed more space, and this time Ellison hoped to find a place where Oracle could stay permanently, no matter how large it got.

He found just such a place next to Highway 101, in a town called Redwood Shores. There Ellison leased two gleaming blue glass towers, one for administration (this was called Larry's building) and the other for research and development (Bob's building). (In an unusual arrangement Oracle leased the buildings but bought the land beneath them for a little more than nine million dollars.) Together these towers offered five hundred thousand square feet of office space, more than enough for Oracle's immediate needs. They also established Oracle in a visible way as a major player in the high-tech business and in the Silicon Valley economy. Oracle Corporation was now a big deal—and looked like one.

Bob Miner was a little uncomfortable about that; even though he was now a huge success, worth hundreds of millions, he didn't like to be seen as anyone important. One day before the company moved into the new headquarters, Miner and Oracle engineer Roger Bamford visited Miner's building. A security guard saw them looking around inside the empty building and—unaware of who Miner was—said, "You can't be in here." Bamford immediately tried to set him straight. "Wait a second," he said. "This is Bob Miner's building. *And this is Bob Miner.*" It was too late. Miner was already apologizing and heading for the exit.

The more software Danny Turano sold, the more money and praise he got, and the more he wanted to sell. He soon found a way. Normally Oracle sold software according to the number of people who were going to use it. A customer who wanted a hundred employees to use Oracle would pay a certain sum for the privilege; someone who wanted a thousand licenses would pay proportionally more. Turano devised another, more lucrative way of making deals. Instead of selling software according to the number of users, he sold site licenses—licenses that allowed an unlimited number of people in a company to use Oracle.

Turano gave his customers strong incentives to buy site licenses. For example, if an insurance company wanted five hundred licenses, he would quote a price. Then he would say something like, "You're going to be needing a lot of Oracle. For just a little bit more I'll sell you a license for an unlimited number of users." The customer would get the additional licenses at a steep discount, compared with the per license rate. Turano was bringing in huge deals, some worth as much as two or three million dollars. Tack on 15 percent a year for a maintenance contract—you had to have a maintenance contract to get enhancements and technical support—and the deals were even huger.

But selling site licenses had a downside: If you sold a site license, you could theoretically never sell to that customer again. Having stocked its shelves with software, what company would buy more?

Ellison came up with an answer: He constantly pushed his developers to come out with new or enhanced products that he could sell to his customers. In the world of high technology this was nothing unusual. Hardware companies expected their customers to junk their old boxes when they bought new ones; that was how they stayed in business. "Larry's mentality was, 'Hey, why can't you do the same thing with software?' " saleswoman Kitty Cullen said.

The trouble was, many customers thought they should get some of Oracle's new products for free. After all, they were paying Oracle 15 percent a year for technical support and *enhancements.* These customers were furious when Oracle came out with something new and then charged them for it. For example, a lot of customers thought they were entitled to the new versions of Oracle that resulted whenever Ellison's engineers made the software work on a new kind of computer, but Oracle always charged for those new versions.

"Many times customers believed that they could just get whatever Oracle had whenever they had it," Turano said. This wasn't because the customers were irrational; rather it was because Oracle's "salespeople wouldn't go out of their way to make it clear what the deal was," Turano said. The terms of these deals were always specified in the sales contracts, of course. But the corporate information guru who negotiated with Oracle usually didn't review the contract personally. A lawyer did that. And the lawyers generally didn't understand software as well as the information people did. Oracle took advantage of that knowledge gap, Turano said.

Oracle Version 6 was the source of a lot of arguments. There were actually two versions of Version 6: the basic one and a sort of turbocharged version that included the transaction-processing option. Some Oracle customers thought they should get the turbocharged version for free, but Oracle demanded payment. "The federal government had a big problem with that," said Cullen, who sold to federal agencies in Washington. "They have limited dollars, and they invest in a product, and they figure it's a partnership."

According to many former Oracle people, the company wasn't as interested in forming partnerships as it was in making deals. "The way Larry's mind would work was he really saw the customer a lot of times as a source of money, not as a user of the product or as a partner in solving a problem," former sales chief Mike Seashols said. Oracle sales had a reputation as a "sell and drop" organization: The salesperson sold you, then dropped you. "The sales rep was given incentive to basically find a new deal, close that deal, and then move on to the next deal," Turano said. "So the guy who just bought from him probably wasn't going to hear from him for a long time, if ever." Oracle user Tony Ziemba said, "It didn't matter [to Oracle] whether you got the software installed and working or not." It mattered only that you bought it.

If customers felt duped when they didn't get the enhancements, why didn't they just find another software vendor—say, Informix or Sybase? One reason was that Gary Kennedy and his salespeople often found a way to mollify them. According to Kennedy, if he believed a customer really deserved a piece of software, he handed it over for little or no cost. He was the boss; he could do that.

But another reason customers didn't defect was that moving to another kind of software would have cost even more than paying for the enhancements. If a company changed vendors, it would have to rewrite all its applications, a hideously expensive task. Oracle's customers may have had options, but none that made any sense. "There's no question that Oracle understood, and Larry understood, and our marketing staff understood, that people who buy databases buy them for at least ten years. Because you really do lock into a company," Seashols said. Or, as Danny Turano put it, "You had them essentially over a barrel." Ellison said he saw nothing unethical about Oracle's behavior. One might disagree with Oracle's business practices, he said, but it was not fair to question the company's ethics.

Whether this was a question of ethics or business practices was just semantics. The fundamental question for any businessperson is, *What are you prepared to do to get ahead?* And whether or not anybody thought about it, Oracle faced that question every day. What would Oracle do to win? Was it willing to deceive its customers—or at least make them *feel* deceived—about what they would get for their money? Was it prepared to loose a sell and drop sales force on the marketplace? To put all the really good people in presales at the expense of customer support? To release software that wasn't ready in an attempt to stay competitive?

Gary Kennedy: "The motto—or the epitaph—could well be written, 'We did what we had to do.' "

In many ways Oracle's vice-president of corporate finance, Stephen V. Imbler, was a typical employee: intelligent, multitalented, and willing to give his entire life to the company. Imbler had been a straight A student at the University of Texas and had made the highest score at the university on his graduate school entrance exams. After finishing his education, he had worked for several years at the accounting firm Peat Marwick, where he had earned as much as $58,000 a year. Oracle had recruited him in 1987 and paid him $90,000, plus a bonus. Within three years Imbler was earning $170,000 a year. Welcome to Larryland.

He worked virtually all the time. "One of the reasons I married my wife is that I was dating her when I started working at Oracle.

After working at Oracle for a while, I realized I was never going to meet anybody else because I was working all the time. So I might as well marry the girl I was dating," Imbler said. "Oracle did tend to be all-consuming in that way." Years later Imbler and his wife were still happily married.

Imbler had the sort of personal background that appealed to Larry Ellison's hiring people and to Ellison himself. The eyebrow-raising entry on Imbler's résumé was that he had once enjoyed a career as a concert pianist. Not long after he was hired at Oracle, he played a concert at company headquarters, which by then had moved to a big building in Belmont.

In the fall of 1989 Imbler began to see some of the fox-in-the-chicken-coop effects of decentralization. Under the standard accounting rules of the time, Oracle's finance people could recognize revenue only when the company had a written commitment from a customer and expected to be paid. If a customer didn't have enough money in the bank to make good on its purchase commitments, the deal shouldn't be booked. But the people working under Gary Kennedy in the finance department of U.S. sales had a different view, according to Imbler. They were trying to get Imbler to book as many deals as possible, even when it was doubtful that Oracle would ever be paid.

For example, in those days Oracle was selling a lot of software to small, minority-owned businesses that did computing work for the federal government. These companies were known as 8As. According to Imbler, Kennedy's people consistently argued that those deals should be booked as soon as the contracts were signed. "But the fact is that eight As were undercapitalized businesses, and they were really only going to pay you if the government paid *them,*" Imbler said. There was no way of knowing which of those companies would ever finish their work for the government and thus get the money to pay Oracle. Some 8As even bought software from Oracle in hopes of getting a government contract sometime in the future. For Oracle, booking such deals was speculative at best. Years later Kennedy acknowledged that Imbler's concerns were "absolutely valid." But at the time Kennedy's staff fought Imbler.

Sometimes Imbler refused to book deals because the sales force wrote outrageous things into the contracts. Once sales wanted Imbler to book a big deal it had made with the U.S. Postal Service. When Imbler read the contract, he wasn't sure the post office had actually made a commitment to buy the product. Without telling anybody in sales, he called the postal service employee in charge of the contract. "Oh, yes, I love my option," the person told him. "It is a wonderful option, and someday I may exercise my option to buy Oracle." Imbler didn't book the deal.

Another time Oracle Europe—for which Gary Kennedy was not responsible—agreed to sell a customer every new release of Oracle's financial applications package through Version 11. (The company was selling Version 6 of that product at the time.) Along the way Oracle was supposed to provide a Turkish-language version of the software. Imbler just shook his head. There *was* no Turkish-language version of Oracle. Sure, maybe someday there would be, but what if there wasn't? The customer would demand a refund. Imbler wasn't about to book revenue in which Oracle had to fulfill some contingency to make good on the deal.

But Kennedy's organization was under constant pressure from Ellison, so it was only natural that U.S. sales wanted Imbler to book every deal as revenue. Every time Kennedy's people got a deal past Imbler, they came that much closer to making their number. At first Imbler's organization reviewed only deals worth more than half a million. But when Imbler began to see problems in the contracts, he lowered the threshold to two hundred thousand dollars. Eventually he and his people started scouring just about every contract that came in. "I found myself probably spending a quarter to half of my time trying to figure out what sales was doing," he said.

Imbler was doing this in defiance of orders. A year or so after decentralization, he said, Jeff Walker told him not to make any more revenue recognition decisions for U.S. sales. Gary Kennedy had his own finance people now; let them decide when to book deals. Imbler disregarded Walker's directive. Imbler and his controller, Tom Williams, were worried that the company was heading for dire financial trouble and thought they had to do something about it. "We took the position that we could overrule any decision that was made in U.S. finance," Imbler said. Soon Imbler was in what he called a "civil war" with Kennedy's finance staff.

Why wasn't CFO Walker more supportive of Imbler? One problem was that Walker, like Ellison, was an entrepreneur, not an accountant; he lacked the inherent conservatism and even suspiciousness of the classic numbers-crunching CFO. According to Imbler, his attitude toward corporate finance seemed to be: What I don't know won't hurt me.

Besides, Walker faced an inherent conflict in overseeing both the applications division and corporate finance. "His bonus was dependent on the salespeople's ability to sell financial applications," Imbler said. "And sales could sell financial applications only in the way they sold all their other products, which was under big quotas and lots of pressure. So Jeff was not really going to be proactive in figuring out ways to limit sales' ability to book revenue."

Imbler went on. "Jeff was like the brakeman on a train. And the brakeman was up there shoveling coal on the fire. Not that he was doing anything wrong; he just wasn't pulling the brakes as the train was going around the corner."

To be fair, Walker had tried to pull the brakes a couple of years earlier, when the idea of decentralization first came up, but Ellison wouldn't let him. One could argue that Walker should have resigned if he felt he didn't have the authority to be a real CFO. But as long as he stayed, there wasn't much he could do.

By early 1990 all the unnatural acts that salespeople were committing to make quota were becoming apparent. At the time Oracle had a policy that allowed salespeople to bring in a contract up to four days after the end of a quarter and still have the deal count for the previous quarter. The idea was to give salespeople a few days to mail or FedEx their contracts to California. But Imbler found that U.S. salespeople were abusing this by using the four additional days to negotiate more deals. (An Oracle salesperson might jokingly refer to March 4 as "the thirty-second of February."[9] Under generally accepted accounting principles, that was not allowable. The rule was that you had to book revenue in the same quarter when you made the deal.

Once an Oracle salesman tried to get credit for a late sale by faxing in a contract with the date cut off. "That's just bullshit, and nobody in their right mind would think that's the way you should do business," Jeff Walker told me. "But it's obvious somebody did. So you find out who it is and you fire them."

But it was rare for a salesperson to be fired, or even disciplined, for such misbehavior. Oracle reserved the right to take back a sales representative's commission if a deal went south, and occasionally it did so. But the threat of a claw-back was not much of a deterrent.

"There was no downside to booking a bad deal. If you didn't book the deal, you weren't going to get your commission. If you booked a bad deal and people later found out about it, you later got your commission clawed back. That's not much of a downside," Imbler said. "In general, salesmen did not get fired, did not get fined, and did not get heavily penalized for questionable sales practices. And that's one of the things that led to the culture that was very difficult for finance. If all you can do is take back a commission, and that's all that's going to happen, they'll try to run three deals past you, and if they get caught on one, hey, they still made money on the other two."

Danny Turano confirmed this. Some sales reps at Oracle even signed customers' names on contracts and sent them to corporate, he said. The salesperson would get his commission right away. But when Oracle billed the customer, it would receive no payment because the customer's accounts payable department had no purchase order for the deal. "It would take months for Oracle to figure it out," Turano said. By then the deceitful sales rep would be long gone.

That wasn't the only outrage. Imbler said he once reviewed a contract in which somebody sold forty thousand dollars' worth of software to a company in Los Angeles. It was a piddling little deal—at the Oracle of the late eighties and early nineties forty grand was lunch money—and Imbler might not have paid any attention if the customer's address had not been a post office box. What kind of company has a post office box as its only business address? A fictional one, in this case. "Apparently a sales rep had just set up a post office box, made up the name of the company, and then got the deal and the commission on it."

Another time Imbler discovered something called a side letter. A second-level sales manager had signed a valid contract with a customer and then had given the customer a secret letter that restated some of the terms of the contract. The contract itself was sent to corporate, but the side letter wasn't. So Oracle finance was in the position of unknowingly booking revenue for a deal that the customer had a right to undo at any moment.

From Imbler's point of view, that was when the behavior of U.S. sales veered into the criminal. "To me, writing a side letter for a public company is like securities fraud," he said. "You are causing the company to misstate its revenues in a public filing, you are getting your money, and you're leaving the company with a horrendous downside."

Imbler mentioned the side letter in an Oracle executive committee meeting that was attended by Ellison, Gary Kennedy, and others. Imbler argued that the sales manager who wrote it should be fired. According to Imbler, Ellison listened to Imbler, turned to Kennedy, and said, "Is Steve telling the truth? Is this really happening?"

Kennedy said it was.

"I think it's about time we changed that," Imbler remembered Ellison saying. Kennedy said he would take care of it. Imbler assumed that Kennedy was using the phrase "take care of it" the way a mafioso might use it, meaning that he would, um, terminate the offending employee. But Imbler later learned that the person did not lose his job. So much for Gary Kennedy's reputation as a hatchet man.

"I took him from second-level manager to a salesman and in retrospect probably should have fired him," Kennedy said. He felt he could rehabilitate the person; the idea of giving up on him "was almost counterreligion." He eventually sent a memo to his sales staff saying that writing side letters was "a no-forgiveness kind of offense.

Don't do it." Still, it's doubtful that the memo had the same deterrent effect that a public firing would have had.

Kennedy said he never saw another side letter, but apparently there were others. Turano, who left Oracle in 1989, said writing secret letters to customers was "in vogue" near the end of his time there. According to Turano, one sales representative made a $980,000 deal with a customer, then provided a side letter saying the customer could back out at any time. The customer did. When Oracle's collections people asked for payment, the customer produced the side letter. End of deal. There were plenty of others that Turano heard about.

Of the many people in Oracle sales, only a small number ever wrote a side letter. But why would anybody do such a thing? People wrote the letters, Danny Turano said, because "they were on the hook for a fairly large commitment to Kennedy," who in turn was on the hook to Larry Ellison.

Ellison never intended for anyone in his company to commit securities fraud. Yet he bore at least part of the responsibility. His example—his bravado, his exaggeration, his punishing demands for revenue, his Genghis Khan business philosophy—created a culture in which side letters were not shocking; they were probably inevitable.

The third quarter of the fiscal year ended on February 28, 1990. By then Steve Imbler had uncovered fifteen million dollars in deals where it was doubtful that whether Oracle would ever get paid. There was a real downside to questioning these deals. Fifteen million dollars were a substantial chunk of the company's revenue for the quarter. Erasing that kind of money from the books would make Oracle's numbers fall short of Wall Street's expectations, and that in turn could trigger a drop in the stock price. But the downside of *not* questioning the deals was even worse. If Oracle booked those deals and never got paid, it would be deceiving the public about its quarterly results, and that could end a finance person's career if the Securities and Exchange Commission found out about it. Imbler had to tell Ellison about the problem.

He did not know how Ellison would react, but he was concerned. He knew that Ellison did not have much use for finance people, even those who could play the piano. As far as Imbler was concerned, Ellison had always treated finance as "a poor-sister organization." The evidence was everywhere. When the fast-growing Oracle began having serious cash-flow problems in 1987 and 1988, people in corporate finance were given morale-boosting T-shirts that said, "Happiness Is Positive Cash-Flow." But the effort to sustain positive cash flow began and ended with the T-shirts; there was never any serious attempt made to deal with the problem.

There were many more examples of Ellison's lack of interest in finance. In Imbler's three years at Oracle, Ellison had never once addressed his finance group, even though Imbler occasionally asked him to do so. Also, while Ellison insisted on hiring only the best people in his sales and development groups, he seemed to care less about the finance department. From the beginning of Imbler's tenure as a vice-president, Oracle forbade him from offering stock options to prospective hires.[10] Imbler did not know whether the directive came from chief financial officer Walker or directly from Ellison, but either way it did not go over too well with the prospects. "I had recruits who would ask about the options during an interview, and I'd say, 'There aren't any options.' They'd say, 'Well, let's just stop the interview right now. Let's not waste your time anymore,' " Imbler said. Meanwhile the finance people working under Gary Kennedy in U.S. sales were still allowed to offer options to *their* recruits.

Corporate finance got short shrift even in the boardroom. Virtually every Fortune 500 board of directors has an audit committee, a group of people who oversee the company's finances and provide guidance. At one point Oracle's audit committee consisted of exactly one person: Don Lucas, who could not have been expected to do the job alone. And of course the chief financial officer, Jeff Walker, was a novice who was also responsible for running one of the company's most important product divisions. It wasn't that financial controls did not exist at Oracle; they just weren't emphasized.

"I've never worried about running finance, even when I should have," Ellison said.

Walker may not have been a strong CFO, but he wasn't ignoring Oracle's problems. On at least one occasion he told Ellison that he was concerned about some of the deals Kennedy's organization was bringing in. Ellison didn't take him seriously. Walker and Kennedy were at war all the time; Ellison assumed this was just another one of their skirmishes. "It was very difficult for me to separate personalities from realistic objective judgment," he said. "But it's my fault; I was the one who created this stew of personalities. They all worked for me, I hired them all, I promoted them all, I put them in all of those jobs."

So it was left to Imbler—one of the "bean-counting idiots"—to tell Ellison that fifteen million dollars' worth of revenue was essentially fictional. The whole prospect made him so nervous that he first consulted with one of Oracle's lawyers about his rights and responsibilities. "I wanted to understand exactly what my obligations were with regard to the company, with regard to our external accountants, and with regard to the SEC. I wanted to make sure that I did the right thing. I wanted to make sure that I protected myself in case somebody tried to do the wrong thing to me." When he began his career in finance, Imbler had never imagined that he would have to consult a lawyer before meeting with his own chief executive officer, but such were the stakes at Oracle.

Nor did he ever suppose that a vice-president of corporate finance would have to justify his revenue recognitions to the CEO. The meeting never should have taken place, he said. "Finance just should have said, 'This isn't revenue, and that's it.' Instead I had to defend my decisions to people who didn't have the technical experience to question them," he said. He was referring to Ellison.

The meeting was attended by Imbler, Ellison, Walker, Kennedy, and one of Kennedy's finance people. Imbler had a worksheet listing all the bad deals, with the worst one at the top and the best at the bottom. He made it a point to include a few marginally acceptable deals on his list so he would appear to be willing to compromise. He discussed the deals one at a time, explaining why Oracle should not recognize the revenue. Kennedy's finance man "acted as the plaintiff's attorney," Imbler said; he argued to book the revenue. According to Imbler, Walker "was largely silent," though Imbler thought Walker deserved credit for helping to arrange the meeting in the first place. The decision about whether to book the revenue was entirely Ellison's. He could recognize millions of dollars in revenue that Oracle might never receive, or he could do the right thing.

He did the right thing. The man who had built his business by pushing for more and more revenue accepted the advice of a bean counter and erased about fifteen million dollars from the books, just like that. Imbler was immensely relieved. Ellison might not have been interested in finance. But once you got his attention—once you sat down and showed him a list of shaky deals—you could count on him to do what really was best for Oracle and its stockholders. Unfortunately Ellison's conservatism was too late and too short-lived.

Ten

THE LIFE OF A BIG CORPORATION IS COMPLEX AND MULTIFACETED; great successes don't happen in a day, and disasters don't either, usually. But if you had to name the date when Larry Ellison's dream began to crumble, Tuesday, March 27, 1990, would do pretty well. That was when Oracle released its financial results for the third quarter of the 1990 fiscal year.

Oracle dropped the bomb after the close of the financial markets, in one of those good news/bad news press releases that corporate communications departments often wrote. The good news came first, of course; Oracle announced that third-quarter revenues increased 54 percent, to a record $236.4 million. The bad news was buried at the bottom of the second paragraph of the press release: Profits increased only 1 percent compared with the same quarter a year earlier, and earnings per share remained stuck at 18 cents, same as the year before. Basically, the sales force hadn't made its number, and losing $15 million in revenue at the last minute was part of the reason. The press release didn't say anything about Stephen Imbler's nerve-racking meeting with Larry Ellison and company, the one at which Ellison agreed to write off $15 million in shaky deals. But it quoted Ellison as saying, "We were disappointed with a $15 million shortfall . . . that went straight to our bottom line."

The results came as a shock to the financial world. Jeff Walker and other Oracle executives had consistently promised another excellent quarter, and as a result, financial analysts expected that earnings would be somewhere between 25 and 30 cents a share. The market was so confident that Oracle's stock price reached $28.38 on March

19, an all-time high. Then, out of nowhere, came the less than satisfying results. Investors felt blindsided and behaved accordingly: They dumped Oracle stock as if it were sludge. On the day after the announcement Oracle's stock price plummeted from $25.38 to $17.50, a loss of 31 percent. Twenty-one million shares changed hands, representing 16 percent of all the company's shares. It was the highest single-day volume ever for an over-the-counter stock.[1] ORACLE STOCK DIVES, said the next day's San Jose *Mercury News.*

The day that headline appeared, two enterprising Oracle stockholders filed suit in San Mateo County Superior Court against Larry Ellison, Don Lucas, Bob Miner, and several other board members and executives. The lawsuit came as no surprise to Oracle. There was a cottage industry in California made up of opportunistic lawyers who filed knee-jerk lawsuits against public companies that lost money or didn't do as well as expected. To build their cases, lawyers followed a simple formula: After a company had announced a disappointing quarterly result, the lawyers did some quick research to find out which insiders had sold stock in the quarter or two before the loss. Those people would be named as defendants. The law firm would also find some optimistic predictions from financial analysts, which would then be offered as proof that the company had misled the market about its prospects. Finally the lawyers would find a couple of stockholders who wouldn't mind being listed as plaintiffs in exchange for part of the hoped-for settlement. And a settlement was often forthcoming. For the defendant companies it was often less costly, in dollars and in negative publicity, to settle than to fight. That is not to say that all shareholder lawsuits were groundless; they weren't. But most of them were thrown together hastily after someone had read that morning's *Wall Street Journal.* The lawyers sued now and asked questions later.

The Oracle lawsuit was hastily prepared and filed. It listed Oracle's business address as "50 Oracle Parkway" instead of 500 Oracle Parkway and said that Jeff Walker had sold ten thousand shares of stock on October 5, 1987, when in fact he sold them in 1989.

Still, the two law firms that filed the lawsuit (one in San Francisco, the other in Philadelphia) told a compelling story. Beginning in 1989, the defendants led the market to believe that Oracle would report "spectacular" earnings results throughout the fiscal year. As proof of this the suit quoted from a report prepared by an industry analyst immediately after Oracle's January 1990 analysts' meeting:

> Oracle's meeting for analysts yesterday was a very positive briefing and it appears that the company's business has shown no evidence of deterioration due to economic or competitive issues.
>
> The company continues to feel that 60–70 percent growth in revenues and earnings is achievable for the current fiscal year and that longer term, 50–60 percent growth is attainable.
>
> Management further reinforced its positive outlook for fiscal 1990 by raising its guidance for revenues and earnings this year due to what appears to be strong sales prospects, particularly for the fourth fiscal quarter . . . Oracle's ability to generate stronger earnings and revenues comparisons over the next year is a key reason for our purchase recommendation.[2]

When Oracle's predictions turned out to be wrong, the stock did a half gainer. According to the lawsuit, the defendants "knew or were reckless in not knowing" that the projections were overly optimistic.

Then came the salacious part. In September and October 1989 several Oracle directors and executives had sold large amounts of stock, for which they reaped "huge profits," according to the suit. It was hard to argue with that adjective. Board chairman Don Lucas sold 60,000 shares and realized $1,467,700. Bob Miner and chief moneyman Jeff Walker did even better for themselves. In three days at the end of September Bob Miner sold 275,983 shares, for a gross of $6,338,117, and between October 3 and 6 Walker realized $5,650,228 on the sale of 234,472 shares. The lawsuit charged that the Oracle insiders made "materially false and misleading statements concerning Oracle's prospects." In other words, they tried to jack up the stock price artificially so their shares would be worth more.

Years later, in an interview, Walker denied doing any such thing. His personal financial strategy at the time was to diversify his holdings to "make sure my family was financially secure." He wanted to invest in tax-exempt bonds but needed cash to do it. He said he sold stock only when Oracle's legal department said he could and only when he would not have to pay a huge capital gains tax for doing so. According to Walker, the fact was that "I sold at every opportunity."

Still, it was also true that Walker sold the stock soon after Imbler started worrying aloud about Oracle's finances. Ellison believed that Walker dumped stock at least partly "because he was concerned. He expressed himself pretty clearly with his financial decision."

Significantly, the lawsuit did not accuse Ellison of improperly selling stock: He had not sold any. Far from making a profit, Ellison took a vicious financial hit. On March 28, 1990, the day that the Oracle stock price dropped $7.88, his personal fortune plunged by $366 million.

Still, the lawyers found something to accuse him of; Ellison still had pretty deep pockets after all. After insisting that the other executives used inside information for their own benefit, the suit said Ellison "knew or was reckless in not knowing of such conduct, yet permitted or acquiesced in such conduct." Apparently the lawyers thought Ellison should have kept Walker from calling his broker.

The Oracle stockholders who initiated the first lawsuit were David and Chaile Steinberg, who lived near Philadelphia. The Steinbergs were private investors who bought and sold stock in many different companies. In November 1996 I asked David Steinberg how and why he and his wife had sued Oracle. At first he could not remember being a party to any such lawsuit. After all, I was calling six and a half years after the filing of the suit. Besides, Oracle was doing so well in the fall of 1996 that Steinberg could scarcely remember a time when it had been in trouble. But the other reason why it was hard for Steinberg to remember was that during his time investing in the market, he had already filed five or six such lawsuits. After a while it was hard to remember the details of any one. The week I called him, he had asked his lawyer to consider suing Veterinary Centers of America, a company whose stock (some of which Steinberg owned) had just cratered.

Steinberg's best recollection about the Oracle lawsuit was that he heard about the company's disappointing third-quarter results and immediately called his lawyer, Stuart H. Savett, of Philadelphia. Savett then contacted someone at the San Francisco law firm of Lieff, Cabraser and Heimann, and soon the lawsuit was filed. Steinberg, whose complaint accused Larry Ellison of breaching his fiduciary duties, said he knew almost nothing about Ellison's behavior as chief executive officer and certainly did not know that Ellison had looked the other way while other people did wrong. "My only knowledge was that something happened that looked like it was not right," Steinberg said.

The lawsuit was pretty thin gruel, hastily prepared and filed. But that did not mean there was no truth in it. The allegation that Oracle had inflated its stock by misleading the public about its financial condition was at least worth looking into, and a lot of lawyers and shareholders wanted to do so. Several more lawsuits followed the Steinberg suit. Later those claims were consolidated into class action and shareholder derivative lawsuits that bedeviled Oracle into the middle of the decade.

Ellison's response to Oracle's woes in the spring of 1990 was characteristic. His childhood friend Dennis Coleman once said of him, "Larry was always the kind of guy who would take it to the limit, and then some. Bet the house, and then bet the house again." That was just what Ellison did in March 1990. Having underachieved in the third fiscal quarter, he told the financial community, "We expect to make it up in the fourth quarter and get back on our annual plan."[3] Stephen Imbler, the vice-president of corporate finance, also used a gambling analogy to describe Ellison's behavior: "It was kind of like he lost a hand in a poker game, so he doubled up his bet on the next hand."

As always, it was going to be up to Gary Kennedy to play out the hand.

Larry Ellison once accused Gary Kennedy of creating "a fear and greed culture" in U.S. sales, but that was not entirely fair. If people were afraid of Kennedy, they need not have been; after all, not even writing side letters would get them fired. One could also argue that Kennedy did not do anything more to foster greed than Ellison did. But certainly Kennedy understood the power of avarice. He used that knowledge to his advantage in the fourth quarter of fiscal 1990, the quarter when Ellison doubled his bet. During that time Kennedy came up with a brilliant and disastrous sales promotion: the Go for the Gold program.

The idea was simple. Instead of paying sales commissions in cash, Oracle would pay in bags of gold coin. There was no extra cost involved in paying people this way; Kennedy even arranged a free redemption program for employees who wanted to trade their gold for cash. "It was a tremendous motivator for the sales force," he said.

It was also a pretty good metaphor for life in the world of high technology. A Silicon Valley software engineer[4] once said that working in the valley was like living beneath a rushing river of gold. High-tech entrepreneurs were constantly trying to reach up into the river, the engineer said. Most grasped only air, but those who got their hands into the river experienced something wonderful and intoxicating: The gold washed all over them. At Oracle the river of gold was not imaginary. It was real.

Years later Ellison said he was appalled that Kennedy paid people in gold. Kennedy insisted that Ellison had approved the idea in advance, but Ellison denied that. "I'm embarrassed to say that I literally didn't know what was going on," he said. "One time a receptionist sent me an E-mail note and thanked me, saying, 'My husband and I wouldn't have been able to go to the Caribbean for Christmas if it hadn't been for the gold that we got.' My God! What is going on! What gold? What gold you got? I was pretty detached." Jenny Overstreet said Ellison got clued in only after he started getting purchase requisitions for gold.

If that was true, Ellison wasn't just detached; he was unconscious. During the fourth quarter the Oracle campus was covered in posters that said, "Go for the Gold!" Overstreet said she and Ellison saw the posters but didn't think there would *really* be gold—not the sort of thing you'd admit casually if you wanted to maintain your reputation as a genius.

Whether or not Ellison knew about the promotion, he had to be pleased with the results. Oracle did $334 million in business—the figure includes sales in Europe, where nobody was going for gold—raising its earnings per share to thirty-nine cents, up from eighteen cents the previous quarter. Not exactly inspiring, but not bad either. According to Kennedy, a lot of people liked their gold coins so much that they kept them for years. Some occasionally dumped them out on the floor and ran their hands through them. "Gold does strange things to people," Kennedy said.

It also did strange things to Oracle. Kennedy said only those people who had really earned gold got gold, but others said the promotion got out of hand. "They gave gold to people who made their goals. They gave gold to people who didn't make their goals. All internal controls were off," Imbler said. "It was like the Republicans' nightmare of a Democratic administration. They just threw money at every problem they could find."

The first quarter of the fiscal year—June, July, and August—was always Oracle's worst. Corporate information managers didn't buy software during the summer. Instead they did what everybody else did: arrived at the office late, looked out the window, went to the beach. The summer blahs were so much a part of the software business that Oracle referred to them in its annual report: "The Company anticipates that the first quarter of each fiscal year will continue to show relatively weak operating results."[5] In 1990, to give just one example, Oracle did twice as much business in the fourth quarter as in the first. Nobody was expecting the first quarter of 1991 to be one big party.

They weren't expecting it to be a funeral either, but that was what they got. When the quarter was over, Ellison and his team made all kinds of excuses. The economy was weak; the software business was in the midst of a correction; people in U.S. sales had not been assigned new territories in a timely way. All true.

But the main reason for the crash was that there was no more business out there. For years, salespeople had been "reaching forward into the future and pulling revenue into the present," in the words of Danny Turano. Now the future had arrived, and there was nothing there.

"There were no deals. That was the problem. Basically we created a desert in front of us that we now had to cross," Imbler said.

Unfortunately the people at Oracle didn't know they were crossing a desert until the sun started blistering their scalps. Years later it was hard to understand how any company could have known so little about its own business. It was even harder to see how a database company—a company that later adopted the slogan "Enabling the Information Age"—could have been so uninformed. But it was. "We were just flying blind," said Tom Siebel, a former sales executive under Kennedy. "The fact of the matter is that Oracle didn't know who its customers were. It didn't know how much product it was shipping. It didn't know if it was shipping Unix product or PC product or Sun product or DEC product. It didn't know. I mean, it was a disaster."

Unaware that a disaster was about to happen, Stephen Imbler was trying to recover from the ones that had already struck. By August 1990 it was apparent that Oracle was never going to collect a substantial part of its receivables. Some customers had already returned the software because it didn't work, some had gone out of business before their payments ever came due, and some had side letters saying they could get out of their deals. Oracle had no choice but to restate its financial results for the previous three quarters—a major embarrassment. Even as Imbler worked on the restatements, he was setting aside another twelve million dollars to cover future uncollected debts.

As it turned out, twelve million was "woefully inadequate," he said. Imbler learned later that about two or three bad deals had gotten by him for every one he had caught. Many of those deals came out of Oracle's consulting organization (which was not run by Gary Kennedy). Oracle had started its consulting business in the mid-1980s as a way of helping customers make good use of their highly complex software. Business boomed. Consulting accounted for 11 percent of Oracle's revenue in 1990, but some of that money was made through fraud. For example, an Oracle consultant would put in a forty-hour week, then bill two different customers for forty hours' worth of advice. "I didn't catch that," Imbler said. The Securities and Exchange Commission did but not until later.

While all this was happening, Oracle was also trying to cope with its cash-flow problem. Just about everyone who advised Ellison on matters of finance—including CFO Jeff Walker and Imbler—had been recommending for months that Oracle raise money by selling equity in the company. Ellison wouldn't do it. He preferred to borrow money instead. At one point, an investment banker from Donaldson, Lufkin & Jenrette proposed that Oracle raise money through a junk bond offering. Ellison liked the idea. He also liked the banker; he was dating her.

"We would have had no financial troubles whatsoever had we done this. But others decided equity was a great idea," Ellison said. It was a stalemate.

Finally Imbler and his treasurer, Bruce Lange, put together a syndicate of thirteen American and international banks, through which they arranged a $250 million line of credit. That would be enough to sustain the company until the next crisis.

A week before the end of the quarter sales vice-president Mike Fields stopped by Gary Kennedy's office. Fields, who reported to Kennedy, commanded about 70 percent of Oracle's domestic sales force, including the entire field sales organization. His quota for that quarter was sixty-two million dollars, and he had already told Kennedy he might not make it. Now he had more bad news. "Things are just not going well. I could be as much as five million off," he said.

Kennedy went and told Ellison, who was furious. According to Kennedy, Ellison said, "I've had enough of this. I want you to fire Mike Fields."

"I'm not going to fire Mike Fields. This is a horrible time to fire Mike Fields," Kennedy said, and Ellison backed down.

About that time Kennedy received more bad news, this time from his finance people. They had made an error when calculating the sales commissions on the "Go for the Gold" program. It was a small error, just a single digit. But that digit translated into a ten-million-dollar mistake. Sales would have to cover the ten million dollars in the current quarter, and it couldn't possibly do it. U.S. sales was now the scene of a horror movie, and the guy with the chain saw was about to arrive.

That Friday afternoon, in the last days of the quarter, Kennedy received a phone call from Fields, who was flying around the country trying to close business. "I gotta tell you, the number is way down," Fields said. Now, instead of missing his number by five million, Fields figured he was going to be twenty million dollars short. In the end even that estimate was low; Fields finished the quarter thirty million under quota.

This time Kennedy took Fields with him when he went to see the boss. Jeff Walker was there with Ellison when they arrived. Kennedy didn't like the feeling he got during that meeting.

"I knew that something was going on, because instead of trying to either blame us or complain or something, Larry said, 'OK, fine, if that's what it is, that's what it is. And I'll see you Monday,' " Kennedy said. "I was pretty convinced that Larry would ask Mike to leave over my objections, ask me to leave, ask both of us to leave, or do something."

The next morning Ellison asked Kennedy to meet him for breakfast at Late for the Train, a restaurant near the railroad tracks in Menlo Park. A lot of business had been transacted there over the years: a lot of start-up companies sketched out on slips of paper, a lot of strategies discussed, a lot of jobs gotten and lost. Gary Kennedy knew he might lose his that day. It could go either way. Kennedy had always thought of Ellison as the older brother he never had; he loved and hated him in the way younger brothers often do. Whatever else was true of Kennedy, whatever his flaws and blind spots, it had to be said that he had taken hold of Oracle when it was a faltering little start-up and whipped it into a high-tech powerhouse, all in the faithful service of his terrible godlike big brother. Now he knew there was a pretty good chance that his big brother was going to tell him to get lost. For privacy's sake, they sat in a booth.

This was how Kennedy remembered the conversation: Ellison told him he wanted to take finance and legal out of the U.S. sales organization and put them back under Jeff Walker. Kennedy had been talking about leaving Oracle, but now Ellison asked him to commit to two more years. "I need you to think about it, because if you can't guarantee two years, then I have to put someone in the position that I can count on," he remembered Ellison saying. Kennedy wasn't willing to do that. He knew there was a chance he would be appointed president of a Mormon mission, an opportunity he did not want to miss. Later that weekend he called Ellison and told him he could not commit to two more years. "That night Larry sent out a mail message announcing my resignation," Kennedy said. He cleaned out his desk a couple of days later.

This was how Ellison remembered the conversation at Late for the Train: "I asked Gary to leave the company." Hadn't Ellison asked for a commitment of two more years? "He's a pathological liar. He's a liar. . . . Pathological liar. Ridiculous. . . . Over my dead body. Amazing. Amazing. . . . It's amazing that people's vanity is so misplaced. I've been fired. Why would he lie about this?" Didn't Kennedy say he didn't want to make a commitment because he might go on a church mission? "Amazing. The only question I'd have is: Does he actually believe it? Amazing. Amazing. Astounding."

The company issued a press release announcing that Michael S. Fields, vice-president of sales, had been named president of U.S. sales. The announcement quoted Ellison as saying that Fields, who had missed his number by thirty million dollars, was "ideally suited" for the job. It said nothing about Gary Kennedy.

Kennedy always said that working for Larry Ellison was like riding a tiger. No matter how wild or dangerous the ride, you had to keep clinging to the tiger's back because if you fell off, the tiger would eat you. Ellison was the tiger, and Kennedy had seen him eat a lot of people. Bruce Scott, the company's first employee, was routinely excoriated by Ellison after he left, Kennedy said. The same thing happened to developer Umang Gupta and sales executive Mike Seashols. Ellison said he was angry at those people only because they left the company at a bad time or went to work for a competitor, but Kennedy saw it differently.

"There are only two kinds of people in the world to Larry: those who are on his team and those who are his enemies. There isn't any other. There's no middle ground."

Even the most valued executives could be made to suffer when they left the company. Tom Siebel was a good example. It would not be going too far to say that Siebel was a superstar at Oracle. After beginning his career as Gary Kennedy's presales guy in Chicago, he was promoted to the Washington, D.C., office, where he achieved 280 percent of quota. He later became vice-president of Oracle's direct marketing division, which finished 1989 at 149 percent of quota. Siebel was named most valuable player in Oracle's U.S. operations. He "personifies the driving vanguard spirit upon which Oracle Corporation is founded and from which it derives its dominance and vitality," a company newsletter said.[6] Siebel's group exceeded its quota again in 1990, the year in which other divisions couldn't make a deal.

When Kennedy left Oracle, Siebel also made his exit. Ellison thanked him for his years of service by stiffing him for thousands of dollars in commissions and bonuses that Siebel claims he earned in his last quarter. "It came as a shock, I have to admit," Siebel said. "It was a real wake-up call. Because if they'd do it to me, they'd do it to anybody."

Why did Ellison do it? "Tom sold a lot of mythological deals. Oracle didn't get paid, so why should Tom get paid?" he said.

But which deals were mythological? Ellison didn't say. The telesales group took part in twenty-two thousand sales calls *a week* in fiscal 1990[7] and made thousands of deals. Some customers didn't pay, but with so many transactions, that was inevitable. And Siebel's deals apparently hadn't been mythological a year earlier, when Larry Ellison and Gary Kennedy gave him his MVP award in Paris. Said Siebel: "Give me a break."

As far as Siebel knew, Ellison never spoke ill of him publicly, though the same could not be said for Gary Kennedy. After Kennedy left the company—after he got off or was thrown off the tiger—Ellison often singled him out as the person most responsible for Oracle's 1990 crash. "Suddenly," board member Joe Costello said, "Kennedy was a jerk. He was overly aggressive and did all these horrible things." Ellison's attacks upset Tom Siebel, who was a friend and admirer of Kennedy's. "He just kind of engaged in a systematic personality and character assassination process that went on for three years. He was just brutal," Siebel said.

When I spoke to Kennedy, he accepted his part of the responsibility for the things that had happened in U.S. sales. He regretted having emphasized victory at the expense of honor. "I wish, in retrospect, that I had said, 'It's better to fail than to cheat.' That I had said, 'Please understand, no matter what happens, it's better to miss your number,' " he said.

After Kennedy left Oracle, he briefly worked at another company, then endured a long illness before serving for two years as a Mormon mission president in Brazil. In 1996 he became president of a San Francisco software company called Keytex, which later changed its name to TenFold. TenFold's founder, a former Oracle executive, was so determined to hire Kennedy that he gave him full voting rights and allowed him to move the company headquarters to Salt Lake City. The founder's name was Jeff Walker.

"Gary was the best sales executive in the software industry. And I felt that from the beginning. He's an incredible guy," Walker said.

A lot of former Oracle people, including Ellison, thought it was funny that the former rivals were now working together. (Dogs and cats were living together in peace.) But Kennedy used the relationship to make a point. In his last months at Oracle, he said, nobody was in a better position to see what U.S. sales was doing than Walker, the chief financial officer. And while Walker at least once expressed concerns about the way U.S. sales was doing things, he didn't hold Kennedy personally responsible. Said Kennedy: "Jeff just turned his company over to me. If Jeff believed that any of the things that were cooked up about me were true, Jeff wouldn't have turned his company over to me."

I asked Ellison about that. If Kennedy was a crook or an incompetent, why would Jeff Walker put him in charge of his company?

"I never said Gary was incompetent," Ellison said. Then there was a long pause. "And I never said he was a crook. To you. Or publicly."

Ellison was good at that: saying things without saying them. He was clever enough to say things laterally, shading his meaning just so. Did he think Gary Kennedy was a crook? Did he tell somebody else that he thought that? Trying to get him on record on these points, or on any point, was like chasing a squirrel around the yard.

Ellison liked to say that he himself was not a crook; he was just inept. "If I was aware of [the things Kennedy did wrong], then I was also doing things that were wrong," he said. "And if I was unaware, I was not a competent and capable CEO. All I'm claiming here is that I was not a competent and capable CEO. I'm still responsible. But it was neglect, not malice." Ellison was too good to himself when he said he was incompetent. What he was guilty of was something less than malice but something more than neglect. He could be accused of a lot of things, but neglecting Oracle Corporation was not one of them. Larry Ellison ruled over his company like a god. All that was woeful and glorious about Oracle grew from his brilliance and determination and numinous will.

He created the company, and then he created Gary Kennedy, who did what he wanted him to do.

"The whole company had built up so much pressure on themselves that you had people cracking and doing dumb things. And unethical things," said Jeff Henley, who became chief financial officer in 1991 and was later hailed as one of the saviors of Oracle. "Larry claimed, 'Well, Gary was doing things I didn't know, but ultimately I'm the boss,' and it's true. Larry wasn't close enough to what was going on. Larry is an aggressive guy, and he put more pressure than he should have on people. He shouldn't have been surprised that they did what they did, because the boss is telling them, 'We really need fifty percent growth,' and if they didn't [achieve that], they knew that they'd get fired."

* * *

Gary Kennedy was history, but that wasn't the end of Oracle's problems. On Sunday, September 2, 1990, Stephen Imbler, Oracle's vice-president of corporate finance, was at home enjoying his Labor Day weekend when the phone rang. Oracle chief financial officer Jeff Walker was on the line, and he had bad news: Revenues for the first quarter of fiscal 1991 were millions of dollars below projections. Nobody knew yet how much the company had lost, but things did not look good. Walker wanted Imbler to be ready to work on the problem when the weekend was over.

Ellison, Walker, and Imbler met to discuss strategy. They knew they had to say something publicly about the loss, or else they might be accused later of deliberately withholding the information. Imbler could not say with precision how much Oracle had lost—he hadn't crunched all the numbers yet—but told his bosses that his best estimate was twenty-five cents a share. According to Imbler, Ellison and Walker wanted to announce an estimated loss of only fifteen cents a share. Imbler was disappointed but not surprised. He thought Oracle should simply tell the truth as best it could. What was the point of publicizing an unrealistically low figure? Everyone would know the truth eventually; Oracle was a public company. "I think that both Larry and to a lesser extent Jeff didn't understand what the CFO's or the VP of finance's job is," he said.

"We just wanted to be accurate," Ellison said, allowing that he had only a vague memory of the discussion. "Finance people will always give you the worst number—that's kind of their job—and sales will always give you the best number. You try to kind of come in between what the finance forecast and the sales forecast is. And the fact is we didn't know, but there seemed to be a reasonable expectation it was going to be fifteen."

Imbler and his bosses "argued and argued and argued"; at one point, Imbler said, Ellison accused him of not caring about the future of the company. Finally Ellison and Walker agreed to announce a loss of twenty cents a share—less than what Imbler would have liked, and more than Ellison wanted, but close enough to the truth to seem reasonable in retrospect. The twenty-cents-a-share estimate was chosen because it "did not imply a high degree of certainty," Imbler said.

Oracle announced the estimated loss on Thursday, September 6. The press release tried hard to couch the news in the most positive way: "Oracle Corporation . . . Thursday announced that for the quarter ended August 31, sales are expected to increase approximately 30 percent.

"As a result, Oracle's total revenue for the last four quarters is anticipated to exceed one billion dollars.

"Unfortunately," the press release continued, getting to the point, "the Company had planned for a revenue growth of 50 percent. As a consequence of this revenue shortfall, expenses grew faster than revenues, and the Company will report the first quarterly loss in its history. Specifically, the Company expects to lose approximately $0.20 a share." The release said Oracle would announce the precise amount of the loss during International Oracle User Week, scheduled to begin a couple of weeks later in Anaheim.

The announcement went on to say that Oracle was reducing its projected revenue growth for the fiscal year from 50 percent to 25 percent, a far cry from the 100 percent growth to which investors were accustomed. The press release also spoke vaguely of a "restructuring and cost-cutting program," something that was sure to mean bad news for a certain number of Oracle employees. "We continue to grow faster than our competition," Ellison said, reassuring the stockholders. "As we adjust to a more conventional growth rate, our company will be stronger than ever." It had to be the first time in company history that Ellison had mentioned the words *conventional growth rate.*

That weekend the San Jose *Mercury News* published a story under the headline HIGH-FLYING ORACLE SET ITSELF UP FOR A FALL.[8] The second paragraph said, "Oracle's brash style has over the years attracted numerous critics, among them analysts, ex-employees and customers. . . . They say Oracle's $1 billion sales level was reached through a growth at any cost strategy, with an over-aggressive sales force over-promising on their products—and that the strategy has come back to haunt it."

One morning about that time Imbler arrived at the Oracle campus early to work out at the thirty-six-thousand square-foot Oracle gym, which had just been completed. Ellison, still a fitness fanatic in his mid-forties, had ordered the place built so his employees could be as fit and trim as they were bright and ambitious. The facility was magnificent; built next to the engineering tower, it was full of weight machines, brand-new stationary bicycles, basketball courts, and so on. As he walked through the doors, Imbler was awestruck—and worried. "It was this wonderful gym with all this equipment. I'm looking around, and I had this fin de siècle feeling. It was like being at an Austrian ball in Vienna just before the war broke out. These were the last days of a proud age, and soon it was all going to change."

Not soon, *now.* The company was hemorrhaging money; it had to start cutting costs right away. Ellison, who had always proudly said that people were the company's greatest asset, now told his top managers to get rid of four hundred people, or 10 percent of Oracle's domestic work force. This was when Imbler had his darkest view of Oracle. He agreed that the company had to lay off some people but was upset about the way it was done. Just a few months earlier Gary Kennedy had gone on a hiring binge to keep up with Ellison's insistent call for revenue growth. Now some of those new hires were going to be shoved out the door. Most of the cuts were made in sales, marketing, and administration; the people who created the products were left alone. Those whom Imbler laid off were given only two weeks' pay—a brutally short severance, in his view. "Forty-five years old? Mortgage? Worked for the company for five years? *Two weeks.* It was across the board," Imbler said. "There's nothing illegal about it. But there was something distinctly distasteful about it. I mean, it wouldn't have made that much difference to our survival if we had given some people four weeks or six weeks." Some of the people who were laid off did not need severance packages because thanks to Ellison's generosity, they had accumulated lots of stock. Others didn't do so well. One of Imbler's people had a large batch of stock options that would have vested in three weeks. But she was given only two weeks' severance, so she never got the options.

Oracle announced the cutbacks on September 17, almost two weeks after the company estimated the size of the loss. This press release, titled "Oracle Announces Cost Reductions," announced that finance had at long last been recentralized under Jeff Walker; the disastrous experiment was officially over. The announcement also said that U.S. sales, which had grown top-heavy under Gary Kennedy, had been restructured so that there would now be fewer managers and more salespeople. "In this way," Ellison was quoted as saying, "we will have reduced sales expenses and increased sales capacity." Again, no mention was made of Kennedy; he had been disappeared, transitively, like the political enemy of a Guatemalan or Peruvian military regime. "We believe that making these changes, as difficult and overwhelming as all of them are, is necessary for Oracle Corporation's return to health and profitability in the second quarter," Ellison said, concluding the press release.

If the layoff was demoralizing, Ellison's next cost-cutting measure was even more so: He canceled the Oracle Christmas party.

International Oracle User Week, a convention for Oracle users worldwide, was held in Anaheim, California, during the week of September 24, 1990. The event was memorable, for all the wrong reasons. As a benefit for the hundreds of customers in attendance, Oracle had installed Oracle*Mail, an electronic mail program, on the convention center computers. The idea was to let people send messages to each other between sessions. But because of the large number of people trying to use the system at the same time, it frequently seized up or went dark. "One user said he had just finished typing a two-page letter when the system apparently 'froze,' " Oracle user Gary Raymond wrote in a newsletter that week. "Unable to quit or escape, the user waited 15 minutes, then turned off the terminal."

For Oracle, the week was about to get worse. That Tuesday the company finally told the truth about the size of the first-quarter loss. Ellison and his team broke the news in a tiny meeting room packed elbow to elbow with customers and financial analysts and members of the press; those who couldn't get seats stood along the walls in the back or crouched in front. The audience was hot and tense and cranky. Oracle's stock, which had climbed above twenty-eight dollars only six months earlier, was now trading in the six-dollar range. The financial analysts, some of whom had advised their clients to buy stock in Oracle, wanted to know what had gone wrong; the customers wanted to know what the first-quarter loss meant to Oracle's future; and the news reporters wanted a good story. They got one.

Ken Cohen, a longtime Oracle marketing man who was advising Ellison during the crisis, stood in the back and watched as Ellison entered the room. "Larry looked like death," Cohen said. "Larry always had a beautiful suntan. But on that day Larry looked lighter than I had ever seen him look. He had been up all night trying to put together this press release, and he had put together forty-two different versions of it. Everybody was role-playing. 'What if this gets asked? What if that gets asked?' He was just absolutely pale. He was like a close relative at a funeral."

Jeff Walker had the job of announcing the loss. For a couple of years now the analysts had been expressing doubt that the company could sustain its pace of growth, and Walker had insisted that it could. Now he had to admit that he had been wrong. "It was kind of like, you know, you gotta go in and tell your dad you did something wrong, and you got caught," he said. "I mean, it was that sort of feeling; you sort of felt guilty."

Walker stood and told the audience that Oracle had lost twenty-seven cents a share—not twenty cents, and certainly not fifteen, but *twenty-seven*—in the first quarter. In terms of dollars, the loss companywide was thirty-six million. There were a few gasps in the room and a lot of cross faces. The analysts were disappointed but not shocked; the big shock had come a couple of weeks earlier, when Oracle had forecast the loss. "We already knew that the patient was dead," Stephen Imbler said. "This was like the autopsy. The question was: OK, what killed him?"

Ellison blamed part of the loss on Kennedy and his people, saying that the reorganization of U.S. sales was handled "very, very, poorly" and was "a very serious fiasco."[9] It was "a day Oracle users will never forget," Tony Ziemba later wrote.[10] "Who would have imagined Larry Ellison as a humble and contrite leader?"

The headlines that appeared later were brutal. THE SELLING

FRENZY THAT NEARLY UNDID ORACLE, *Business Week* hollered. CEO BLAMES MANAGERS FOR ORACLE WOES, said the San Francisco *Chronicle.* ORACLE "FIASCO" RESULTS IN $36 MILLION LOSS, shouted *PC Week*. ORACLE LOSS IS BIGGER THAN EXPECTED, the San Jose *Mercury News* said.

After the announcement, marketing man Ken Cohen went to the airport to catch a plane back to Oakland, where he lived. He was sitting in the terminal, reading a newspaper, when he saw Larry Ellison walking by. Ellison, who had changed out of his business suit and into jeans and a work shirt, was soon to board a flight to San Francisco. He appeared to be in a daze, as if he had just awakened from a heavy sleep. Cohen was surprised to see him alone; Ellison was almost always accompanied by Jenny Overstreet or someone else from Oracle.

Cohen approached him and said, "Rough day, huh?" Ellison responded with a faint shake of his head, as if he had barely heard the question.[11] Cohen saw that he had tears in his eyes; it was one of the few times he had ever seen Ellison in emotional pain.

"So," Cohen said, "you headed back up to the Bay Area?"

"Yeah, I have a lot of work to take care of," Ellison said.

"Maybe you need a day or two on the beach."

"No," Ellison said, shaking his head again. "You know, Ken, we ought to be able to fix this. . . ."

Soon Ellison began to talk about how. The more he talked, the more animated—the more like himself—he became. Ellison was a man with an inexhaustible supply of words, and now he was tapping into his supply with his usual joy and abandon. Even when he was feeling his worst, Ellison remained an optimist, a man who couldn't help looking forward. He lived in the future, it was true; part of him believed that Oracle was already fixed. But the part of him that Cohen could see didn't look so good.

"Man, you look terrible," Cohen said.

"I don't suppose you'd believe it if I told you I have a cold."

"I think it's time for straight shooting on this one, Larry," Cohen said.

"I don't think the analysts would have believed it either."

"I think some of them wished you had worse."

"Yeah," Ellison said, "but they'll see. They'll see."

Ellison asked Cohen how he thought the announcement had gone. Had it gone as badly as it had seemed to go? Were people really as angry as Ellison thought they were? Yes, Cohen said, it had gone terribly, and people were furious. "I was afraid of that." Ellison sighed. To Cohen, Ellison seemed like a little boy who regretted that he had made his parents so angry. At one point Ellison asked him, "Did I say anything really wrong?" Cohen did not think so, but Ellison was not comforted. "He was just a crumbled mess," Cohen said later.

Still, as Ellison made his way to his gate, Cohen did not doubt that he would build Oracle back up again.

When Tony Ziemba learned that Oracle stock was trading around six dollars, he saw an opportunity. A year earlier Ziemba, president of the New York Oracle Users, had personally seen to it that Oracle received bad press for releasing a buggy Version 6 and for ignoring customer complaints. Now that Oracle's questionable business practices were costing it dearly, he figured it was time to cash in. Ziemba was no financial wizard, but even he could see that at six bucks Oracle was undervalued. So he called a broker.

"Last Tuesday, September 18, I bought 100 shares of Oracle at the opening bell for 6¾. . . . It was all I could afford," Ziemba later wrote in a user group newsletter.[12] "This investment won't pay for my trip to Tierra del Fuego in 1991, or the Range Rover in 1992, but it may be my chance to get in on the ground floor of what could be the hottest information services stock of the 1990s. I hope that over the next few months I may be able to add to this holding."

Ziemba's purchase turned out to be a wise one. He was also not the only person who made a killing by buying Oracle at a scratch-and-dent price. About that time Stephen Imbler bought thirty-eight-thousand dollars in Oracle stock. As of February 1997, he still had it. It was worth one million dollars.

But Ziemba wasn't interested simply in padding his bank account; he also wanted to change Oracle—from the inside. He pointed out in the newsletter that the entire company could now "be bought for less than the amount of its expected sales in fiscal 1991." If Oracle's customers took all the money they planned to spend on Oracle products and services and used it instead to buy stock, "they would own the company." Once they owned it, he wrote, "we could scale back the grandiose marketing plans of yesteryear [and] redeploy human resources to customer support and services." The result would be "a perfect software company." It was a customer's dream—a wild and farfetched one, but still a dream.

Even Ziemba knew that a takeover wasn't realistic. So he urged all Oracle users to do as he had done and buy one hundred shares of stock. If they did so, "we could easily end up with over one percent of the stock and maybe we could eventually get someone on the Board of Directors to expressly represent the user community."

Ziemba eventually gave up even the idea of getting a representative on the board of directors. Instead he and the leaders of several other user groups sent Larry Ellison an open letter in which they each pledged to buy one hundred shares as a gesture of support for the company.

"We all share a common desire that Oracle, the product in which we have all invested our 'blood & guts,' be the very best product to meet our diverse information needs," the users wrote to Ellison.[13] "We all knew we were hitching our future to your bandwagon when we first bought your product and invested our training time, our careers, our companies' futures, and our very own human capital in your vision. We knew we were taking you in as a partner. We are still confident that you can lead us all into the future."

Eleven

ON OCTOBER 31, 1990, ORACLE'S STOCK DIPPED TO A SKELETAL PRICE of $5.25 before closing at $5.38. But if the stock price was deathly on Halloween, it was even sicklier the next day, when it rattled and hacked to an all-time low of $4.88.[1] Oracle wasn't alone in its misery. That same day *New York Times* business waiter Floyd Norris filed a story that began, "If the market for big stocks is depressed, then the market for small stocks seems almost lifeless."[2] The article quoted an expert as saying that only a few technology companies—Apple, Intel, a couple of others—were still holding the interest of investors. Still, Oracle's stock was sinking only partly because of a down market. It was also falling because investors had lost confidence.

If Oracle had been in any business besides software, it would have been vulnerable to a hostile takeover. Certainly the low stock price would have made it a fat target. But as things were, a coup was never really a threat. If a raider had tried to depose Ellison and Bob Miner (who together held a formidable stake in the company), it almost certainly would have driven away all the top managers and engineers, leaving Oracle with nothing more than a clever name.

"I don't think we were terribly concerned," Ellison said. "I don't think we could figure out who would do it other than Microsoft. And Microsoft doing a hostile takeover was just out of character with their management team and with their strategy." Maybe so, but Bill Gates apparently was following the news about Oracle. According to Ellison, Gates took "a huge position" in Oracle stock. "It was fairly common knowledge in the industry that he had made the investment," Ellison said. Gates would not confirm this.

As Oracle's fortunes fell, so did Ellison's. His stock, worth $954 million that spring, was worth only $164 million on November 1. Good-bye, $790 million.[3] Years later Ellison made a joke about the loss. "I'm a man of modest needs," he said. "I could live on a few hundred million dollars."

At the time the loss was no joke. The reason was that Ellison's assets were not liquid. Far from it. In the four and a half years since Oracle had gone public, Ellison had not sold a single share of stock on the open market. While most other high-tech insiders regularly sold stock as a way of diversifying their portfolios, Ellison clung to his Oracle shares the way a hermit might guard the trinkets in his house in the woods. His advisers did not think this was a good idea. In early 1990, before Oracle got into trouble, the securities firm Donaldson, Lufkin & Jenrette, which worked closely with Ellison and the company, strongly recommended that he diversify. "It's not like he was born to wealth. If Oracle went bust, he went bust," said Scott Smith, then with DLJ. Ellison refused to sell.

"How is that going to affect my life?" he said to me. "Let's say [the stock falls and] my net worth goes from, at that time, one billion to three hundred million, or now, from eight billion to two billion. I don't think it's going to affect what I have for dinner, or how I raise my kids, or anything. It might affect my ego. I might feel like I didn't do a good job and let a lot of Oracle shareholders down. But it's not going to affect my personal life."

Ellison said similar brave things at the time. Stephen Imbler was in one meeting in which Ellison ruminated that Oracle's hard times did not make him sorry for himself. What he really regretted, he said, was that some of his employees and shareholders had lost what was, to them, really big money. Imbler couldn't tell how sincere Ellison was. "He was saying these things, and he probably half believed them. But I actually think that Larry was still searching for what he really felt about all this."

Ellison's refusal to sell any of his own stock was partly an expression of faith in the company and in himself. To him, selling stock would have been a tacit admission that something might actually go wrong, which was unthinkable to him. "I think he deserves a tremendous amount of credit for putting his personal fortune where his mouth was," said DLJ's Smith. In Ellison's view, Jeff Walker had not been willing to do the same a year earlier, when he sold more than five million dollars' worth of stock.

Ellison had another, more peculiar reason for holding on to his shares. He felt attached to them. "I identify very closely with the company. It's part of me, and I just don't want to get rid of it," he said. He believed that keeping his shares as a way of expressing confidence in the company was "rational" and that keeping them because he loved them was "emotional." "And I'm not quite sure how those two things interact. . . . It's hard to know which one is the dominant force," he said.

But if he wouldn't sell stock, how did Ellison finance the lifestyle of a multimillionaire playboy? How did he afford the Japanese-style home in Atherton and the vacations to Hawaii and Japan (he liked to be in Kyoto every year at cherry blossom time)? Where did he get the money for the fast cars and tailored suits? He also invested heavily in other companies, spending millions. But how?

Ellison's 1990 salary and bonus totaling $1.6 million were part of the answer. According to the San Jose *Mercury News,* he was the second-highest-paid executive in Silicon Valley, behind John Sculley of Apple Computer.[4] But even that generous salary wasn't enough to cover Ellison's expenses. To finance his way of life, he was spending the best kind of money of all: somebody else's.

What he had been doing for years was borrowing money from a brokerage house, with his Oracle stock as collateral for the loans. "Larry had this view that he could borrow money against his stock, and the appreciation of the stock would more than offset the interest rate on the borrowing," Jeff Walker said. Things had always worked out precisely that way. No matter how much Ellison borrowed, his stock was always worth far more, and his Oracle salary paid him more than enough to repay the interest.

But when Oracle bottomed out, Ellison was in a crisis. He would not discuss his personal finances in detail, but certain things were clear. Most brokerage houses called in their margin loans as soon as a client's stock fell below five dollars a share. They did this because they wanted their clients to repay them while they still could; after all, a stock worth five dollars a share is only one piece of bad news away from being worth fifty cents. Oracle's stock hovered around five dollars for a few weeks in the fall of 1990. At any moment Ellison's brokerage house could have called in his loan, meaning he would have had to dump his Oracle stock at bargain basement prices to come up with the cash.

"There was the possibility of losing absolutely everything I had," he said. How did that feel? "It's exciting. It's a rush, man," Ellison said, laughing the easy laugh of a man who almost lost everything—and ended up with six billion dollars instead.

Ellison did not get a margin call, but he still had money problems. With the stock twirling down the toilet, he wasn't in a position to borrow any more money on margin. "Larry had big cash-flow problems when the stock dropped," Bob Miner's daughter Nicola said.

In December 1990 Ellison finally sold some stock—but not Oracle stock. At the time Ellison served on the board of directors of Cadence Design Systems, whose chief executive, Joe Costello, was on the Oracle board. (The person who arranged all this was venture capitalist Don Lucas, who sat on *both* boards.) While on Costello's board, Ellison had amassed 149,361 shares of Cadence stock. He sold 126,584 shares in three days in December, grossing $2.97 million. He also arranged a line of credit with the securities firm Morgan Stanley.

"Morgan Stanley came to my rescue," he said. "Morgan Stanley said, 'We think you're bankable. . . . Tell me how much you need.' Well, that's kind of an extreme statement. If I had said, 'Twenty-five billion dollars,' they would have said, 'No, no, no, no, no, no. You need to need less.' But Morgan Stanley expressed confidence that I still had enough equity in Oracle stock to continue to be worthy of credit."

Ellison never lost confidence in his company and never became emotionally detached from his stock, but he finally quit hoarding his stock. He sold stock in 1991, 1992, 1993, 1994, and 1996, raising many millions.

"I'm sure Larry saw the bogeyman in those days," Jeff Walker said, "and didn't ever want to see it again."

Larry Ellison had always liked working from home, where he could eat his sorbet while looking out on the ponds and the pool and redwood trees. But late in 1990 and early in 1991 he seemed to be even less visible than usual. Ellison said he wasn't hiding out. He was busy rebuilding his company. "When things get bad, when things get really tough, I get a little bit reflective for a while," he said. "And then all you can do is go back to work."

It was unlike Ellison to brood—"You never saw the guy get down about anything," one employee said[5]—but he was apparently more than a little bit reflective after the crash. Barbara Ellison said he was terrified that he would fail, confirming all his father's darkest predictions about him. "Black days. It was awful. The kids and I never saw him. We couldn't see him. He couldn't take the time to see us," she said. "It wasn't that he didn't want to. But everybody had said that he was going to fail. And there were people waiting for him to fail, people who didn't like him. I remember talking to him, and he'd be real worried and scared and wanting to turn it around. . . . There was a note in his voice that you didn't usually hear with him—just scared, worried."

He had reason to be. Even after the layoffs and other cost-cutting measures, Oracle was short on cash. By the end of December the company had borrowed $170 million from the international banking syndicate, and though the credit limit was supposed to be $250 million, the banks had decided not to lend Oracle any more money. "We were bleeding cash like crazy. The banks were just up in arms," one executive remembered.[6] The company had to get cash somehow or go under.

It began by making a serious effort to collect money owed by its customers. For a long time Oracle's collections people had been writing off some debts as uncollectible just because they wanted to shorten their list of receivables. Jeff Walker put a stop to that. He began calling some customers personally and saying, "Hey, this money's due. What's going on?" At one point Walker heard a rumor that an Oracle customer in Santa Clara was about to go bankrupt. The customer still owed Oracle some money, so Walker called and threatened to quit providing technical support unless the customer paid up. Somebody from Oracle drove to Santa Clara later that day to pick up the check.

But collecting from deadbeat customers wasn't going to be enough to solve the cash-flow problem. When Oracle was desperate for money, Ellison looked to the country he thought of as his spiritual home, Japan.

Nippon Steel, a nineteen-billion-dollar industrial company,[7] already had a sizable high-tech division and now wanted to invest in Oracle. Ellison hated to sell even a tiny part of his company, but he had little choice. Early in 1991 he set out to make a deal. The goal was simple: Get two hundred million dollars from Nippon Steel under the best possible terms. The person he chose to conduct this critical negotiation was not a wizened business person, not even a senior executive. Instead he chose a thirty-year-old Oracle employee named Ron Wohl, who had two important attributes: He was smart, and he would do whatever Ellison told him to do.

"I wanted to do the negotiation myself," Ellison said. "And if I wanted to control the negotiation, Ron is the guy I want carrying my message. I don't trust anybody else to—with fidelity—carry out my instructions." Wohl was Charlie McCarthy to Ellison's Edgar Bergen.

Nobody really knew what Wohl's title was or how senior he was. Imbler thought of him as a "minister without portfolio," but if you'd had to give him a title, secretary of state would have done nicely. When Ellison wanted to know all the dirt about something going on inside Oracle, Wohl would go on a fact-finding mission. "He didn't say no very often," Overstreet said. Said Imbler: "If Larry wanted to know something, he would send Ron to ask it. Then Ron would go back and tell Larry what I had said."

Wohl conducted the Nippon Steel deal with the help of Oracle's auditors and lawyers. At first the Japanese were taken aback that Oracle had sent such a young man to make the deal. But in the end Wohl's youth did not prove to be a problem. In mid-1991, after a negotiation that lasted several months, Nippon Steel agreed to lend Oracle two hundred million dollars in exchange for debt instruments that Nippon Steel could later convert into Oracle stock. Oracle would get cash, and Nippon Steel would get Oracle—or part of it anyway.

Ellison agreed to the deal—but never went through with it. By the middle of the year Oracle was no longer as desperate as it had been in the fall. Money was beginning to flow in. The economy had bounced back. Oracle's stock was climbing again. Instead of going through with the original plan, "We went back to Nippon and changed the deal on them, basically," said one executive involved in the negotiations.[8] Now, instead of borrowing two hundred million, Oracle would take only eighty million dollars. And instead of getting stock in Oracle, Nippon Steel would get part of Oracle *Japan*.

"I think Nippon Steel felt that time was on their side. The slower they moved, the more desperate we would become for cash," Ellison said. "I felt that just the opposite was true: that time was on our side. . . . We were able to drag out the negotiations long enough to change it from a debt instrument on Oracle stock to a debt instrument on Oracle Japan stock. We actually pulled this off without really compromising Oracle shareholders' position one iota, and we raised all the money we needed from Nippon Steel at fantastic interest rates."

According to Wohl, there was at least one uncomfortable moment in the negotiation. With the deal almost done, the Nippon Steel people wanted Ellison's initials on the draft of the agreement. But nobody could find Ellison. He had done one of his periodic disappearing acts, supposedly to go shopping.

Joe Costello came to the study of Lawrence Joseph Ellison in the late 1980s, when Don Lucas invited Ellison to sit on the board of directors of the company where Costello worked. The company, SDA Systems, made software that could be used to design semiconductors, computer systems, and electronic equipment. At the time Costello was running sales. When the job of president and chief executive officer became available, Ellison insisted that the board hire Costello. Either give him the job, Ellison said, or I'll quit the board and hire him at Oracle.

"The whole board was ticked off at me," Ellison said. But Costello got the job.

Costello thought Ellison was a brilliant but unpredictable board member. Once, in a meeting of SDA's compensation committee, Ellison delivered a long, impassioned speech about why a certain thing should be done a certain way. Then he went to the regular meeting of the board and argued the other side of the same issue just as vociferously. "He won the day—both ways," Costello said.

Another time Costello called a meeting of the people involved in taking SDA public: lawyers, investment bankers, executives, and so on. Forty-five minutes into the meeting he heard a loud noise. "Voom! The boardroom doors open. Swoosh! Larry walks in," Costello said. "I looked at my watch and I said, "Jeez, Larry, you're either six weeks too early or six weeks too late." Ellison thought, incorrectly, that this was a board meeting. "He'd missed by six weeks—and forty-five minutes."

But Ellison was more than just a funny character. He was an electric figure, and his presence on the board was mostly positive, Costello said. "It's that native curiosity, energy, intensity, enthusiasm. He gets enthusiastic. He's wildly enthusiastic. I mean, it's infectious. He's just like, whoa. When he's into something, man, there's energy around it. And he drags people along in his wake. That's superpositive. . . . He'll get into a sport or a hobby or a woman—it doesn't matter what it is. He's wildly enthusiastic, in all of the above." And when Ellison felt he was wrong, Costello said, he would back off "in the following sense: 'You idiots. Do what you want. I know I'm right. I've spoken my piece.' "

The year Oracle began to come apart, Don Lucas arranged to have Costello take a seat on the Oracle board. Costello was desperately needed. At the time the board consisted of Lucas, Ellison, and

Bob Miner—and Miner wasn't even interested. Once Lucas proposed that Oracle enact a continuing stock grant program to attract better people to the board. "Bob Miner's response—and Bob Miner was the nicest guy, so this was almost out of character for him—was to say, 'I'd rather give my lowest developer stock than any of you guys. You do nothing for the company,' " Stephen Imbler said.

Costello was about to do something for the company: bring in a new chief financial officer. Jeff Walker had decided it was time for him to go, and he surely was right. He had almost no support in the company, and Wall Street was "beating the hell out of me," he said. Costello and Lucas suggested Jeffrey O. Henley, a former Silicon Valley CFO who had spent several years running companies for a wealthy private investor. Henley, whom Don Lucas called "a straight arrow," had a brief meeting with Ellison in his office and later met him for dinner.

"This was a guy that was extremely bright, amazingly bright. You can get on to subjects about finance or anything, and he just has an amazingly quick uptake," Henley said. But that wasn't the only thing that made him want to work for Ellison. "What I liked about him was that he definitely seemed like he had been humbled. I read all the stories about how arrogant he was and all that. And clearly he had been taken down a few pegs."

He was about to be taken down a few more. By the mid-1990s Ellison was often applauded for hiring Henley, and justifiably so. But he actually didn't want to hire Henley at first. "We hired Jeff Henley almost over Larry's dead body. He didn't like him that much. He was OK. But he didn't like him that much," Costello said. "We—Don and I—kept pushing this. [We said,] 'Let's get going. This is what we need. We gotta do this.' "

According to Costello, "Larry said, 'He's just a goddamn finance guy, he's not that smart, he's a fucking plodding finance guy, and he just knows how to plunk numbers,' and all this shit. And I said, 'Yeah, right, we just need a fucking finance guy so we're not two hundred and fifty million dollars in the hole; how about that, you fucking idiot?' These were the kinds of conversations we had."

Ellison said he hesitated to hire Henley only because he believed

Lucas and Costello were offering him too much money. "I thought that the size of the option we were giving Jeff Henley was astoundingly large," he said. It certainly was. If Ellison broke down and hired him, Henley would receive options to buy five hundred thousand shares of Oracle stock at an average price of $9.13. If the stock went anywhere at all, Henley was just about going to drown in that river of gold. He would also get $74,000 in salary for the couple of months remaining in the fiscal year. And if he turned out to be a loser and Oracle pushed him out, he'd float to earth under a golden parachute of $25,000 a month for up to a year.

According to Costello, he and Ellison kept at each other. "Larry didn't like [the offer to Henley]. He said, 'It's too much. Too rich. Cut it.' And Jeff was going to say, 'Well, screw you.' " Finally Ellison said he would quit if he was forced to make the hire, Costello said. "We just stared back and said, 'OK.' We called his bluff. That was the moment of real truth. And he did back down."

Ellison said Costello's version of the story is "absolutely not" true, adding that "Joe is not one of my favorite guys." Don Lucas recited the public relations version of the story. "To Larry's credit, when I suggested Jeff Henley, Larry hired him on the spot," Lucas said. Not even Ellison thought *that* was what happened.

Though it may be hard to imagine given the money he stood to make, Henley had some reservations of his own. In some ways Oracle looked strong. It was the leader in the relational database market, with a huge installed base of customers who would continue to want upgrades and support. But there were questions, most of them having to do with Ellison. Henley had just finished working for an entrepreneurial character and didn't know if he could put up with another one. And maybe Ellison wasn't the right person to turn Oracle around. "The question was, Did we need to blow out Larry? What did we have to do? Usually when these things happen, the CEO is gone," Henley said.

Was Ellison in danger of losing his job? He thought so. Some people certainly wanted his head. At an analysts' meeting sponsored by Oracle in January 1991, according to Jeff Walker, one person asked Ellison if Oracle would be looking for a new chief executive. Ellison deflected the question with humor. Without saying a word, he patted himself down to make sure he was still there. Yet even Ellison wondered if he would be—in the vernacular of Silicon Valley—shot.

"There had to be serious conversations at the board level about replacing me," he said. Years later Ellison had a sense of humor about it: "I think getting rid of me would not have been the right thing to do. What a surprise that I hold that view to this day."

Stephen Imbler also believed Ellison should stay. He considered Ellison "a great man," though a flawed one, as all great people probably were. "Yeah, we had lost hundreds of millions of dollars of shareholder value," Imbler said. "But who was the primary person responsible for creating that shareholder value in the first place? There are very few people who can lose hundreds of millions of dollars because very few ever create that much."

There were only three people on the Oracle board, so it wasn't hard to figure out where everyone stood on the issue of firing the CEO. Joe Costello was willing to consider firing Ellison. Bob Miner apparently wasn't. That left Don Lucas, who had always supported Ellison and whose personal fortune had multiplied as a result. "We did have some discussions regarding the subject," Lucas said. "I took a very strong position, and it was never officially discussed. . . . I never had any doubt that Larry was the right person to run Oracle. No doubt."

In the end Henley didn't either. He decided to become chief financial officer only after he was sure that Ellison would remain as CEO. The half million stock options he got didn't hurt either.

Henley worked on the Nippon Steel deal for a few weeks before taking over as CFO. He soon realized that a lot of essential work had already been done: Oracle's creditor banks were getting paid, the company had positive cash flow, and finance had been recentralized under Jeff Walker. Whatever Walker's shortcomings as CFO, he had done a good job of putting Oracle back together, something few people outside Oracle ever knew. "Later we read in the press,

'Jeff Henley has done such a great job of turning this around,'" Walker said. "And so obviously our pride makes us say, 'No, we turned it around, and Jeff Henley got the credit.'"

This was not to say that Oracle was completely out of trouble. Henley was to become CFO on a Monday in March. Over the weekend Ellison called "in a panic," he said. Oracle's receivables were much higher than Ellison had thought. The board was going to hold an emergency meeting that Saturday night to discuss the problem. He wanted Henley there.

The numbers were astounding. Oracle had about ninety million dollars in receivables—roughly half in bad deals, and another half in uncollected technical support payments. Henley took an ax to the problem. He recommended that Oracle write off the entire ninety million. Just forget about it. The company was never going to see any of that money anyway, he said. "Let's get that behind us," he said. The board agreed, and that was that.

That first meeting established a theme for Henley's tenure at Oracle. He was a sensible, conservative guy, and his conservatism instilled confidence in people. The people at Oracle "were nervous kittens," he said. "What the company needed was somebody who looked like he knew what he was doing."

Most of the changes Henley made were commonsense ones; they only seemed revolutionary because Oracle had been so lacking in common sense. No longer would the company sell a maintenance contract and recognize all the revenue up front; instead it would book the revenue a month at a time, as the payments came in. No longer would Oracle have a tiny bad-debt reserve; from now on it would always keep enough cash to cover any deals that fell through. And no longer would customers be given a year to pay for their software. "I said, 'I don't care if [the accounting rules say] you can recognize the revenue or not; we're not going to do that,'" Henley said. "So let's start doing thirty-day terms, and if people aren't willing to buy, it means they really don't need the products anyway."

All these changes slowed revenue growth, of course, and that was why Larry Ellison had not made them years earlier. Oracle's revenues grew only 12 percent in 1991 and 15 percent in 1992—a long way from the 102 percent growth of 1989. But of course that kind of growth had come at an extremely high cost, both in dollars and in reputation. Ellison and Oracle were no longer in a position to be cocky and aggressive. They had to be conservative or die.

"We just basically flushed out all the aggressive practices," Henley said. "We had to kind of get the company back to a normal way of doing business."

Larry Ellison and his then girlfriend, Kathleen O'Rourke, spent the Christmas break in Hawaii that year. In the past year or so he had watched his beloved Oracle become the object of pity and ridicule, been sued repeatedly by people he had never met, withstood the rage and fury of investors and financial analysts, nearly lost a billion-dollar fortune, and seen Oracle post a loss of twelve million dollars for the 1991 fiscal year, its first-ever annual deficit. Now seemed like a good time for a vacation.

One stormy day Ellison went body surfing on the Big Island of Hawaii. Ellison was no expert body surfer, but he was athletic enough to enjoy almost any sport he tried. "I had no intention of body surfing in a storm like this. But there were three Hawaiian guys—probably teenagers, eighteen, nineteen years old—who were out, so I thought I'd go out. I tried to dive through these huge, huge waves. Suddenly I found myself in a perfect position to catch a wave. So I said, 'OK.' I didn't think. I was in a perfect position and just jumped on the wave.

"The first half second I thought: This is amazing. I couldn't believe the acceleration off this wave. The second half second I knew I was in deep, deep trouble. Holy smokes. I knew I had to bail out of this wave before I got onshore or I was going to be maybe killed. So I duck. I go into a tuck to try to get out from underneath the wave. The wave took me down. It was taking my head right for the sand and would have snapped my neck like a twig, but I finally just rolled and got onto my back and right shoulder.

"The hydraulic pressures just started building. Within a couple of seconds I started hearing a sound in my ear that sounded like someone with a handful of shredded wheat, just crushing all this shredded wheat. Those were hairline fractures forming all over my body. And the first thing that gave was my clavicle, which exploded into three pieces and—unbeknownst to me—went into my neck and broke my neck in two places. Then I started hearing snap, snap, snap; it sounded like green twigs. Those were my ribs breaking one after another. One of them went into my lung."

Later Forbes tried to turn this clumsy accident into high drama with a print advertisement for the magazine. "The three billion dollar man," the ad began, referring to Ellison. "His life story reads like a piece of pulp fiction. Man grows up as math whiz on Chicago's tough South Side. Man masters computer. Man builds world's largest database software company, *Oracle.* Man drives business to billion dollar growth. Business hits wall. Man breaks neck bodysurfing in Hawaii. Man nurses both body and business back to health." Man, where did they get these copywriters?

"The wave finally receded, and I got up. I wanted to call one of the Hawaiian guys to get some help. My shoulder was no longer here; my shoulder was about at my waist. I couldn't move my right arm. I also couldn't breathe because the rib had gone into my right lung and collapsed it. And I couldn't inhale or exhale. I could kind of breathe a little bit, but I couldn't make any noise. I couldn't make a sound. It was up to me to get to shore and I just kind of staggered to shore. Everyone kind of cleared out. I guess I looked pretty bad when I came out."

Ellison was taken to the hospital, where he was placed in a brace so his collarbone could start healing. Beyond that there wasn't much the doctors could do. After all, they could not put his ribs in a cast. Ellison rested for a couple of days in his hotel, then flew back to California. Only when he saw his doctor there did he learn that his neck was broken in two places. Fortunately it was a stable fracture, and Ellison was in no danger. Within a few days he was back at Oracle, sore but basically all right. Still, he said, "It had been a very bad year."

* * *

The recriminations lasted way beyond 1991. For one thing it took a couple of years for Oracle to get its books straight. In the beginning of fiscal 1991 the company restated its financial results for the last three quarters of 1990—a major embarrassment. What was even more embarrassing was that the numbers still weren't right. Early in fiscal 1992 Oracle restated its results for the first three quarters of 1990 and for the 1990 fiscal year as a whole. The company was forced to admit that it had overstated its net income in one quarter by 159 percent and understated in another by 21 percent.[9] The company's books had not been so fouled up since the days of the dyslexic bookkeeper.

There were more humiliations to come. On September 24, 1993, after a long investigation, the Securities and Exchange Commission filed a complaint against Oracle in federal court in San Francisco. Between June 1989 and February 1991 the company had committed a long list of sins, according to the document. All the things Stephen Imbler had been concerned about—and a few he hadn't known anything about—were mentioned. "Oracle double-billed customers for products, double-billed customers for technical support services, invoiced customers for work that was never performed, failed to credit customers for product returns, booked revenues that were contingent, and prematurely recognized revenue," the complaint said.

It mentioned deals that were booked after the end of a quarter, deals with side letters, deals that could be canceled at will by the customer, and deals for software that Oracle hadn't even created yet. It spoke of consultants who billed customers who had not signed contracts and consultants who sent out bills for work they hadn't done. In sum, the complaint described a company overrun by crooks and incompetents, a place where the management was so busy trying to rule the database world that it either didn't know or didn't care what anybody else was doing. "Defendant Oracle, unless restrained and enjoined, will continue to engage in the transactions, acts, practices and courses of business described in this complaint," the document said. Oracle settled with the SEC by paying a token hundred-thousand-dollar fine without admitting liability and promising to behave in the future.

There was one more matter to dispose of. In the years since stockholders David and Chaile Steinberg first sued Oracle, other shareholders had banded together to sue just about everybody associated with the company: executives, directors, even auditors. It would have cost Oracle millions to defend itself in court (and God knew how much more if the jury didn't like what it heard), so it decided to settle. Without admitting wrongdoing, the company agreed to pay $23.25 million to the shareholders and as much as $750,000 to their very happy lawyers. On July 5, 1994, after all the payments had been made, a judge dismissed the lawsuits. Oracle's stock that day sold for roughly the same price as it had before the stockholders filed their opportunistic lawsuits.

There were definitely consequences to what Oracle had done, and Jeff Walker knew them by heart. "The company became the dominant supplier of databases. That was a consequence," he said. "The company has generated a return for its shareholders that is almost unprecedented in the history of capitalism. Customers have gotten value for their money. As for the people in the financial community, you know, those people who try to bet on the weather—I mean, the ebb and flow of market valuations—sometimes they made a lot of money and sometimes they lost a lot of money. . . . There were many, many winners, and many losers. The losers probably complain. I'm not sure the winners are complaining."

Walker was right, of course. In the end none of it mattered. Within a few years people were referring to what happened at Oracle in 1990 as a "hiccup" or a "stumble" or a "dip," just something that happens to a fast-growing company. The rule bending, the cheating, the arrogance, the shabby treatment of customers—none of it made any difference because Oracle was making money again, and that was what counted. Oracle did a billion dollars in business in 1991 and was up to two billion in 1994. The stumble or dip or whatever mattered so little that Ellison actually turned it into a positive. In 1990 Ellison almost ran his company into the ground. But that was history, and Ellison always had a special way with history. In the future the story would be different. In the future he would be portrayed as Oracle's savior—a mystical and powerful figure who saw the company perish and then raised it, godlike, from the dead. This, like many of the things he said, was largely true.

"Man nurses both body and business back to health," said the advertisement. "Man emerges as major player in the biggest venture of all. The race to rule the world's information superhighway. *Stay tuned.*"

Twelve

LARRY ELLISON WAS A MAN OF MANY INFATUATIONS. HE EXPERIENCED the world with the excitement and wonderment of a little boy and had the attention span to match. Thus he could fall quickly and completely in love with almost anything: a new technology, a piece of Japanese art, a certain kind of car (in the late 1980s he had a fling with Ferraris), whatever. Some of these attachments lasted, and some didn't. Ellison was well known for becoming enamored of the professional abilities of certain Oracle executives, though the feeling usually wore off after a while. And he was positively head over heels about Steve Jobs, to whom he felt so close that "sometimes I can't distinguish between him and me." But perhaps Ellison's greatest infatuation, in the early 1990s, was with a company called nCube.

nCube, founded in 1983 by four former Intel Corporation engineers, built what were known as massively parallel processing computers. While personal computers had only one processor, massively parallel computers packed hundreds or even thousands of processors inside a single machine—"like taking a . . . network of a thousand PCs spread throughout a skyscraper and shrinking it to fit inside a filing cabinet," as *Fortune* magazine elegantly put it.[1] All those processors could do some nifty things. nCube's machines were capable of taking complex mathematical problems, exploding them into thousands of parts, and assigning the parts to individual processors to solve. Best of all, nCube computers could perform these calculations hundreds of times faster, and far more cheaply, than the fastest mainframe computers.

Ellison became aware of nCube in the late 1980s. In the early years nCube had made sales mostly to the commercial and scientific markets. Ellison saw a larger market for the technology. If he could make Oracle software run on a massively parallel machine, he could dramatically shorten the process of retrieving information from a database—without raising the cost. In 1988 he put Oracle's developers to work on the problem, and in 1989 he and Bob Miner bought a controlling interest in the company. And while nCube's fortunes were to rise and fall, Ellison's enthusiasm remained high.

Still, there was a darker side of the nCube story that was never told: Ellison stubbornly and cavalierly refused to pay the debts associated with buying the company, even though he could easily afford to do so. His selfishness infuriated his creditors and threw some of their lives into chaos.

When Ellison and Miner bought into nCube in February 1990, they signed papers agreeing to make quarterly payments totaling $16.9 million to the company's principal shareholders.[2] Ellison was supposed to pay 67 percent, and Miner 33 percent. Miner always made his payments on time. Indeed, when Oracle's stock was depressed and Ellison was strapped for cash, Miner once wrote a check to cover his portion *and* Ellison's.

If Miner was dependable, Ellison was, well, something else. He missed $1.4 million payments on September 30, 1990, and June 30, 1991. The creditors hounded him, but he didn't pay, nCube investor Charles Masters said. Finally Ellison came up with an idea. He asked Bob Miner to write a check for all the money that he, Miner, still owed. That way the nCube people would get some cash, Miner would never have to make another payment, and Ellison would get off the hook, at least temporarily. Ellison promised that he would resume paying his part—without Miner's help—beginning September 30, 1991. Everyone agreed on the plan, and the parties signed papers restructuring the deal. Miner paid the shareholders $2.7 million.

It all would have worked out fine, except that Ellison didn't make the September 30, 1991, payment. "As soon as it was time for Larry Ellison to perform again, he just stiffed us," Masters said. "It became totally clear that he was not a man of good faith at all."

That November four of the nCube shareholders, including company cofounder Stephen R. Colley, sued Ellison to get the money he had not paid them. Six other shareholders sued him separately about the same time.

The Colley group's attempt to serve Ellison with the lawsuit was a story in itself. About noon on November 8, 1991, a process server named Erasmo Romo went to Ellison's Japanese-style home in Atherton. Everything that happened after that was in dispute. Romo said Ellison's gardener told him to follow the long driveway to the house (the gardener denied this). He said he saw Ellison near the front door, identified himself as a process server, and left the papers.

Ellison said he was sitting in his dining room when "I noticed a prowler walking along the outside patio deck. I immediately yelled through the window that he must leave my property immediately and proceeded to pick up the telephone and dial 911. The prowler threw some papers in the direction of the window I was standing near." The "prowler" then went on his way.

Five minutes later, alerted by Ellison, the Atherton police stopped Romo's car on nearby El Camino Real, drew their guns, ordered him out of the car, forced him to kneel on the ground, handcuffed him, and hauled him to the station. Apparently the Atherton police had a zero-tolerance policy toward paper-throwing prowlers.

Romo was never prosecuted. The Colley group said Ellison "had full knowledge of the reason for Mr. Romo's presence on his property." They said Ellison had been trying to dodge service for days.

In January 1992 Ellison wrote checks totaling more than three million dollars to the Colley group to cover the payments he had missed so far. But he did not include any money for the shareholders' attorneys, as he was required to do under the contract. While the parties argued about that, Ellison started missing payments again. In a sworn deposition in June 1992 a lawyer for the Colley group reminded Ellison that he had missed a quarterly payment to the four nCube shareholders on March 31.

Q. That principal payment is now outstanding; is that correct?

A. Absolutely.

Q. Do you intend to pay it?

A. Absolutely.

Q. When?

A. I don't know.

Q. Interest, I assume, is accruing since March 31, 1992, on those four payments?

A. Absolutely.

Q. Do you intend to pay that interest?

A. Absolutely.

Q. Do you know when you intend to pay that?

A. I do not.

Ellison interpreted his agreement with the shareholders in a unique and interesting way. He said the contract offered him a choice: He could make the principal payments on time or skip them and allow interest to accrue. When could the nCube shareholders expect him to pay the snowballing debt? "Assuming 100 years would be ridiculous. Two years would not be ridiculous," he said.

To the shareholders, the whole situation was ridiculous. Ellison owned more than thirty-three million shares of Oracle stock, which was trading around fifteen dollars at the time of the deposition. The nCube people believed he was stalling in hopes that his stock would rise quickly enough to cover the interest he was paying them. Ellison was employing the same strategy he had used when borrowing money on margin. There was one difference. "We were unwilling lenders," said Masters, who was part of the Colley group suing Ellison. "He could have paid what he owed, but it was cheaper to renege."

Masters, for one, was not in a position to lend money to people in the Forbes 400. Years earlier he and another investor had gotten nCube started with a few hundred thousand dollars in seed money. When Ellison bought the company, Masters received a promissory note worth just under two million dollars. He immediately made various investments in the belief that Ellison would be sending him checks. When Ellison didn't pay, Masters had to sell a house and some of his stocks in order to meet his obligations.

Masters, who owned some Oracle stock, once approached Bob Miner at a shareholders' meeting. He had always liked Miner, not least because Miner paid his bills. "I said, 'Bob, how can you stand to work with a guy like that? You've met all your obligations, and this guy thinks he's God almighty. He just thinks he can do anything he wants.' And Bob said, 'Well, you know, he made me a lot of money.' "

It galled Ellison's creditors to know that he could pay them but just didn't choose to. "This is not a case involving a legitimate dispute between two parties which ultimately resulted in a lawsuit," a lawyer for the Colley group wrote.[3] "Rather, this case involves a fabulously wealthy defendant who, while acknowledging the debts he owed, casually dismissed and ignored such obligations because it was not financially advantageous or convenient for him to honor them."

Eventually the courts ordered Ellison to pay his debts to both groups of creditors. The judgments against him included principal, interest, and attorneys' fees. The Colley group got a $2.4 million judgment against Ellison in February 1993. A month later Ellison still hadn't made a payment. The shareholders' lawyers asked the judge to sell Ellison's house out from under him. "We were in the process of moving in on his Japanese art collection when he decided he'd cave," Masters said.

Paying the judgment was no problem for Ellison. In May 1993 he sold 230,000 shares of Oracle stock, grossing $8.8 million. He still had $1.25 billion worth of stock left over.

Now that nCube was bought *and* paid for, it was up to Ellison to do something with it, and he soon did.

When an executive headhunter asked Raymond J. Lane if he would like to interview for a job with "a software company in California," Lane said thanks, but no, thanks. It didn't matter which software company it was. At the time Lane was a Texas-based senior vice-president at the management and technology consulting firm Booz·Allen & Hamilton. He had worked there for twelve years. If he stuck around, he might someday become chief executive officer, a title held by only a half dozen people in the firm's eighty-year history. And now somebody wanted him to go work at a software company? No way.

The headhunter eventually persuaded Lane to go to the interview anyway. Maybe Lane could sell some consulting services to the software company; who knew? Early in 1992 an unenthusiastic Ray Lane flew to California to meet with Larry Ellison. He had no idea what he was getting into.

Ellison was looking for someone to replace Mike Fields as head of Oracle's domestic sales. But he wanted more than just another salesman. Ellison wanted someone who could handle all of the company's diverse business interests: sales, consulting, and Oracle user education. He wanted a mature manager. He wanted a respected businessperson. He wanted Ray Lane and said so. He offered Lane a hundred thousand stock options to join the company.

Lane was flattered. "I don't think so," he said. He lived in Texas and wanted to stay there.

"You can run things from Texas," Ellison said. All Lane had to do was make his way to Redwood Shores for the Monday management committee meetings.

Lane demurred. His real ambition was to be the head guy at Booz·Allen. He had invested too much in the firm to leave now.

Ellison upped the ante. Instead of a hundred thousand options, he offered Lane three hundred thousand options at $15.50 apiece. Three hundred thousand options? Lane couldn't believe this was happening. "If the stock moved five points, it would be more money than I could ever make at Booz·Allen," he said.

All this was happening at a critical moment for Oracle. For three years the company had been talking about Oracle7, the new version of its database management system. But the software was never actually released. The product became one of the most infamous pieces of vaporware in the business. Other database companies believed that Ellison deliberately tried to forestall competition by implying that Oracle7 would be released at any moment. (This was known as creating fear, uncertainty, and doubt.)

Ellison's response: Who, me? "We always told people when we believed things would be ready," he said. "I never intentionally deceived anybody. I certainly was wrong on my dates. But Bill Gates has been wrong on his dates. We're always wrong in the same direction. We always take longer."

Oracle tested an early, or alpha, version of Oracle7 at fourteen customer sites beginning in August 1991. The product had all sorts of snazzy features that were sure to brighten the lives of corporate information czars. Some of these features could be understood by normal people. Let's say you wanted two hundred order takers at your company to have access to certain tables in your database. With Version 6, you had to enter two hundred commands—once for each individual. With Oracle7 a single adjustment would do the job. "V7 appears to be the database we've all been waiting for," one Oracle user wrote late in 1991.[4] "The question is, how long do we have to wait?' "

Six more months, as it turned out. The company released Oracle7 in June 1992. Raymond J. Lane joined Oracle that day. "That was a hell of a day," Ellison said.

It turned out to be. Within a couple of years Lane would be hailed, along with chief financial officer Jeff Henley, as a savior of the company, someone who helped it grow from a reckless and irresponsible adolescent to a reasonably dependable (if still somewhat arrogant) young adult. "Lane is calm, reasoned and practical. . . . He has the presence of the old-style gentleman executive," an industry magazine said.[5]

Lane succeeded partly because in Oracle7 he had something excellent to sell, a product that did pretty much everything Ellison said it would do. Oracle's days of sacrificing quality at the altar of market share appeared to be over, at least as far as the core product went. "They finally did go in and really make major changes to their underlying architecture. So if you look at the product today, it's an excellent product," industry analyst Jeffrey Tash said.

The customers' satisfaction was reflected in Oracle's bottom line. In Lane's first four years on the job, companywide sales grew from $1.5 billion to $4.2 billion and earnings per share quadrupled.

Oracle gave him the moon in return. Lane's salary and bonus in his first year totaled $986,721, not including 350,000 stock options (300,000 when he joined and another 50,000 at the end of the year). After fourteen months he was promoted to president of worldwide operations, and in June 1995 he was appointed to the board of directors. By the summer of 1996 he was chief operating officer, the undisputed number two man in the company behind Ellison. His salary and bonus that year totaled $1,525,127, and he owned or had options to buy 894,931 shares of Oracle stock. Once, when Ellison decided to sell his Ferrari, Lane said, "I'll buy it," and wrote him a $55,000 check.[6]

At first Lane wasn't sure things would work out at all. He arrived at Oracle to find that someone else—a longtime employee—thought *he* was in charge of domestic sales. This was Ellison's fault. Instead of telling the other person that Lane had the job, Ellison had left it to Lane to do so. The same thing had happened to an Oracle marketing vice-president years earlier, when Ellison hired Mike Seashols as his sales manager. Early on Lane came to a conclusion about his boss: "You get Larry as a whole package. You don't get the good without the bad."

Lane's arrival and Henley's continued stable presence changed Ellison's role in the company. With his top guys running the day-to-day business, he was free to do the things he enjoyed—brainstorm with his engineers, reread Poe's "Annabel Lee," visit Japan, entertain women, buy things, give interviews, and dream. Ellison didn't have to behave like a grown-up businessperson anymore (not that he ever really had). Now he could really indulge the twelve-year-old inside him. Sometimes the twelve-year-old would show up at the Monday management committee meetings and pitch ideas that made Lane and Henley "fall off our chairs," Lane said.[7] But they could always get back into their seats and talk sense to him. "I can always tell him, 'Larry, I think this is fucked up. Larry, I think this is crazy,' " Henley said.

Although Ellison and his top managers worked effectively together, they weren't exactly a close-knit group. Ellison's quirky ways could be maddening; one high-ranking executive said his chronic lateness was "flaky" and "disrespectful."[8] And while Henley and Lane generally liked and admired Ellison, they didn't pretend to know him. "Larry's a pretty alone, cold guy, you know," Henley said. "I'm not sure he could ever get real close to anyone."

If bringing in seasoned managers changed things for Ellison, it changed them even more for Oracle's customers. Lane quickly became known for listening to customers and trying to meet their needs—something Oracle had never really done before. There was a story that was well known at Oracle about a customer whose entire business had ceased to function because his database had crashed. The desperate customer called Oracle support and explained what had happened. Said the support technician: "Bummer, dude." Under Ray Lane, no customer would ever again be addressed as "dude."

The work Oracle did with R. R. Donnelley & Sons Company, the giant Chicago-based printing and distribution company, was another example of Lane's influence on the company. One of the things R. R. Donnelley did was print books and distribute them to bookstores. For years the stores would guess how many copies of a certain title they would need, order the books, sell as many copies as they could, and return the rest. This was not an efficient or cost-effective way of doing things. No matter how well the booksellers predicted their customers' tastes, they often ended up with a lot of unwanted and unneeded books on their shelves. And of course publishers wound up with stacks of books they couldn't sell.

R. R. Donnelley used Oracle to address this problem. With the help of an Oracle consultant, the company developed an overnight order fulfillment system. Now bookstores could order books as they needed them, and not before. Oracle "really has gone above and beyond the call of duty to help me be successful," R. R. Donnelley's technology guru said.

This guru was Al Guibord, who had lost his job at a New Jersey company in the mid-1980s after Larry Ellison sold him a copy of Oracle that didn't work. Much of the story of Oracle—the trickery, the treachery, and the fancy recovery—was written on this guy's forehead. After Guibord left Timeplex (the company whose Oracle MVS software never worked), he landed a job as an information manager for a company in Boston. There he once again bought Oracle—not because he enjoyed pain but because salesman Danny Turano promised to make him happy. Turano did so.

Later Guibord moved to R. R. Donnelley, which used the Oracle database and Oracle financial applications to run its business. There Guibord became a Ray Lane devotee. Lane treated him as royalty, appointing him to the panel of customers that met with Oracle executives twice a year and advised the company on products, licensing, marketing, and so on. Lane even gave Guibord his home phone number. Once Oracle had made Guibord a goat; now the company "has actually made me a hero in this company," he said. "Years ago Oracle was just basically a bunch of guys running around taking orders. It's a different company now. It's being run like a business," he said.

Not that he ever forgave Larry Ellison. As far as Guibord was concerned, Ellison, who rarely met with the customer advisory board, had nothing to do with Oracle's new customer-friendly approach. He gave all the credit to Lane. "If Larry was still running Oracle, I wouldn't be doing business with Oracle," he said.

Guibord's attitude was not uncommon. A lot of people didn't like to give Ellison credit because he had been brash and egotistical and cavalier about customer needs. But of course Ray Lane didn't breeze into Oracle through the air-conditioning ducts. Ellison hired him. He gave him three hundred thousand stock options and a mandate to run the company in a professional way—that is, in a way that Ellison had not been able to run it himself. This required Ellison to put aside his ego, something he normally guarded as vigilantly as his stock. "I think Larry did not get as much credit as he deserves for having turned that thing around," former sales executive Tom Siebel said.

You could even argue, somewhat perversely, that Guibord had to suffer through his first experience with Oracle in order to enjoy the second one. Oracle might never have been in a position to hire the likes of Ray Lane if it had not first won a considerable part of the relational database market share. And Oracle probably would not have got so much of the market if it had not constantly hyped and released new products, including some that weren't ready. Guibord and a lot of other customers got trampled in the 1980s. What they didn't know—this is the perverse part—was that it was for their own good.

Oracle's top managers met in a conference room at 500 Oracle Parkway to talk about product schedules. Larry Ellison and Ray Lane were there, and so were a few others. Bob Miner was also present, though he probably would have liked to be someplace else. After fifteen years as partners Ellison and Miner had grown apart. There was a time when Ellison had called Miner at home after dinner every night to go over the day's events ("None of us could understand what they'd have to talk about if they worked together all day," Nicola Miner said); now they hardly communicated at all, except to argue. The years, the money, the profound differences in character and lifestyle—all these things had put distance between them. Now the gap was going to become a canyon. Ellison demanded to know the progress of certain products Oracle was developing. Miner either didn't know or didn't care.

"He said, 'If you want schedules, I'll give you schedules,' " Ellison said. "And he threw a blank piece of paper on the table. There was dead silence in the room." That was the moment when Larry Ellison knew that his partner was finished as head of development.

The day had been coming for some time. After fifteen years Miner had nothing left to prove, no reason to keep pushing himself or his people. Certainly he didn't have to work. "We all felt, and he especially felt, that he had too much money, that you get to a point in life where you have too much money, and you don't need any more," his daughter Nicola said. He had already established a family foundation to give money away—always anonymously, so he wouldn't be embarrassed.

While Ellison's ego and compulsive personality drove him to seek ever-greater wealth and fame, Miner didn't care about those things. He rarely gave interviews; one of the few he ever granted was to a homegrown cable television show called *High Tech Heroes.* Miner was so low-key that some journalists who wrote about Oracle didn't even know who he was. That was OK with him. By 1992 he was working mostly for fun.

The trouble was, he wasn't having any. Miner had founded Oracle because he knew he would never be happy working as a drone in some giant corporation. Now Oracle was a giant corporation. It employed 8,160 people in the United States and forty other countries. Miner had so many people working for him that he didn't know who a lot of them were—a fact he considered almost unbearable. "He had created what he wanted to escape," Nicola Miner said.

Working with Larry Ellison wasn't any fun either. After a decade and a half Miner was tired of trying to make good on Ellison's promises. Once Miner was riding in his car, listening to a tape of a speech Ellison had given in which he said Oracle would have a certain technology ready by a certain time. Miner laughed out loud, according to his daughter, who was with him. There was no way Oracle could ever make the deadline Ellison had set. "He was angry because Larry had set him up to look like an idiot," Nicola Miner said. Ellison was not one to apologize. "I pissed a lot of people off that way. Still do," he said.

Miner's frustration with Ellison, his dislike of corporate life, and his growing appreciation of the good things in life (tennis, golf, a winery he had bought in Napa) all showed in his work. He just didn't care anymore. "Bob needed to go. I liked Bob a lot, but Bob was holding the company back because he just didn't want to work," chief financial officer Jeff Henley said. "The company had greatly outgrown him. He was head of development—I mean, it was a joke. . . . In the last year or two here he just wasn't driving development. He played chess in the afternoons. It was clear to me that he was not equipped to manage hundreds of developers. The products weren't coming out fast enough."

When Miner produced a blank sheet of paper instead of the

much-needed product schedules, Ellison wondered if he would have to fire his own partner. He didn't wonder for long. Soon after that meeting Miner called Ellison and said, "I'm going to make this easy for you." He said he was going to open a small office in San Francisco and tinker with some new and speculative technologies. To make sure he enjoyed himself, he was going to take a few of his favorite engineers with him. Miner later referred to this group of techies as "elitist weirdos," which he meant as a great compliment.

Miner opened the San Francisco office late in 1992. His time among the weirdos would be all too short.

Larry Ellison met Adelyn J. Lee in an Oracle elevator, a means of transportation that provided a metaphor for their eighteen-month relationship. At the time Lee was a thirty-two-thousand-dollar-a-year marketing coordinator. She and Ellison had never seen each other before, but that wasn't unusual; Oracle had thousands of employees, and Ellison didn't spend much time at the office. During the elevator ride Lee, a tall Chinese-American woman in her late twenties, mentioned half-jokingly that she would like to go for a spin in Ellison's Ferrari. Then the doors opened, and they parted.

Later Ellison, who was forty-seven at the time, received an electronic mail message from Lee. "Please forgive me if I was being too forward in asking" about the Ferrari, she said.

Too forward? Not at all. "Not only will I be happy to take you for a ride," Ellison wrote, "you are welcome to take a turn at the wheel." You could almost hear his engines revving.

According to Ellison, Lee asked him out five times before he broke down and said yes. (Suddenly he was hard to get.) Their first date was, appropriately, purely automotive. Ellison's Ferraris kept catching fire on him, so he took Adelyn Lee with him to test-drive an Acura NSX sports car, a low-slung two-seater capable of hitting seventy before you got out of the garage. He liked the NSX so much that he eventually bought one, and another after that, and another.

He also liked Lee, whom he continued dating a couple of times a month.

Ellison was not available to see Lee more often because he was also dating two other Oracle employees: Kathleen O'Rourke, the longtime girlfriend who had witnessed his accident in Hawaii, and a woman named Andrea Zeman. How did Ellison complete this hat trick without getting caught? He didn't. "Kathleen came over to my house one night, and Adelyn's car was there. She got furious," he said.

In spring or summer of 1992 the relationship between Ellison and Kathleen O'Rourke was at a turning point. "Kathleen and I were working on our relationship. . . . Can't figure out if we're going to get married or break up," he said. They went to Napa to ride bikes and talk about it. They were pedaling along when Ellison got upended going over some railroad tracks. He landed on his left elbow, shattering most of the bones in his elbow and arm. "The doctor who finally fixed it said it looked like a war wound, like I'd been shot with a high-powered rifle in the elbow," Ellison said. A surgeon at the University of California, Davis, put the pieces back together. Ellison was so pleased with his work that he later pledged five million dollars for a new musculoskeletal institute at the university.

He was still on the mend when he told Adelyn Lee that he wouldn't see her anymore because he wanted to work things out with O'Rourke. She sent him a message saying she would like to keep in touch "on a platonic level." "I would love to stay in contact," he answered.

They stayed in contact. Within a couple of weeks Lee sent Ellison a message saying there was a man in her jujitsu class who had an "uncanny resemblance to him." Ellison's reply: "Hmmm . . . Is he really that good-looking?" Eventually they began dating again.

In some of the electronic mail messages she sent Ellison, Lee made it clear that she liked him for more than just his good looks. On January 10, 1993, she sent him a message marked "Happy Thoughts."

> Hi Larry, would you really consider giving me a loan for $150k (at a fair interest rate of course)?
>
> In keeping with my New Year's Resolution, I need to move, not necessarily to a larger place—just a better location. Fyi, I will be using these funds to invest in the market (I am currently investing successfully, however practicing on a larger scale would prove to be faster), therefore I should be able to repay you in a short period.
>
> Please consider my request and give me your thoughts.
>
> Adelyn

Ellison—who received hundreds of E-mails every day and replied to dozens—made a brief but pointed reply. "How about dinner instead?"

If Lee noticed the word *instead* in Ellison's message, she did not pay any attention to it. Two days later she sent another message asking about the loan.

> Hi Larry, dinner would be great! Any day this week except Wed & Thurs.
>
> Adelyn
>
> ps. could you work with me to try and figure something out in regards to a loan for me.

Apparently Ellison never responded to the request for a loan. When I asked him about this later, he was incredulous that she had even asked. "Adelyn would say, 'Gee, I'd love to make some investments of a hundred fifty thousand dollars in cash, and I'll pay you back the loan. Could you give me a loan? A hundred-thousand dollar loan?' She said, 'Will you think about it?' I said, 'OK, I'll think about it,' " he said. "I tried to brush it off as a joke."

I said I would have serious doubts about the long-term prospects of the relationship if I thought the woman was just in it for money.

"Sure," Ellison said.

"Then again, maybe you weren't concerned about the long-term prospects," I said.

"There's that possibility," he said.

By February 23 Lee apparently had forgotten about the loan. Now she had other things on her mind.

> Hi how are you doing? Couple of things,
>
> —are we on for HI
> —is nCUBE going public anytime soon
> —the lady from the watch store keeps calling, what can I tell her
>
> Adelyn

"HI" turned out to be Hawaii; apparently Ellison and Lee had discussed going there. nCube was the massively parallel computer company Ellison owned. In his reply Ellison brushed off all three questions. He pretended he didn't know what "HI" meant and asserted that nCube would not go public "for at least a year." As for the watch lady, he said, "tell her next week."

Next week never came. Three weeks after that exchange, Lee asked again about the watch.

> Hi Larr, when are we going to go watch shopping.
>
> Adelyn

Ellison said Lee's requests for money and gifts "didn't bother me that much because I had no intention of giving [them] to her. . . . If I think someone is dating me . . . to get stuff, I will automatically give them nothing." He may not have given her a watch, but he certainly gave her the time of day. "She was a very intriguing woman," he said. Translation by Jenny Overstreet: "He thought she was really pretty, and they had a great sexual relationship."

Unfortunately Adelyn Lee was not making her immediate supervisor as happy as she was making the CEO. That January Lee had become the administrative assistant to Craig Ramsey, an Oracle sales vice-president. Ramsey was already dissatisfied with Lee's work. She was supposed to be at work each day by 8:30 A.M. but showed up late at least once a week and sometimes twice. When she finally got to the office, she was often unpleasant to deal with, Ramsey said. From time to time Ellison needed to speak with Ramsey. But when Jenny Overstreet placed the call, Lee was often abrasive or abrupt.[9] Once, when one of Ellison's top executives called for Ramsey, Lee put him on the speakerphone, as if *she* were the busy executive and he were the secretary. "She didn't understand how to treat people in business, especially not her boss's superiors," one Oracle employee said.[10] By late March 1993 Craig Ramsey was in the extremely unenviable position of wanting to fire his boss's girlfriend.

Enter Jenny Overstreet, who knew almost everything that went on at Oracle and who inevitably got word of Ramsey's problem. No admirer of Adelyn Lee herself ("She was horrible. Always horrible"), she asked Ellison what Craig Ramsey should do if he wanted to get rid of a certain administrative assistant with whom Ellison happened to be sleeping.

"Treat her like any other employee," Ellison replied. He later explained, "If this had been a woman that I was engaged to be married to, that I loved dearly, the answer would still be the same." If ever there was a clear statement of Ellison's priorities, that was it. Nothing—not friendship, not sex, not even supposedly eternal love—was going to get in the way of what was best for his company. He said later that he wasn't especially interested in the reason for Lee's dismissal.

Overstreet said she had that conversation with Ellison on— appropriately—April Fools' Day. Soon people at Oracle were laying the paper trail necessary to terminate Adelyn Lee. Craig Ramsey discussed the situation with Oracle lawyer Juana Schurman and human resources chief Phil Wilson; at one point Schurman wrote Wilson a memo saying, "We are generally agreed that Adelyn must go." Now Ramsey had to find the right time to tell her.

Meantime Ellison kept dating Lee but never told her she was about to be fired. His reason for doing this had a certain odd logic. From his point of view, his personal relationship with Lee had always been completely separate from their working lives. Their sexual relationship had nothing to do with her employment at Oracle, and her firing had nothing to do with their romantic life: That was how he saw it. Ellison also said he kept dating her because he wanted to be "a little bit supportive under those circumstances." Ellison was like a death row priest comforting a condemned woman—except that Lee had no idea she was going to be executed. It was hard, in the end, not to conclude that Ellison had been at least a little bit callous: Sure, Lee was going to lose her livelihood, but why should that spoil his good time?

The two kept up their E-mail banter during the first part of April; Lee kept asking Ellison for things, and he kept asking her to settle for less. In an April 12 message she described a car she wanted Ellison to buy for her. It was a 1991 Acura NSX sports car (they were addicted to these things), black with a tan leather interior. The asking price was $50,600. Along with the description, Lee sent this message: "how about this instead of a watch?"

Ellison's reply: "hmmm . . . are you sure that you are tall enough to drive an nsx?"

Lee immediately wrote back: "yes, I am tall enough, I found this out when I did the test drive. per our conversation, do you still promise to buy me an NSX!"

Again Ellison appeared to avoid the question. "yes—of course I will buy you an nsx and anything else you want: a home in woodside, a gulfstream jet, the hope diamond, the general electric corporation (which is quite expensive), even dinner on Friday."

Lee responded the next day.

> ok. I would have settled for just the nsx, but since you offered:
> atherton rather than woodside (we can be in the same hood!)
> leair jet (did I spell it right?) rather than gulfstream
> any collector diamond around 2 karats would be fine
> GE hummmm, why did you pick this particular company
> and dinner fri at the Barbarossa @ 6:30 would be fine.
>
> Adelyn
>
> ps. the question still posed is a new or used NSX?

Ellison apparently never answered the question. On Thursday, April 15, she sent Ellison a message asking about a new job within the company. Lee had begun her career in Oracle's worldwide marketing division, and she was now pressing to go back there. Her message to Ellison had a much more professional sound than the ones she had sent earlier in the week.

> If we are truly going to work toward expanding our market share worldwide . . . I believe that we first need to be able to work within and help ourselves by helping each other.
>
> My recommendation, would be to reinstate my position as Inter'l Mktg Mgr w/a commensurable compensation package to that of . . . the person who took over my responsibilities when I transferred positions.
>
> Please feel free to contact me to further discuss this issue or if you have any questions. I look forward to your reply.
>
> Best Regards, Adelyn

Ellison, who often read his E-mail late at night, sent his reply just after midnight. She wasn't going to get the marketing job or even keep the job she had. But Ellison didn't tell what he knew. "aj—I would like to stay out of this. larry."

Lee and Ellison kept their date that Friday evening, but they never made it to dinner at Barbarossa or anywhere else. Lee met Ellison at his home in Atherton. They stayed there awhile, then went out to see the movie *Benny & Joon,* about an auto mechanic who cares for his mentally ill sister. Everything else that happened that evening became a subject of legal dispute.

On Monday morning, three days after Lee and Ellison went to the movies, Craig Ramsey fired Adelyn Lee. He gave the following reasons on her termination form: "Inability to interact with her peers or my direct reports. Lack of sensitivity to business issues. Attendance problems." Lee did not take it well. She dumped Sprite on her computer keyboard, knocked the books off her shelves, stole Craig Ramsey's calendar, and locked herself into an office and tried to reach Ellison by phone. No such luck: He was through with her. Finally, security escorted her out of the building.[11]

The next evening Larry Ellison sat down in his home office in

Atherton to read his electronic mail. He noticed a message that Craig Ramsey apparently had sent that morning.

> Received: 04-22-92 08:19
> From: cramsey. US1
> To: lellison
> Subject:
> I have terminated adelyn per your request. cdr

Ellison stared at the message for a moment to be sure it said what he thought it said. *Per your request?* Without thinking, he started to type. "Craig, are you out of your fucking mind? I never fucking told you to fire Adelyn, you fuck—"

Then he stopped. "Wait a second. Wait a second," he remembered telling himself. "Holy shit. Oh, my God, oh, my God, oh, my God. *Court case.*"

He deleted everything and started over.

> Received: 04-23-93 10:59
> From: Larry Ellison <LELLISON.US>
> To: cramsey. US1
> Subject: Re:
> Cc: rocampo joverstr
> In-Reply-To: cramsey. US1's message of 04-22-93 08:20
>
> PRIV AND CONFIDENTIAL
>
> craig, are you out of your mind! I did not "request" that you terminate adelyn. jenny told me you wanted to terminate adelyn. I decided not to veto your decision. I did not want to get involved in the decision for obvious reasons. this is the most amazing note. wait a second . . . craig, did you send this note? larry

"I pretty much reconstructed [the note] without the profanity and wrote a new note that I knew would be part of the evidence," Ellison said. He sent copies of the message to Jenny Overstreet and

Oracle's general counsel, Ray Ocampo. "It was clear that a lawsuit would be coming."

On October 18, 1993, Adelyn J. Lee sued Oracle, Larry Ellison, and Craig Ramsey in San Mateo County Superior Court. The accusations: wrongful termination, failure to prevent discrimination, and negligent mental distress.

A couple of weeks later Bob Miner suffered a collapsed lung. He did not know why, but after going to the doctor, he had his suspicions. At one point he told his kids that "he hoped he would make it,"[12] confusing and alarming them. What was the big deal? "We didn't know that [the collapsed lung] was a symptom of cancer," Nicola Miner said. Miner did not share his fears with some of his colleagues either. When Jeff Henley, Oracle's chief financial officer, saw him at the company Christmas party that December, Miner did not look well. "I've got this thing, and I just can't shake it" was all Miner said.

For a while the doctors had a hard time arriving at a diagnosis, partly because Miner had so much fluid in his lung. After two months they found what they expected to find: a tumor, evidence of cancer of the lung.

By the time the doctors discovered the tumor, Bob Miner and Larry Ellison were no longer close. Miner and a few of his favorite developers were in San Francisco, working on their pet projects, while Ellison was down the peninsula at Oracle headquarters, talking up the information highway. He and Miner still spoke on the phone once in a while, but since Miner wasn't in the thick of things anymore, Ellison did not see him as much. Also, Miner had long since sold Ellison his interest in nCube because he didn't think the company would ever make money. He and Ellison had even made a bet: If nCube became profitable by a certain time, Ellison could place a statue of himself on Miner's property in Napa. If not, a statue of Miner would be erected outside one of Ellison's houses in the Bay Area.

The news of Miner's illness was not delivered to Ellison by Miner himself but by Ellison's assistant, Jenny Overstreet, a close friend of the Miners'. Ellison's reaction, according to Overstreet: "They've already cured that." As always, Ellison was optimistic and forward-thinking, if not quite realistic. As he often did, he dealt with this highly personal crisis in a purely intellectual way, the way that made him most comfortable. The only way for him to cope with the idea that his longtime partner had cancer was simply to say that he did not have cancer or wouldn't have it for long.

Miner also dealt with his illness in a way that was characteristic of him: He downplayed it. Though he had enough money to hire the best cancer specialists money could buy, he decided to see doctors at a hospital in San Francisco because it was a short walk from his home. When Ellison heard that, he was furious. "[I told Bob,] 'I'm so glad it's convenient for you, Bob. You know, walk from your house. That's how I'd pick a hospital. Walk from your house. Goddamn right. Makes sense to me. Shit, yes. Closest hospital. Walking distance would be great.' He said, 'Larry, don't give me any of this crap.' " Ellison said he offered to help Miner find better care, but Miner refused.

The hospital near Miner's home put him through a grueling six months of radiation and chemotherapy. Then the doctors changed their diagnosis. He did not have smoker's cancer after all, they told him. He had a much less common form of cancer called mesothelioma, or cancer of the lung lining. Sixty percent of the people with the illness got it from exposure to asbestos; Miner, who, so far as he knew, had never been exposed to asbestos, was among the 40 percent who got mesothelioma from God knew where. This new, accurate diagnosis was the worst possible news. The radiation and chemotherapy had been a waste of time. There was no effective treatment.

Miner—who, Nicola said, was "very stoic but very scared"—called Ellison and asked for his help. "[I told Bob,] 'I'll send someone over to get all of your records.' 'But Larry, this thing is inoperable.' I said, 'Bob, the guys [at the San Francisco hospital] should have little round red noses that glow in the dark and big shoes with bells on the end. Give me the records.' "

Ellison had connections in the world of cancer treatment. One of his close friends was Dr. Marguerite Lederberg, a psychiatrist at Memorial Sloan-Kettering Cancer Center in New York. (Lederberg's husband was Nobel laureate Josh Lederberg, a molecular biologist in whose laboratory Ellison once spent a couple of weeks assisting with experiments.) Miner was eventually referred to Dr. Robert Ginsberg, the chief of Sloan-Kettering's thoracic service. Ginsberg removed one of Miner's lungs, to no avail. The cancer had spread into his back and stomach. Years later Ellison was still upset with the decisions Miner made about his treatment.

"If we had gotten him to Memorial [Sloan-Kettering] earlier, I think his life could have been saved," Ellison said. "Now it was a pretty aggressive cancer, and maybe it couldn't be [cured]. But sometimes you get lucky."

Nicola Miner saw things differently. She was not upset with her father's decision to be treated close to home because "when you're scared and you're facing the end of your life, you don't make the most rational decisions." She also believed the mesothelioma would have killed her father no matter what; the cancer was that virulent. "There was nothing Larry could have done. The only difference that it made by Daddy not going to Sloan-Kettering first was that he got a lot of treatment that made him feel sick," she said.

Even after Sloan-Kettering could no longer help him, Miner was willing to try anything. One speculative treatment required him to paint a purple substance on his feet; people who saw him on those days couldn't imagine what had turned his feet that color. Another time a healer prescribed a potion of Chinese herbs mixed in a bag. Miner was told to soak the bag of herbs in water and drink the resulting brew. Under no circumstances was he to open the bag. He opened it, of course, and found hundreds of black bugs inside. "That was the end of that particular therapy," his friend Condon Brown said.

In the fall of 1994 Bruce Scott, who had worked closely with Bob Miner in the early days of Oracle, wrote him a letter. The timing was coincidental: Scott had no idea Miner was sick. He wanted to get in touch with his old boss simply because he missed him. Scott believed he had grown a lot since his uneasy parting from the company. In the old days he had complained about the exhausting pace of work at Oracle. Miner had always told him, "You'll never have a boss who will be as good to you as I am." That had proved true.

Scott thought about calling Miner at Oracle, but a letter seemed safer. In his brief note to Miner Scott said, "If you aren't still angry, I'd like to have contact."

That Saturday Scott's phone rang. It was Miner. "Ten years just vanished," Scott said.

"I got your letter," Miner told Scott on the phone. He was crying. "And it came at a really bad time. Because I just found out I have about three weeks to live."

Scott visited Miner in San Francisco. Miner was still able to take short walks, so the two strolled together through the city. Scott tried to apologize for the way he had left Oracle—he resigned by E-mail and left Bob to finish Version 3 by himself—but Miner stopped him. "The only thing you did wrong was that you were young," Miner said. Still, Miner seemed perplexed, even hurt that Scott had left him so many years before. "He asked me, 'Why did you leave?' He still didn't know why I left," Scott said. The two talked for a while, then parted, their relationship healed. Scott never saw Bob Miner again.

It had been all too clear for some time that Miner would not be returning to Oracle. He would have been happy to have someone pack up his San Francisco office and send his things home, but Ellison would not hear of it. Miner might die, but Ellison was not about to let him resign. "Larry is so funny. He kept my dad on the payroll," Nicola Miner said. "He was like, 'You're coming back to work, you're not quitting.'" Miner's reaction, according to his daughter: "It's ridiculous, but I'm so sick I don't care."

"Dad and Larry were both into living forever, the fountain of youth kind of stuff," Nicola said. "I think that my dad's illness really freaked Larry out. He didn't see him nearly as much or even call him nearly as much. He was really scared and really uncomfortable, which is how a lot of people get when people have cancer. My dad always said that Larry had a hard time facing his own mortality."

In the days before he died, Miner invited a number of close friends to his home to say good-bye. Ellison was not among them. Yet Ellison was willing to be there if Miner wanted him to be. On November 11, 1994, Jenny Overstreet received a call from Miner's doctor. Bob had only a few hours to live; if Ellison wanted to see him, he should come right away. Ellison and Overstreet were thinking about whether to go—they were willing, but didn't know if they should—when Nicola Miner called them on the car phone and asked them not to come. "We just wanted to be a family," she said. Bob Miner died with his wife and children by his side. He was fifty-two years old.

Though Ellison knew Miner was going to die, he didn't believe it when he got the news. "How do I know he's dead? I haven't seen the body," he told Overstreet. Ellison did not mean to be uncaring, and Overstreet did not think he was. His reaction, she knew, was just a defense. Miner's death, like the death of Ellison's adoptive mother, Lillian, was not something Ellison could cope with on an emotional level; the news inflicted a pain he could not bear to feel. For the closed, remote Ellison, being rational was a means of survival. Just as he had once asked a marriage counselor to explain love to him in objective terms, he now sought a rational understanding of Miner's death: He wanted proof.

"It does seem odd how someone can be here and then be gone," Ellison told me. "It doesn't make any sense to me. Death has never made any sense to me. How can a person be there and then just vanish, just not be there? Clearly the reason they're not there is they're off doing something else. Like Bob was. I thought, nothing has really changed; he's just off playing tennis, or he's at a restaurant in the city. That's why he's not here. . . . Death makes me very angry. Premature death makes me angrier still."

Miner died as he had lived—quietly, humbly, and anonymously. The press missed the story. The San Francisco *Chronicle,* Miner's hometown paper, did not run an obituary until five days after his death. It appeared in the back pages, next to a couple of other obituaries and a story about the accidental death of a five-year-old boy. The *Chronicle* referred to him, correctly, as the "architect of the world's first commercial relational database management system," but otherwise the story left a lot to be desired. It did not mention that Miner started with four hundred dollars and built a company worth billions, that he was on *Forbes* magazine's list of the four hundred wealthiest people in America, or even that he enjoyed tennis and chess and golf. The obituary in the San Jose *Mercury News,* which covered technology well, was even briefer: only 183 words. It said Miner "assisted in the development of [Oracle's] data-base project," which seriously understated the case. It called Miner a "major shareholder" in Oracle but didn't say how major, even though the information was readily available. To be fair, the news media did not get much help in preparing their stories; at the request of the family Oracle released only a sketchy biography. Still, the news accounts seemed a poor send-off for a man as accomplished, and beloved, as Miner.

A few weeks later Oracle held a memorial service—purely secular, of course—in Miner's honor at the Herbst Theatre in San Francisco. Miner had asked Jenny Overstreet to put something together—not because he wanted people to say nice things about him, but because he thought that a gathering might help them feel better. Overstreet, the consummate organizer, spread the word about the event, made an audiotape of Miner's favorite music, brought in wine from Miner's Napa winery, and arranged for food from Vivande, Miner's favorite San Francisco restaurant. She also produced a videotape consisting of images of Miner, his friends, and family. The last image on the video—a picture of the Miners enjoying a family rafting trip—lingered on the theater's big movie screen as the service began. About 150 people were present. Larry Ellison sat in the third or fourth row, alone.

"Look at us all," Overstreet said, then drew a deep breath to keep her composure. "We and our memories are the program today." She invited people to come to the front and share their stories about Miner. Miner's friend Roger Bamford, one of Oracle's chief programmers, would speak first; after that anybody could say anything he or she wanted. "We can be here as long or as short as we all feel like being here," Overstreet said.

Roger Bamford had known Miner for ten years. "I think I was his longest direct report. He was my manager the day I joined Oracle, and I wouldn't let go until he had to quit to look after his illness," Bamford said. When Miner left Oracle headquarters in 1992 to work on special projects in San Francisco, he asked Bamford to go with him; Bamford said Miner referred to the group of technologists as elitist weirdos. "He was very fun to be with. He took a personal and completely unpretentious interest in everyone that worked for him," Bamford recalled.

Bamford then recited a litany of things that Miner enjoyed. The list included throwing darts, shooting pool, cracking jokes (especially in serious meetings), playing tennis (of course), driving fast, and gleefully rubbing people's noses in the fact that he had beaten them at something. "Most of all, Bob cared about his family, and that was really clear at times when things weren't going well," Bamford said. "The only time Bob seemed down was when he had problems at home."

Bamford also spoke of Miner's generosity with money. There were many times when Miner lent—or gave—money to people who needed it. Bamford knew it was true because the people who received the money told him so. "He could brag about winning a chess game and about being so cool, but he never said anything about the money he was giving away," Bamford said. When Miner gave to charity, "They'd thank him by calling up and asking for more." He was capable of saying no. Bamford could remember a time when a certain school had to look elsewhere for the money to buy toilets for its gymnasium.

"It was really special being around Bob, and I'm really going to miss him," he said.

People came to the microphone one after another. They often found it hard to put their feelings into words. Brian Cassidy, an Irishman who worked for Oracle in Europe for many years, said the occasion reminded him of an old Dublin custom. In the 1920s, before Guinness began bottling its ale, one son from every Dublin family was sent to the local pub each night to fetch a jug of ale for his father. The boy would place a swatch of Irish linen over the jug, then carry it home as fast as he could. Inevitably some of the ale would spill along the way. When bystanders saw this, they would always shout, "Hey, you're spilling your message." That was how it felt to talk about Bob Miner, Cassidy said. No matter how hard his friends tried to describe him, "You're just spilling the message. The message isn't somehow quite the flavor that you wanted to convey."

Yet even at the memorial service people got most of the message home. Kirk Bradley recalled a time when Miner, who was worth several hundred million dollars by then, won $280 from him in a poker game. After the game the two went to a record store so Miner could blow his winnings. "I spend maybe four minutes and I pick out about thirty CDs. He spends twice as long—I have to go find him—and he has two CDs in his hand, and he's hovering over a third," Bradley said. "I said, 'Bob, what are you doing?' He said, 'This is fourteen ninety-five.'" Miner did not want to spend even fifteen dollars foolishly. "I just don't understand that and never will. But it was so consistent with his way," Bradley said.

The greatest tributes came from people who had worked for Miner. Bill Friend, an early Oracle employee, said people occasionally asked Miner to make software do impossible things. Miner would always say, "No problem—that's an SMOP," which meant "simple matter of programming." Once Oracle executive Derry Kabcenell went to Miner to discuss a work problem. Miner listened carefully, smiled, and said something that sounded like "Kameab yoyo." This turned out to be an acronym: "Kiss my ass, baby—you're on your own." An Oracle engineer said he and Miner often played chess when they were supposed to be working; the person marveled that Miner always managed to get some work done anyway. Ed Oates, one of three cofounders of Oracle, said Miner "took Oracle as a big game, for the most part. . . . That kept him human."

Some people didn't seem to be talking about Miner at all. Geoff Squire, who had run Oracle Europe until he had a falling-out with Ellison, said Miner "was one of the few people in the world who were both successful and nice. It's a very, very rare combination." Oracle people took the remark as a potshot at Ellison.

Larry Ellison spoke early in the service. There was an awkward moment when he and another person moved toward the microphone at the same time, but the other person quickly yielded. Ellison stood behind the podium and looked for a moment at the audience. The moment was full of possibility. Ellison had been Miner's partner and at times one of his closest friends, and everyone in the room expected him to say something brilliant and touching. It wasn't fair; Churchill himself might have withered in that situation. Besides, Ellison was not one to make public displays of deep emotion—or private ones either. Whatever he really felt about Bob's death was bound to stay locked inside him, a thundering heart in a stainless steel cage. Indeed much of what Ellison said was, if you listened carefully, really about Ellison: his wit, his brilliance, his charm. Yet the speech was not a disappointment. In the end he found some tender words for the man who, in many ways, had made his success possible.

Ellison, dressed appropriately in a dark blue suit, spoke softly at first. "My name is Larry Ellison," he said, though of course he didn't need to. "Let's see, I knew Bob for about twenty years, and I worked with Bob for about twenty years," he said, then paused. "I had a long written speech, and I just left it in my seat. I'm not sure what I'm going to say. . . . I met Bob twenty years ago at a place called Ampex Research on the peninsula, and Bob was my manager." Ellison told of how he had asked to have Miner as his boss because he felt Miner was "far and away the brightest of the people who were available to be my manager." Then his tone changed. "Freud said there's only two things that are important in life—love and work, and not necessarily in that order. I worked with Bob for twenty years. I certainly loved Bob, though I didn't realize it every day that I worked with him. Some days I didn't realize it more than others, I suspect. Bob and I didn't always agree on everything. Twenty years is a long time."

He concluded his remarks with a story. It was a typical Larry Ellison anecdote, in that it was plausible, brilliantly funny, and extremely difficult to check. (Nicola Miner later said to me, "Who knows if it's true?") This is what Ellison said: "When you have an initial public offering, not only does the company sell stock to raise money, but there's also the opportunity for the founders to go ahead and take some of their stock and also make that a part of the public offering. And of course the public then buys your stock and you get a rather large check as a result. Bob and I both sold stock in the IPO. The contrast in how we both reacted to receiving these huge checks was really quite stunning. I made an appointment with the bank manager to pay off my mortgages. The bank manager met me; it was a big deal.

"Bob, finishing his tennis that afternoon, decides to wander into the bank to deposit his check. He's got this three-million-dollar check," Ellison said. "It was Friday. There were a variety of workers in line. This was the day everyone got paid. There were rather long lines in front of each of the tellers. So Bob inserts himself into one of these lines. So there he is, wearing these baggy tennis shorts and old tennis shoes and a T-shirt and this conspicuously unfashionable tennis hat."

People started to laugh. Ellison, hearing them, went on for a while about the hat. "I don't know where he got this hat, but this was not a handsome tennis hat. It was one of these things that goes all the way around, loose brim. And of course he has just played tennis. He was a little bit sweaty; his shirt was sticking to him in a few places.

"He fills out this deposit slip that basically says, 'I would like to deposit two million nine hundred ninety-nine thousand eight hundred dollars into my savings account, and I'd like two hundred dollars in cash." The audience roared. Ellison held his hands in front of him and said, "I'm not making this up. Because he always took two hundred dollars out of his check for cash for the weekend." Miner handed the deposit slip and the check to the teller. "She gasps. She is just stunned. She gasps; then she screams. She thinks she's being robbed." As the auditorium shook with laughter, Ellison conceded that the teller's reaction did not make sense. "Think about it: You don't make a deposit when you're robbing. You make a withdrawal."

According to Ellison, the bank manager rushed over to see what was happening. When the manager saw Miner's check, he called Miner "Sir" and told him it wasn't necessary for him to stand in line.

"Bob said, 'I don't understand. What's changed? What's different? Why don't I have to wait in teller lines anymore?' And the manager said, 'You just don't have to.' The manager missed the point. He missed the point," Ellison said. "Bob really didn't want to be treated any differently because he had a lot of money. That was the worst thing Bob could imagine, was being treated any differently. Bob treasured being one of the guys. That was his most valuable possession. He never wanted to give that up, and he never did. Thank you."

Then Larry Ellison sat down.

Thirteen

LARRY ELLISON'S INVESTMENT IN NCUBE DID NOT GENERATE THE RE-turn he had hoped for. The machines built by nCube and other companies were devilishly hard to program; getting thousands of processors to work together on a problem was—well, a real problem. And it was never easy for a new technology to gain wide acceptance, as Ellison well knew. Maybe the day would come when massively parallel computers would run the world, but that was sometime in the future. Still, Ellison was "absolutely convinced that massively parallel processing has a huge place in the next generation of computing."

Soon he got a chance to prove it. In the early 1990s everybody in high technology was talking about the information highway or the information superhighway. These terms meant different things at different times. In the beginning the "highway" was going to be little more than an expanded network of fiber-optic cables that could deliver five hundred or more television channels into a single home. (Just what America needed: *Gilligan's Island* playing on sixty-three different channels.) Later, as technology improved and people raised their sights, the definition changed. No longer would folks have to sit passively as the cable TV companies pumped programming into their homes. Instead television would become *interactive*. People would use their television sets to shop from home, communicate with others on the network, and watch movies and TV shows when they wanted to watch them, not when the cable companies told them to.

This last idea, video on demand, was going to be big. Never again would people drive someplace to rent a movie. Instead they would push a couple of buttons, and—zzzt!—*Rocky IV* would materialize on their sets. Video on demand "will enable people to summon movies at any time and use their interactive TVs as 'virtual VCRs' that can stop, restart, rewind, or fast-forward at will," *Fortune* magazine promised.[1]

If there was going to be an information highway, somebody would have to build it. First the cable TV or phone companies (the line between them was blurring) would have to build a high-bandwidth network capable of handling thousands of movies at once. But that was only one of the challenges. Somebody would have to create a database that could store all those movies, and somebody would have to build a computer that could distribute all those videos to the people who wanted them.

That was where Oracle and nCube came in. When Ellison saw the potential for the information highway, he told his software engineers to build a system that would supply video on demand. According to lead engineer Andy Laursen, "He said, 'I got nCubes. Make an nCube into a video server.' " Within a few weeks Laursen and his team had figured out how to get bits streaming off the nCube. There was just one problem: The TV monitor was blank. "We weren't actually displaying anything. You couldn't see the video," Laursen said. Back to the drawing board.

Then Ellison raised the stakes. In two weeks five thousand Oracle users would be in Orlando, Florida, for their annual conference. Ellison told his team that he wanted to demonstrate the video on demand system there. "He really likes to challenge his technical people," Laursen said. Another team member, Evan Goldberg, said, "When you work at Oracle, generally it's a hundred percent commitment."

The engineers worked day and night. Finally, at eleven one evening, they got something—nobody could remember what—to appear on the television monitor. Goldberg called Ellison at home and told him about the breakthrough. Ellison said he would be there in fifteen minutes. "Larry comes over in fifteen minutes—with a date in tow. And we're having this geekfest where we show him this streaming video," Laursen said. Ellison's date may or may not have been impressed, but Ellison was. "He thought it was awesome. He was ecstatic," Goldberg said. "He was saying, 'You've done it! You guys are geniuses!' When Larry likes something that he sees, you basically come out of there feeling like you've reached a higher plane of consciousness."

Oracle demonstrated video on demand that fall in Orlando. The audience of database managers and software developers saw music videos by Jon Bon Jovi, Pink Floyd, and Peter Gabriel and a clip from the film *Demolition Man.* According to Goldberg, the demonstration "wowed them, though they had no idea what to make of it." That wasn't surprising. The trouble with high technology demonstrations was that they sometimes didn't look as amazing as they were. What Oracle did in Orlando was use highly complex computers and software to play several streams of video on demand, a real breakthrough. But all the audience saw was television, hardly a new or exciting experience.

No matter. Oracle was more than just a database company now; that was what counted. Ellison was so excited that he dubbed Oracle's relational database management system Oracle Media Server and pronounced it capable of managing video, audio, and text. This was another bit of clever marketing. The new name might have created the impression that Oracle's relational database system could handle video, but it couldn't. Oracle's video server really had "nothing to do" with the database system, according to Oracle's new media division chief, Farzad Dibachi; they were two independent products packaged together. Later Ellison changed the name again, to Universal Server, but Oracle still had not solved the tricky problem of how to store video in a relational database. The term "Universal Server" infuriated Ellison's old rival Mike Stonebraker, who was now at Informix, working on a relational database that could handle video. "What Oracle has is a relational database system, a turnip, a carrot, and a strawberry. They're just packaged together and called the Universal Server," he said.

In the next couple of years Ellison was to evangelize the information highway as if the salvation of every soul in the world depended on it. "Skeptics may disagree, but I believe the sheer impact of the interactive network into the home will rival that of the electric light, the telephone, and the television," he once said. "We won't just talk or shop on the information highway; we will live on it."[2] Ellison was right in a way. An information revolution was indeed about to happen. But it wasn't the one he, or anyone else, was expecting.

The story Adelyn Lee told in her lawsuit against Larry Ellison was worthy of the tabloids or the afternoon soaps. It focused almost exclusively on the events of April 16, 1993, the night Lee and Ellison went to the movies to see *Benny & Joon.* According to the lawsuit, Lee went to Ellison's home that night with the intention of ending their relationship. While she was there, Ellison "struggled with [Lee], pulled her into his bedroom and pushed her onto his bed in an effort to force [her] to have sexual intercourse," the lawsuit said. She refused. Then he told her to give in "or else." Or else what? Ellison told Lee that if she ever left him, "you will leave Oracle," according to the lawsuit. Finally she agreed to "manually stimulate" Ellison. A couple of days later she was fired, even though she had always received positive job reviews.

As evidence of Ellison's vindictiveness, Lee offered Craig Ramsey's "I have terminated adelyn" message to Larry Ellison, of which she had somehow obtained a copy.

The sensational allegations were big news in Silicon Valley. ORACLE CHIEF SUED OVER FIRING: WOMAN CLAIMS REPRISAL FOR ENDING RELATIONSHIP, the San Jose *Mercury News* said. The San Francisco papers also carried the story. News of the lawsuit later found its way into *The New York Times, Wall Street Journal,* and other national publications. When reporters called Oracle, a spokesman declined to comment because Ellison had not yet seen the lawsuit. Thus Adelyn Lee's side of the story was the only one printed or broadcast on the first day.

The second day was another story—literally. Having read the lawsuit, Oracle went on the offensive: It released a one-page statement calling the lawsuit an attempt at extortion. Without offering evidence, it accused Lee of being the true source of the "I have terminated adelyn" message purportedly sent by Craig Ramsey. "Oracle believes that the facts of the case will clearly show that the lawsuit has been a carefully planned attempt to extort money from Oracle Corp.—efforts which included Ms. Lee's forging electronic mail under the name of her direct supervisor, Mr. Craig Ramsey," the statement said. It said the details of Lee's story were "fictitious inventions designed to smear Mr. Ellison."

Oracle's statement said "CEO Lawrence Ellison was not involved in either Ms. Lee's termination or its approval." This wasn't exactly true; everybody had been afraid to fire Lee until Ellison told Jenny Overstreet that his girlfriend should be treated the same as every other employee. The statement went on to say, accurately, that Lee was fired by Ramsey, who "had repeatedly counseled Ms. Lee concerning her unsatisfactory performance and [had received] numerous complaints about her performance from other Oracle managers and from her co-workers." The statement also accused Lee of "incompetence and lack of integrity," qualities for which Ellison had been willing to fire her but not to break up with her. Oracle also included a quotation from a partner in a big California law firm, who called Lee's lawsuit "the most blatant setup that I have seen in my years as an attorney handling cases of this type." Lawyer Cecily Waterman continued, "This type of trumped-up case does a gross disservice to those women with legitimate complaints in this area." The San Jose *Mercury News* published excerpts from Oracle's statement on October 21. This time the headline said, ORACLE CEO CALLS SUIT AN EXTORTION TRY: HE DENIES FORMER EMPLOYEE'S CLAIM SHE WAS FIRED AFTER BREAKUP.

In a statement issued later, Oracle released copies of the E-mail messages in which Lee had asked Ellison for a $150,000 loan, a watch, and an Acura NSX. And for the first time Oracle addressed the issue of the incriminating E-mail note from Ramsey to Ellison.

"Ramsey has vehemently denied sending this note," Oracle said. "Did Lee write and send this note? This note was printed only seconds after it had been sent. Who printed it and why? Lee has a copy of his note in her possession even though it was printed a day after she was terminated. How did she get it? The only plausible explanation . . . was that Lee sent the note herself and immediately printed a copy of the note as a part of her attempt to frame Oracle Corporation, Ellison and Ramsey."

According to Oracle's statement, one of Ellison's old girlfriends had come forward to say that Adelyn Lee once suggested to her that they both could "get a million dollars" from Ellison. The statement didn't give her name, but the old girlfriend was Andrea Zeman, one of the three Oracle women Ellison was dating late in 1991.

There was another point that Oracle could have made in that statement but didn't: The process of firing Adelyn Lee had begun long before her April 16 date with Ellison. That fact would make it harder for her to argue in court that she had been terminated because of what happened that evening.

Still, there were some holes in Oracle's story. How did Lee send an E-mail note to Ellison under Ramsey's name? And how did she get a copy of it later? Those questions became increasingly important.

The lawsuit came at what should have been a glorious time for Ellison. Soon after it was filed, *Fortune* magazine put his picture on the cover next to the headline SOFTWARE'S OTHER BILLIONAIRE. While Ellison must have been annoyed at being identified as the "other" anything, he had to enjoy the exposure. The article (which did not mention the lawsuit) described Ellison as "tall, thin and urbane, turned out in double-breasted English suits or black silk Japanese sports shirts with black slacks." Who would argue with a description like that one? More important, the piece detailed Ellison's plans to bring video on demand to the masses. It also drew the battle lines for a high-tech war between Oracle and Microsoft, with Ellison and Bill Gates as the generals. "Gates and Microsoft are perhaps the only CEO and company in the industry that match Ellison and Oracle for aggressiveness, drive, tenacity, chutzpah, and unbridled self-confidence," the article said.

The *Fortune* piece changed everything. No longer was Larry Ellison just an anonymous nerd trying to squeeze a few more transactions per second from his database software; for the first time he was a national business figure dealing in sexy consumer technology. He was also seen as a threat to the untouchable Bill Gates.

The moment was all but ruined by the lawsuit. Oracle's denials aside, a lot of people were inclined to believe Adelyn Lee's story. After all, she was telling it only a couple of years after law professor

Anita Hill had made sexual harassment a hot national issue with her accusations against Supreme Court nominee Clarence Thomas. Those who believed Hill probably also believed Adelyn Lee. Besides, Lee's story was plausible. Ellison was a randy, self-centered, tough-dealing billionaire who admitted to dating attractive female employees by the batch. Was it so hard to believe that he had gotten angry at Lee and threatened her?

"The worst thing that happened to me was just going up and down the elevators," Ellison said. (His only habitual contact with regular people was in elevators.) "I remember the way one woman looked at me in an elevator. . . . There was nothing I could do, nothing I could say." Ellison's family was upset too. His son, David, got some grief about the lawsuit from his classmates, and one day reporters showed up at the home of his former wife Barbara, in hopes of getting an interview about Larry. She was so rattled that she hired security guards to protect her and the children. The whole ordeal made Larry Ellison miserable—and furious.

His anger was apparent in Oracle's court filings. The company's lawyers filed papers officially denying everything Lee had said. Lee's claims that she wanted to end the relationship with Ellison, that he coerced her into having sex, and that he vengefully ordered her firing: These were all lies, according to the Oracle lawyers. "[Lee]'s motivation in initiating and soliciting a personal relationship with defendant Ellison was tangible gain," the documents said, adding that she had "encouraged the attention of defendant Ellison for improper reasons unrelated to the retention of her employment at Oracle."

The lawyers began taking sworn statements from the witnesses the following May. In her videotaped deposition Adelyn Lee described the moment that Ellison threatened her in his bedroom. "He took my hand and placed it on his penis and he whispered in my ear, 'Fuck me or else,' " she said. She added details to the story, saying she had dug her fingernails into Ellison's chest when he got rough with her.

Oracle's lawyer quizzed her at length about the messages she had sent to Ellison while they were dating. Why did she ask Ellison for a $150,000 loan? "He offered to assist me in any way possible." Why did she ask him if nCube was going public? "Out of curiosity."

Why did she ask so many times about the Acura NSX sports car? "It was more of a joke at the time," she said.

Lee was asked if she had somehow forged the "I have terminated adelyn" message. She said no. How did she get a copy of it? It was in a stack of messages that a friend at Oracle found in a printer the day after she was fired, she said. The friend had delivered them to her at lunch that day.

Lee's lawyer, Lawrence E. Viola, took Larry Ellison's deposition a couple of weeks later. Even with his reputation at stake, Ellison could not resist putting on a performance. He was funny, sarcastic, impatient, commanding—anything but worried. At one point Viola asked Ellison why he didn't call Lee to console her after he learned that her firing had finally been carried out.

ELLISON: Well, because I got a very bizarre mail note on the 22nd of April which I believe Ms. Lee forged.

VIOLA: And which mail note is that?

ELLISON: It's a mail note that says it came from Craig Ramsey to me saying that "I have fired Ms. Lee as you have ordered me to. Signed, Craig Ramsey."

VIOLA: I'm sure I have that note somewhere here, but I don't believe it says "you have ordered me." I believe it says "per your request."

ELLISON: I'm sorry. "Per your request," right.

VIOLA: That's OK.

ELLISON: "I have shot the station wagon full of nuns. They are all dead. Signed, Craig Ramsey."

VIOLA: I take it that you at no time asked Mr. Ramsey to terminate Ms. Lee.

ELLISON: Of course not.

Ellison dropped a bombshell during the deposition. He told Viola that some technical people at Oracle had conducted an investigation into who had sent the incriminating message—"I have fired adelyn per your request"—to Ellison. "Basically we have traced the call where this note originated to a modem in Adelyn's neighborhood," he said. "So we have absolute technical evidence that the call was made from a house . . . in Adelyn's neighborhood, in a very small area where she lives, to Oracle Corporation."

Viola was alarmed, with reason. If Oracle could show in court that Lee had somehow sent a message using Ramsey's E-mail account, the whole lawsuit could be exposed as a fraud. In a letter to Oracle lawyer Ray Ocampo, Viola asked whether the company planned to call any expert witnesses to testify about where the message had originated. Ocampo's response indicated that Ellison had been mistaken and that Oracle did not have "absolute technical evidence" after all.

"We have concluded that we are unable to trace the source of the call. Mr. Ellison's deposition testimony will thus be amended to reflect this conclusion," Ocampo wrote. Then Ocampo went on the offensive. Given Craig Ramsey's sworn testimony that he was in his car when the incriminating message was sent, there was "only one person with both the motive and opportunity to place the call and initiate the E-mail in question—your client." Ocampo strongly suggested that Lee drop the lawsuit.

That autumn Ellison sat for a long interview with Julian Guthrie, of the San Francisco *Examiner*. In the interview, which was published as a cover piece in the *Examiner*'s Sunday magazine section, he offered his first public explanation of what happened at his home the Friday night before Adelyn Lee was fired.

"She said I forced her to have sex with me. That's just not true. She came over to my house, an hour late. We were supposed to go to dinner and a movie. There was no time for us to have dinner, so I wanted to have sex. So we had sex. The thing was, she said she didn't want to have sex. She wanted to go to dinner, but finally she said 'OK.' This, apparently, is where I forced her to have sex. You know what we did after I 'forced' her to have sex? We went to the movies, to see *Benny & Joon*. I don't think that a woman who has just been assaulted would want to see a movie. It's just unbelievable that someone could make charges like this."

The lawsuit dragged on. In early February 1995 the two sides met for a mandatory settlement conference but were not able to come to an agreement. The trial was set to begin on, of all days, St. Valentine's Day. About that time Ellison told the Los Angeles *Times,* "Settle? I'd rather be robbed at gunpoint."[3] Then he settled. On February 21 the two sides filed a stipulation for dismissal; there would be no trial.

"The parties have been able to work out their differences in a professional manner. The differences arose from a misunderstanding in a personal relationship," the stipulation said. "Accordingly, all allegations in the complaint are withdrawn." Oracle Corporation (not Ellison) agreed to pay Lee a hundred thousand dollars "as consideration for particular elements of her lawsuit," whatever that meant.

Ellison had a pretty good case going into the trial. He could show that Craig Ramsey was on his car phone when the incriminating message was sent, making it virtually impossible for him to have sent the message. Oracle's technical people were still trying to prove that Lee had somehow sent it. Ellison could also prove that others had decided to fire Lee a couple of weeks before the April 16 date at Ellison's home. So why did he settle?

"The trial was going to be a media circus," he said. He had heard rumors that Barbara and his kids would be called to testify, though he didn't know about what.

"My lawyers were telling me I should settle, that it would cost more money to go to court and try the case than it would to settle. Plus, with any jury there's no guarantee. We thought there was a ninety-eight percent chance that we would win, but look at the O. J. Simpson [criminal] jury. Oh, my God!" Ellison said. Even if he had prevailed at trial, "She could just say, 'Mr. Ellison is an incredibly wealthy person, and justice has not been served here. The man attacked me, and now they've let him go.' " There was no choice but to pay her off, he said.

After paying her lawyers, Adelyn J. Lee went home with fifty-eight thousand dollars,[4] almost twice her annual salary at Oracle. She probably thought that was the end of the story.

Joe Costello did not know much about the database business when Don Lucas asked him to join the board of directors of Oracle Corporation. But the database business wasn't what attracted Costello to

Oracle anyway. Rather it was the applications division, the one founded by Jeff Walker, that interested him. Costello believed Oracle could make a killing on applications—software that automates business functions such as accounts receivable, payroll processing, and accounting. Every big business needed financial applications, and only a few software companies were building and selling them. "There was no one there; the market was wide open," Costello said.

But all in all, Oracle's applications business had been a disappointment. For a long time Oracle shipped software that didn't work right. That was partly because applications were new, and new software always had bugs. But it was also because Oracle was more interested in developing technologies than in perfecting them. George Koch, who took over the applications division after Walker left, arrived to find the division cranking out new releases of the financial applications software every six months. "[The attitude was] code it, ship it, wait till they complain, and then fix it. There wasn't any testing in there," Koch said. He improved quality by slowing down the process but ended up being moved out of the job because Ellison thought he wasn't moving fast enough on a certain project. "There's probably an element in Larry that is so market-driven that he will risk reputation in order to be first in the market," Koch said. "But there's probably at least as much a part of Larry that is Pollyannaish. If he hopes hard enough that something will be true, then it will be true." All these problems hurt sales; other applications were growing more quickly than Oracle, stealing the business that Joe Costello had hoped to get.

Costello blamed many of the problems on the person Ellison chose to succeed George Koch. That person was . . . Ron Wohl, the same young executive who had negotiated the Nippon Steel deal in 1991. Costello did not think Wohl was an inspired choice. "A zero," Costello called him. Although few doubted that Wohl was bright and dedicated, he was no Larry Ellison when it came to charisma. Costello didn't think Wohl stood a chance against the high-powered executives running SAP America Inc., the applications market leader.

"I kept saying, 'Look, Larry, think about it this way: It's him against the SAP guys. That's what's reality. Him against them,' "

Costello said. He thought it was like asking Mister Rogers to fight Mike Tyson. "Yeah, probably Rogers will win. You know, he might go down a few times, but I think he's got staying power. That was what it looked like. It was stupid."

Costello was not the only person in the company with reservations about Wohl. Ray Lane had his doubts, as did other top people. Wohl was "a bit of a lightning rod" for criticism, according to one person familiar with the controversy. Said Costello: "Larry just didn't want to face the problems, was my opinion about it."

Costello and Ellison had frequent arguments about Wohl. Finally, at a board meeting in 1995, Costello blew up. He shouted at Ellison, "Larry, you look exactly the same way you did in 1990, only it's worse. Your ears aren't working anymore. You believe so much in your own bullshit about your success that you're not paying attention to the problems. You're not. This kind of arrogance is going to lead to another disaster. That's the only end. . . . *Fucking wake up.*"

Ellison—who believed Costello was after his job, an accusation Costello laughed off as "preposterous"—screamed profanities back at him. "Joe kept saying, 'You're fucking it up, just like 1990, you're fucking this thing up again. You're totally fucking everything up.' And all the other board members were sitting around in total disbelief," Ellison said. "I think I got rather upset with Joe. Unfortunately I was just back from Europe and a little bit sleep-deprived."

For Costello, that was the beginning of the end, and he knew it. His carping about Wohl and the applications division wasn't doing anybody any good, and now that he and Ellison had had a screaming match, the situation was always going to be uncomfortable. "I didn't see that I was going to be able to provide much more value after that one defining moment," he said.

Costello's tenure on the Oracle board ended, not surprisingly, with an explosion. His company, Cadence Design Systems, was trying to choose an applications package on which to manage its corporate finances. Among the contenders were Oracle, which Cadence had used in the past, and SAP. As an Oracle board member Costello might have been inclined to stay with the Oracle applications. But as CEO of Cadence he had a duty to buy the package that would be best for his own company. So he let his technical gurus make the decision, and they chose SAP.

Kaboom! Ellison demanded Costello's resignation from the Oracle board and got it. At the same time Oracle's Ray Lane quit the Cadence board; he could hardly serve a company that had just rejected the products he sold. But that wasn't the end of it. Ellison was so "pissed off" at Costello, he said, that he told Don Lucas to resign as chairman of the Cadence board, only to change his mind later. "I said, 'Don, look, if you leave Cadence, it's a personal financial sacrifice, and I feel like it's not right to ask you to make a personal financial sacrifice. I'm just going to back off on this.' " What did Ellison mean by "financial sacrifice"? Lucas received a seventy-five-thousand-dollar-a-year retainer for serving on the Cadence board (upped to a hundred thousand dollars in 1996), and had options to buy 39,583 shares of Cadence stock at a favorable price (60,000 shares in 1996).[5] Even if Lucas had quit the Cadence board, he could have muddled through: He served on eight other corporate boards at the time, for which he received millions of dollars' worth of retainers and stock options.

Oracle's applications division began to do better after Costello left the Oracle board. One small but satisfying example was that Oracle's applications package was named best applications product of 1996 by *Datamation* magazine, which said: "Oracle's Applications [leave] its competitors gathering dust."[6] Still, many in the company believed Ron Wohl should be replaced with someone more dynamic and more experienced. Late in 1996 I asked Ellison if Wohl was the best person for the job.

"Best person at Oracle," Ellison said. "And the best person I've found so far. And Ron has a lot of experience right now. Ron is not the same person that started in this job. Ron is an incredibly bright guy. Very organized and knows a lot of the details of the technology and I think has done a terrific job. I don't want to start discussing my executives' weaknesses, but he's not perfect. None of these guys are perfect. There are trade-offs. You might be able to find a more charismatic leader than Ron. But I'm thrilled to have Ron. I think he's doing a great job, and I think we're going to cream our competition."

As for Costello and Cadence, they got what they deserved, Ellison said. "Their SAP implementation is going dismally," he said. "Cadence is still using the Oracle applications because they can't get the SAP applications up and running. Ha-ha-ha."

Not so, Cadence said early in 1997. The company knew the SAP applications would be difficult to put in place. But the implementation was "right on track," spokesman Mike Sottak said, "We're not unhappy, and we don't think buying SAP was a mistake by any means."

Costello knew that Ellison was angry. After he left the Oracle board, he said, Ellison "said some really nasty things about me, things he was going to do to destroy me and ruin my personal life and destroy my career in the future. You know, B movie shit. C movie shit. Just crap."

But what really bothered Costello wasn't what he considered Ellison's vindictiveness; it was the way he ran the company. "He still doesn't set a great culture for a company in terms of: What do we stand for? How do we operate? What are our principles of behavior? What's good behavior and bad behavior? What's this thing all about?" Costello said. "You don't get that from him. It's good in this high-tech age to not have too much baggage. But he went a little too far."

Ellison was masterful at motivating people toward certain goals: killing Ingres, doubling revenue, delivering video on demand, and so on. "You can always get people to do that: go after money and success. But those things lack long-term substance," Costello said. Ellison's energy and intellect were impressive, but leadership, Costello believed, was not just about being right.

What Costello thought was missing was any sense of where Oracle was going or how it would get there. "You can't actually sit there and have a rule book about everything. But more and more, as companies get larger and more complicated, as things move faster, you've got to set a magnetic north," he said. This was especially important in a crisis. Without any sense of direction, how would people in the company make decisions? "When you go into some companies, you get a feeling, a theme. I didn't get that strong a sense about who they were [at Oracle], what they stood for, what they represented, and where they were heading. . . . Oracle feels opportunistic. It's a massive opportunity. But it feels opportunistic."

He continued. "There's something fundamental missing in Oracle, I think. And I've felt this way all along. There's this lack of an underlying current or substance to it. And it comes from Larry, I think. What does he really believe in, really fundamentally, that's going to keep guiding him? He's got all this childlike enthusiasm and all this and all that. But what's the set of principles and values, the base-level stuff that he's always going to come back to? And how does that manifest itself in the culture of the company? I think he gives short shrift to a lot of that discussion. He's uncomfortable, actually, with a lot of that discussion."

Oracle's video server solved only half the interactive television problem. Yes, massively parallel computers using Oracle software theoretically could turn movies into bits and send the bits over a network into people's homes. But what would happen after the bits arrived? Consumers would need computerlike devices that could turn the bits back into movies. Because it was expected that these devices would sit on top of TV sets, they were known as set-top boxes. Though lots of companies were building set-top boxes, Ellison believed one company was perfect for the job: Apple Computer.

Like a lot of people who used computers, Ellison was crazy about the Apple Macintosh, the famously easy-to-use personal computer. "I've been an Apple bigot since 1984, when the Macintosh first came out. . . . I love the Mac," Ellison once said.[7] From his point of view, the Macintosh operating system was everything that Microsoft's Windows was not: simple, straightforward, efficient, clever, likable. (When you wanted to delete a file from the Macintosh, you "picked it up" with the mouse, dragged it to a picture of a trash can, and dumped it inside.) Microsoft had labored for years to come up with something nearly as good, and only in 1990, with the release of the third version of Windows, had it come close. Even after Windows became popular, the Macintosh retained a zealous following. Sometimes Mac enthusiasts went a little overboard: In the book *Insanely Great* author Steven Levy describes the Macintosh as "the most important consumer product in the last half of the twentieth century." That was extreme, but a lot of people agreed with Ellison when he described the Mac as "beautiful." He thought that it (or something like it) could serve just as beautifully as a set-top box.

Ellison eventually got a chance to get Apple involved in interactive television. At the height of the information highway craze, British Telecom asked Oracle to participate in an interactive TV trial in Ipswich, a hundred miles northeast of London. Ellison jumped at the chance. He already had everything he needed except a set-top box. That was when he got on the phone to Apple Computer. "Oracle—specifically, Larry Ellison—single-handledly forced Apple into the [interactive TV] business," Oracle's Farzad Dibachi said. "He called [CEO Michael] Spindler and said, 'You have to get into this business. This is perfect for you guys.' " Apple had little trouble putting together such a device, and soon a small number of British Telecom customers were cruising the information highway, banking from home, watching educational videos, and ordering movies when they pleased.

"The important thing . . . is that [the equipment our customers have] doesn't look very different from the sort of equipment they're already used to. They use the normal television set. There's a set-top box that's been made for us by Apple, which looks very similar to our satellite decoder box, and is driven by a remote control which is very similar to a normal VCR type of remote control," British Telecom's Alec Livingstone told an interviewer.[8] "What Apple have done for us is to build something which doesn't look like a computer. But that's what it is. It's a very powerful computer we're putting in everyone's home. Inside that set-top box there is, as well as the computing power, all the picture decoding, television driving capability, infra-red decoders and whatever else."

Although British Telecom was happy with Apple's set-top box, Apple "showed a lot of reluctance" to make a product line out of the devices, according to Oracle's Dibachi. At the time Apple had other things on its mind. Microsoft's Windows operating system was commanding a larger and larger share of the personal computer market, reducing Apple to a sort of boutique company that supported its loyal longtime users but wasn't attracting many new ones. Also, Michael Spindler was on his last legs as CEO, and the company was having trouble doing much of anything. It was no wonder Apple didn't want to build set-top boxes. At first Ellison considered licensing the Apple operating system for use in set-top boxes that others would build. But that idea soon yielded to a much grander one. If Ellison really wanted the Apple operating system, there was only one sure way to get it: Buy Apple Computer. But he wouldn't dream of doing that without including his good friend Steve Jobs.

Jobs was one of the most compelling figures in high technology. He founded Apple Computer at nineteen and introduced the Apple II (he called it the first real packaged computer) at twenty-one. "I was worth . . . over a million dollars when I was twenty-three and over ten million dollars when I was twenty-four, and over a hundred million dollars when I was twenty-five," Jobs once said.[9] "And it wasn't that important because I never did it for the money." Like Ellison, Jobs had a moon-size ego and a searing need to know exactly what he was capable of. The people who worked for Jobs, who were caught in his gravitational pull, quickly found that he cared much more about his own standards than he did about their feelings. "When I wasn't sure what the word 'charisma' meant, I met Steve Jobs, and then I knew," Apple's chief scientist, Larry Tesler, once said. "He wanted you to be great, and he wanted you to create something that was great—and he was going to make you do that."

The Apple Macintosh, introduced in 1984, was that great thing. The Mac endured for years, but Jobs didn't: He was ousted by Apple's board of directors after a clash with the company's chief executive, John Sculley.

Larry Ellison met Jobs about the time he left Apple. Jobs's magnificent Mediterranean home stood on a hill in Woodside just above where Ellison lived with his wife Barbara; his pet peacocks would sometimes wander onto the Ellisons' property. As they got to know each other, Ellison and Jobs learned that their lives were remarkably similar: Both were reared by adoptive parents; both considered themselves, not unjustly, to be unconventional thinkers; both had wide-ranging interests and sophisticated tastes; and both tended to attract and repel others with great force. "We both joke around. I say Steve's my best friend, and Steve says I'm his best friend," Ellison said. "And he says, 'Well, you're my *only* friend, so you must be my best friend.' That was Steve's joke. I could use the same joke."

Ellison tended to describe the things in his life in grandiose terms: He owned the fastest boat in the world, the best surgeon in North America had operated on his elbow, and so on. He talked about his relationship with Steve Jobs the same way. This wasn't just a friendship; it was a pairing for the ages, a meeting of the minds, the greatest love of all. For example, when Bob Miner died, Ellison thought about how much he cherished Jobs. But for Ellison, one of the most important moments in his relationship with Jobs was the night the two men attended the premiere of the film *Toy Story* in San Francisco. The movie was produced by Pixar, of which Jobs was chief executive.

"I am very competitive, and sometimes, when somebody does something really great, I get upset because I just feel like, *that isn't me,*" Ellison said. "And my reaction toward Steve wasn't competitive at all. I felt what he had done was so wonderful, and I was so proud of him, and I love him so much, it was almost as if I had done it. I didn't feel the least bit competitive. I was very, very happy for him. Which was a bit of a revelation for me. The wonderful thing about loving somebody else is that it can expand your ego in the best sense. If they do something great, you feel terrific about it. And so here we were at this premiere, and I'm giving Steve a big hug and telling him all of this. It was a pretty emotional moment for both of us. . . . It was a great, intimate moment."

Jobs also felt warmly toward Ellison, though he used different language to describe the feeling. He described Ellison as bright, funny, and tasteful. Also, "He's done a phenomenally good job of running Oracle, in my opinion. And yet he's not like a Bill Gates, where that's all he knows." (Gates baiting: That was another thing they had in common.) And was Ellison indeed one of Jobs's closest friends? "Absolutely," he said. "Some of these things, it's hard to put into words. And some of them shouldn't be put into words."

Now, in the heyday of the set-top box, the two men were considering reclaiming Apple in the name of egotists everywhere. If they could figure out a way to buy the company, which had a market capitalization in the seven-billion-dollar range, Jobs might get another chance to make Apple great. And Ellison would get to own a legendary company, one that would make his name as well known in the industry as Bill Gates's. Jobs did not know exactly what he and Ellison would do with the company or what Jobs's role would be. "Who knows?" Jobs said. "We would have tried to fix it. But it might have been irreparably broken."

For Ellison, scheming to buy Apple was fun; he was like a kid figuring out how to get the money together for a new bike. At one point Ellison, Oracle's new media division chief, Farzad Dibachi, and another Oracle employee, Evan Goldberg, went to Apple for a meeting about set-top boxes. As they were walking along Infinite Loop, the road leading to the impressive Apple headquarters, Ellison smiled, gestured grandly, and said, "Someday this will all be mine." Dibachi laughed. "You know, Larry, someday somebody is going to write *Barbarians at the Gate Two,*" he said, referring to the classic business book about the fight for control of RJR Nabisco. "And I'm going to tell the author that you said this." Prediction fulfilled.

Ellison could be excused for being excited. "This was not a thing that was at all weighty. He was enthusiastic about it," said Joe Costello, then an Oracle board member. "And it was very childlike. . . . Larry and his buddy Steve Jobs were kind of up in the clubhouse one day, in the treehouse, and they said, 'Gosh darn it, they really let our old clubhouse go to rot. We ought to take that baby over. We could make something out of that still. It's a good property.' Steve's perfect for that kind of thing."

During that time Ellison sought advice from his friend Michael Milken, the junk bond dealer who went to federal prison for his part in the securities scandals of the 1980s. Costello imagined how Ellison, still in the treehouse, might have talked about Milken. " 'He'd be neat to have in the club, don't you think? Everybody thinks he's a scoundrel, but you know, he's kind of a cool scoundrel. I don't think it would hurt our club's image too much to have that scoundrel in it.' " In reality Ellison did not think of Milken as a scoundrel. Far from it. "Sometimes I do things that are conspicuously not self-serving, almost self-destructive. And people have said that my vigorous defense of Michael Milken, and my association with Michael Milken, is something where I'm taking a risk [because of the stigma associated with Milken]," he said. "But the Michael Milken I know—I didn't know Michael in the Drexel days—is one of the most humane, most gifted human beings I've ever met in this lifetime. And there are so few of those. I just can't afford to sacrifice a relationship with any one of them. They're what makes life wonderful."

Despite what Ellison said was useful advice from Milken, Jobs continually tried to discourage Ellison from buying Apple. "Larry has hardware envy. Which is really stupid, I keep telling him," he said. Jobs was right. As a business proposition, buying Apple did not make much sense. Computer hardware was hideously expensive to produce, and Apple's market share was steadily declining. Ellison would do better to stay in the software business, with its low overhead costs and wide profit margins.

But that was not the point. Ellison's interest in Apple was not about money but about glory, and Jobs understood that as well as anyone. In time Ellison came up with a plan: He would get a couple of consumer electronics companies to join Oracle in buying the company. Oracle would take the Macintosh operating system, and the other companies would divide up the hardware. The partnership would relieve Ellison and Oracle of the burden of coming up with the purchase price on their own. In time Ellison enticed two household name companies into the deal: Philips Electronics and Matsushita. He was hopeful. And when Katherine Bull of the computer trade publication *Information Week* broke a story about the deal, the industry really began to buzz.

In the end Ellison backed off. "It was never an idea that made any sense," Oracle chief financial officer Jeff Henley said. Buying Apple would have been too expensive, even if others had invested, and would have led Oracle too far astray from its core business, database management systems. Joe Costello, the board member, thought that Ellison was just following the idea to see where it might lead him. "Sometimes he's just enthusing, you know? He's got a thread, he's on to something, he's got a scent, he's on the track of something. He's not sure what the hell it is, and sometimes it's nothing. And he'll drop it," Costello said. Costello thought Ellison's childlike curiosity was one of his best traits.

Over the next couple of years Ellison and Jobs often mused about buying Apple, which continued to struggle against Microsoft's dominance. Ellison said he was ready several times to go through with it—Milken even helped him with a financing scheme—but Jobs always balked. "That was a very personal decision for Steve. And I thought Steve should have primacy. You know, Apple was his company, so he should make the decision," Ellison said. Early in 1997 Jobs returned to Apple as a consultant, but a lot of observers felt he had arrived too late to turn things around. That March Ellison again made noises about buying the company, only to announce a month later that he had changed his mind.

During one of their discussions about buying the company, Jobs said something that made Ellison laugh every time he thought about it. The two men were sitting at a table on the deck behind Ellison's Japanese house, the ponds and pool and redwoods stretched out before them. After a long talk they arrived at a financing plan that they thought might work. But Ellison had a question: If they raised the money the way they had discussed, how would Ellison make money on the deal?

Jobs leaned across the table and grabbed Ellison by the shoulders. "Larry, listen to me; I'm your friend," he said. "*You don't need any more money.*"

Oracle remained a major player in the era of the information highway, right up to the day the era ended. After the successful British Telecom trial, Oracle participated in a similar trial conducted in Virginia by Bell Atlantic, another phone company looking to horn in on the cable companies' turf. Evan Goldberg was one of the California-based Oracle employees who crossed the continent to work on that project. Goldberg used the film *The Fugitive,* starring Harrison Ford, to test the system; by the time the trial was set to begin, Goldberg knew the script almost by heart.

Bell Atlantic customers in Fairfax County, Virginia, were supposed to get much more than *The Fugitive* from the so-called Stargazer interactive TV system. "From day one of the trial, approximately 42,000 minutes of programming—refreshed monthly—will be available to customers on demand, at any time of the day," a Bell Atlantic press release said in March 1995. The service was to begin that May. "The initial Stargazer product mix will consist of approximately 200 new, recent and classic movies; more than 120 episodes of recent and classic television sitcoms, dramas, talk shows and news specials; and 45 sports, comedy and music performance specials."[10] The information highway apparently was going to be hundreds of lanes wide.

Larry Ellison was bubbling over with enthusiasm at the possibilities. "Bell Atlantic is planning to have a million homes online by the end of 1995," he told *Red Herring* magazine.[11] "By the end of 1997 or 1998, there will be tens of millions of households hooked into the interactive highway. It will be a huge business long before this decade is over."

As was often the case, Ellison was getting ahead of himself. Bell Atlantic never had more than a thousand paying households taking part in the Stargazer trial. Then, late in 1996 the company shut down the experiment, saying it would begin offering interactive TV service again in 1998, in Philadelphia.

So what had happened? As a Bell Atlantic press release put it, the Stargazer system was "surpassed by the fast pace of technological change." During the interactive TV trial Bell Atlantic transmitted video over conventional copper telephone wires that had been specially adapted for that purpose. But copper wires wouldn't be sufficient to handle the thousands of simultaneous streams of video that the company someday hoped to sell. That would require an entirely new network of high-bandwidth fiber optic cables, a true information superhighway. Building such a network was going to be damnably expensive. Despite the predictions of technofuturists such as Larry Ellison, the phone companies weren't sure they would ever see a return on their investment. Would large numbers of people actually subscribe to a video on demand service? Or would they just keep making the trip to the neighborhood video store? Nobody really knew. And nobody wanted to bet a few hundred million dollars on the answer. So the phone companies, which once had been sprinting toward the future, now moved forward on tiptoes. The information highway was going to be built sometime between several years from now and never.

Some of Larry Ellison's detractors thought he should have been embarrassed about having been so optimistic about the information highway. "Larry Ellison was touting video on demand as the be-all and end-all. And of course he was totally wrong," his old rival Mike Stonebraker said. These detractors should have known better than to think Ellison would apologize. One of his bedrock personality traits was that he was almost completely unembarrassable. Asked why he had been wrong, he simply blamed the phone companies. "What Oracle doesn't do, and can't do, is run the wire from the server to the subscriber household," he said.[12] Ellison *would* have been right if only the phone companies had followed through; that was the idea.

Why be embarrassed anyway? The hullabaloo over video on demand had landed Ellison on the covers of *Fortune* and *Business Week* magazines; both articles had portrayed him as a pretender to the throne of Bill Gates. More important, Oracle had become, if not a household name, at least a recognizable one. No longer was the company known only by savvy investors and a few thousand techno dweebs. "In a 1995 survey, 40% of respondents recognized the name Oracle. In 1992 only 2% did," *Fortune* magazine reported.[13] Video on demand may not have happened, but it was not a failure. "In the end not that much has come out of it, but you can argue that Oracle got a lot of publicity in a new area, and the whole thing made Oracle look like a pioneer," Oracle's Goldberg said.

There was another reason why Larry Ellison wasn't embarrassed about interactive TV: He didn't have time to be. By the time everyone figured out that interactive TV wasn't going to happen, the whole industry had moved on to something new. The next big thing wasn't a highway after all. It was something much more intricate and intriguing—a web.

Early in 1993 two young men at the University of Illinois at Champaign-Urbana—coincidentally, one of the institutions from which Larry Ellison did not graduate—finished writing a computer program that would change the world of high technology. Its name was Mosaic.

To understand the importance of Mosaic, it helps to know a little history. For many years government officials, university professors, and others had been exchanging information through a little-known network of computers called the Internet. Then, in 1989, an English computer scientist named Tim Berners-Lee created a subset of the Internet called the World Wide Web. The Web was, essentially, a global network of documents, some containing text, others containing images or sounds. People got around the Web by using software tools called Gopher, FTP, Telnet, and so on. These tools were functional but clumsy; the average person didn't like using them.

Then Mosaic came along. Mosaic was a "browser." It let computer users jump from place to place on the Web merely by pointing a mouse and clicking. Let's say you were reading a document about Shakespeare stored on a Harvard University computer. And let's say this document contained a link to more information at Britain's Oxford University Web site. All you had to do was click, and—

presto!—a picture of the bard would appear on your screen. With Mosaic, computer users could theoretically get a world of information without ever getting out of their chairs. The implications were immense. Now anybody with a computer, a modem, and a copy of Mosaic could travel the planet, virtually. Millions of people downloaded free copies in 1993, the same year Larry Ellison made the cover of *Fortune*. The program was clearly a huge success, but coauthor Marc Andreessen was skeptical about its future.

"The Internet was a toy then," he told *Fortune* magazine. "That was when the whole interactive TV thing was peaking, one of those frenzies the industry indulges in when it loses sight of what's happening in the real world. So I wasn't sure about the browser."[14]

He eventually became sure. Andreessen, still in his early twenties, formed a company called Netscape to commercialize his browser, now called Navigator. When Netscape went public in 1995, Wall Street went crazy. The stock, which was offered at twenty-eight dollars a share, opened at *seventy-one* dollars and traded as high as seventy-five dollars the first day. The emergence of Netscape had far-reaching consequences: Thousands, and then millions, of people began surfing the Web, looking for everything from pornography to poetry; on-line information services, such as America Online and CompuServe, signed up hordes of excited customers; Silicon Valley entrepreneurs formed countless start-up companies in hopes of capitalizing on the new medium; Bill Gates had to rethink completely Microsoft's business strategy, which was like trying to make a crisp ninety-degree right turn in the *QE* 2; and of course Marc Andreessen became a fabulously wealthy young man. Good-bye, information highway; hello, World Wide Web.

Larry Ellison didn't see the Web happening any sooner than Gates did. But once it did happen, Oracle quickly figured out a way to capitalize on it. Ellison had always thought video was the "killer application," the thing people really wanted, for multimedia computing. He believed it in the days of the information highway, and he continued to believe it in the era of the World Wide Web. It was clear that the highway—the high-bandwidth fiber-optic network—would not be built anytime soon. But why couldn't Oracle software deliver video and other content over copper phone lines to people's personal computers, just as it had delivered them to set-top boxes? Maybe the picture quality wouldn't be great, but it would be passable.

Then Ellison took a small logical step: Why couldn't a personal computer *be* a set-top box? Oracle knew from experience that set-top boxes could be built for a few hundred dollars. And set-top boxes were really just computers that rested on top of TV sets. So why not turn the set-top box into a small, inexpensive computer, one that would be easier to use than a personal computer? No reason.

Ellison watered this little idea until it grew into a big one. Here was a way to get at Microsoft. This was his thinking: The personal computer, once a wonderful innovation, had seen better days. It had become overweight and unwieldy. At two or three thousand dollars it was affordable only by a privileged few; it was hard for everyday people to use and impossible for them to repair; and its dominant operating system, Microsoft's Windows, was clumsy and inelegant and disagreeable. Owning a personal computer was like having a mainframe on your desktop. It was just too much.

The world had changed. The personal computer was no longer the point. Now the network was the point. It didn't matter how much power you had on your desktop; being connected was what counted. What people needed was a cheap, easy-to-use computer, one that would derive its power not from a hard drive and the Windows operating system but from the network itself. No more would people drive to a store, buy a box of software, drive back home, install it, do their work, and then store the work on their hard drives. Instead they would use a cheap computer—the device would eventually acquire the name network computer—to connect to the network. They would use this cute little machine to download software. They would do their work, then send their work across the network to a server maintained by private enterprise—say, a phone company. Servers would be everywhere. And every one of them—you must have known this was coming—would be running Oracle software.

Larry Ellison looked into the future and saw . . . himself!

Later, after Ellison sent the whole industry into a tizzy, people at Oracle disagreed about who was the first to have the idea for the network computer. Ellison of course said he did; others credited his technical people. Who knew? The fact was that the idea was the by-product of an organic process, one involving video on demand and servers and set-top boxes; the idea was so right that somebody at Oracle would have had it eventually. Indeed Oracle was not the only company thinking about small computers. An Atlanta company, ViewCall America Inc., was working on ways to provide both hardware and content to consumers, and a Silicon Valley company called WebTV Networks was trying to turn television sets into World Wide Web monitors. Probably Ellison's most important ally was Scott McNealy, chief executive officer of the server maker Sun Microsystems. Sun had come up with a new programming language called Java, which could be used to write Web-based word processors, spreadsheets, and so on. Like Ellison, McNealy wanted people to use network computers (his would be called JavaStations) instead of Windows-based PCs. The combination of Sun's servers and Oracle's server software would wipe Microsoft off the face of the earth—or so the companies hoped.

Ellison may not have been the one to dream up the network computer. But as always in his career, whether he had the idea first did not matter. What he did with it was what counted.

At 12:01 A.M. on August 24, 1995, computer store managers across the United States unlocked their doors and jumped out of the way as herds of customers stampeded their shelves. What these bleary-eyed people were battling to get was a software upgrade—namely, Windows 95, the latest release of Microsoft's ubiquitous PC operating system. The cost: about ninety dollars. Through an extraordinary worldwide marketing campaign, Microsoft had succeeded in making computer users believe that Windows 95 was a great improvement upon Windows 3.x, and that they would be missing something if they didn't get it right away. This was how Bill Gates had made the largest personal fortune in the world: by selling his customers more or less improved versions of what they already had.

But hey, people wanted Windows 95. Some went to the computer store that night for fear that they would not be able to get the upgrade if they waited, but that was not a rational fear: Microsoft was stamping out Windows 95 CD-ROMs the way Nabisco punched out Oreo cookies. Most people who bought the program that evening—er, morning—did so because they believed that the introduction of Windows 95 was an authentic cultural event. A business student in Auckland, New Zealand, was the first person in the world to buy a copy of Windows 95 (because of the placement of the international date line, the new day dawned first in New Zealand); hours later Tim Liao, a student at San Jose State University in California, bought his copy at a CompUSA store in Santa Clara. "How many times in your life can you be a part of something like this?"[15] he said. Once every couple of years, if Bill Gates had anything to say about it.

A few hours later, in the light of day, thousands of Microsoft employees, industry types, journalists, and curiosity seekers swarmed the Microsoft campus in Redmond, Washington, for the official Windows 95 product launch. Jay Leno, host of *The Tonight Show,* was the rent-a-host for the big event. "Ladies and gentlemen, welcome to the launch of Windows 95. Yes, welcome, Microsofties, nice to have you all here," he said, pacing a grand stage. Leno was an appropriate choice; NBC, on which *The Tonight Show* aired, had just agreed to provide news content for the Microsoft Network, or MSN, Microsoft's new on-line service, and its first attempt to capitalize on the World Wide Web. "But now let's welcome the chairman of Microsoft. Listen to this: This is a man so successful his chauffeur is Ross Perot, ladies and gentlemen. . . . Please welcome Bill Gates." Gates, dressed in a golf shirt (sartorially he was no Larry Ellison), said, "We wanted people to be able to appreciate how Windows 95 makes computing faster, easier and more fun. . . . This moves the whole PC industry up to a whole new level." The proof of the product's importance was right there in front of him: The Microsoft campus was dotted with booths at which technology companies were displaying Windows 95–compatible games, spreadsheets, word processors—and relational database management systems. Larry Ellison was no fan of Microsoft, but he knew that most Oracle users worked his software from a Windows-based personal computer. Microsoft was the IBM of the 1990s; it was the drum major, and the rest of the industry was marching in step.

The release of Windows 95 was the cold boot heard round the world. In Britain subscribers to the normally staid *Times* of London received a brochure for Windows 95 inside their morning papers; Microsoft bought out the entire press run to make sure of it. In New York the multicolored Windows logo was projected on the side of the Empire State Building, and in Toronto the eighteen-hundred-foot Canadian National Tower was decorated with a Windows 95 banner three hundred feet high. Computer users in China, Hong Kong, and Eastern Europe were also talking about Windows 95: They had been using bootlegged copies for months.

At the end of the day a market research firm reported that the first day's sales of Windows 95 met or exceeded its expectations, and Wall Street analysts predicted that Microsoft would ship fifteen to thirty million copies by the end of the year. Meanwhile CompUSA, America's biggest computer retailer, exulted over "the single largest day in our company's history," according to executive vice-president Larry Mondry, while thousands of journalists began to ponder how Microsoft had gotten itself so much free publicity. Thanks to the media, people everywhere were whistling "Start Me Up," the Rolling Stones tune that served as the Windows 95 theme song, and repeating the Microsoft advertising mantra, "Where do you want to go today?"

Where did people want to go? Where *could* they go, except Microsoft? As computing columnist Dan Gillmor put it, "Wherever you want to go today, or tomorrow, you're going to pay Microsoft to get there."

Not if Larry Ellison had anything to say about it. On August 24, 1995, Ellison was not in Redmond, Washington. He was hundreds of miles away, in California's Silicon Valley, developing a scheme to achieve the only real goal he had left: to knock Bill Gates off his pedestal and get up there in his place. When the hubbub over Windows 95 subsided, there was going to be a great silence in the computer business, and Larry Ellison was going to fill it.

Fourteen

ELEVEN DAYS AFTER THE WINDOWS 95 LAUNCH ELLISON WAS SCHEDuled to speak at an annual conference in Paris called the European Information Technology Forum. All the powerful people in the computer business would be there to hear him or would read his words in the trade press. What was more, Bill Gates was also scheduled to give a keynote address. Gates, already enjoying widespread, favorable news coverage, was about to publish *The Road Ahead,* a book describing his vision of the future of computing. The forum would be a perfect opportunity for Ellison to offer his own vision for the future of computing, his own *Road Ahead.*

A day or two before he was scheduled to leave for Paris, he and Farzad Dibachi sat down at a computer terminal to go over Ellison's presentation. They always prepared Ellison's speeches the same way: Dibachi would write a draft, and Ellison would rework it with Dibachi present. "He would sit at the terminal, type away, and basically redo everything," said Dibachi, who said he did not mind being edited. This time they came up with a simple framework: Ellison would demonstrate how a big computer at Oracle headquarters in California could send video, sound, and text across phone lines to a personal computer at the conference center in Paris. That was about it. When the two men finished their work, Dibachi did not know whether Ellison was even going to mention the idea of cheaper, easier-to-use computers.

Ellison made his presentation in Paris on the morning of September 4, a Monday. The talk began badly. The technical people could not get the slide projector to work. "As you can see, technology is going to dominate the world," Ellison quipped, and everyone laughed. He was most likable when he was ad-libbing; all he ever had to do to win over his audiences was make sure something went wrong.

Finally the slide came up. "Oracle does one thing, and only one thing, well," he began, speaking without notes. "We manage vast amounts of data and deliver enormous amounts of data across the network." In the beginning Oracle software moved information back and forth between VAX minicomputers and so-called dumb terminals—bare-bones machines that were used mostly for entering data. With the dawning of the personal computer era, "client-server" computing became the fashion. In this model big "server" computers shared information with PC "clients," which, unlike dumb terminals, had minds of their own. Whatever the computing paradigm someone might be using, Oracle was there.

Now, in the Internet age, Oracle software could handle more than just database tables. It could also deliver sound, picture, and even movies to a variety of "clients"—personal computers, set-top boxes, hand-held computers such as Apple's Newton, and what Ellison called Internet appliances. "While Internet appliances don't exist yet, we think they'll be around. In fact we think they'll be around as early as next year," he said.

Ellison said he believed high technology was in the midst of a paradigm shift (an idea that everybody in Silicon Valley liked because it meant that somebody was going to make a lot of money). There had been a time when mainframe computers ruled the information world, Ellison said. Then personal computers came along, and suddenly everybody had what amounted to a small mainframe computer on his or her desk. He noted ironically that the introduction of Windows 95 received more coverage than the war in Bosnia. "It was staggering, and it does mark the high point of the personal computer," he said.

Now another seismic shift was taking place. In the Internet age it didn't matter what kind of machine you had. What mattered was that you were connected to the network because that was where the information was.

"When you suddenly have a network-centric point of view, you don't need a device anywhere near as complicated or as expensive as a personal computer," Ellison said. "You can build a multimedia Internet terminal and deliver it for about four to five hundred dollars. . . . You just plug it into the wall to get power—to get electrons—and plug it into the network to get information—bits—and you're done." This device would use an operating system much less complicated (and much less capable) than Windows 95. Instead users would simply download a small operating system every time they turned on their machines.

The audience could see two versions of Ellison as he spoke: the life-size one standing before them and the giant, Oz-like image projected on the screen above. Larry and Lawrence, ego and superego.

Ellison continued on the subject of inexpensive computers. Then, finally, he introduced an idea that preoccupied the computer industry for months to come. This idea got him onto the pages of *Newsweek, The New York Times,* and *The Wall Street Journal* and onto broadcasts of every television news show from *The NBC Nightly News* to *Oprah* to CBS's *Coast to Coast.* As a result of Ellison's speech, millions of everyday people became at least passingly familiar with Oracle (until then most regular folks had never heard of it), and Ellison became more famous than ever before. In the trade press Ellison's idea drew praise and wonderment and ridicule and contempt. At Microsoft it inspired amusement and derision and—finally, amazingly—a measure of acceptance. And many, many companies, including Apple and IBM, invested untold time and energy in creating a new kind of computer—all because Larry Ellison thought that doing so would be a good idea.

He described an Internet appliance that would be more than a dumb terminal because it would have its own local memory and a fast microprocessor. But it wouldn't have a disk drive. "This is a full, multimedia, personal—" Ellison started to say "personal computer" but stopped himself. "It's not a personal computer. It's a *network computer.*" That was the first time he ever publicly spoke the words.

He began his demonstration. Ellison stood before a personal

computer, a picture of its screen projected large for all to see. Using a special high-speed telephone line, he connected to a server in California. He clicked the mouse, and the PC displayed a TV news story, with pounding music, about the new Rock and Roll Hall of Fame and Museum in Cleveland. Another click, and the audience heard NBC's Brian Williams introducing a report about the abortion drug RU-486.

"The interesting thing about this technology, what you see here, is that it will run on a four- or five-hundred-dollar Internet terminal," Ellison said. Although he was using an expensive multimedia computer for the demonstration, there was no need for disk drives and loads of expensive application software.

After playing some music by Mariah Carey, Ellison turned to the audience and said, "That's it. Thank you very much."

But he had only just gotten started. After the speech he sat before the audience and took questions from an interviewer, International Data Corporation analyst John Gantz, who asked Ellison to talk more about the importance of the network, which was like asking water to run downhill.

"The PC is a ridiculous device. The idea is so complicated and so expensive," Ellison said. "People in the PC world make fun of the mainframe. They say it's much too complicated and much too expensive, and therefore it has to die. You can make the same statement about the PC." Ellison estimated that big organizations spent five thousand dollars a year to maintain each of their personal computers (he didn't say where he got the number). "These appliances"—network computers—"will cost four to five hundred dollars and give you video and audio. And if they break, unplug it and throw it away."

Disposable computers! This *was* a paradigm shift.

Gantz asked Ellison if he had "a time frame for the death of the PC." Ellison said the personal computer would not die, just as the mainframe computer had not died when the PC came along. But he said the PC was "no longer going to be the center of the universe. The network is going to be the center of the universe." He predicted that network computers would outnumber personal computers within ten years and that people would have them at work, at home, and in their pockets.

That was that. In less than an hour, speaking off the cuff, Ellison had placed himself and his company at the center of a great debate. The idea of the network computer had been so compelling that he might even have to get somebody to build one.

Bill Gates spoke next. He did what any sensible multibillionaire would do when confronted with a farfetched and radical idea: He ignored it. "I think it's becoming clear now that the PC, together with Windows, low-cost communications, and Internet standards, forms the core of not only what will be a new way of doing business, but a new way of learning and entertaining ourselves," Gates began. He said fifty million personal computers had been sold in the past year and predicted that worldwide interest in the Internet would drive more sales in the future. He mentioned in passing that "the PC will take on new forms," such as the wallet PC. But whatever form it took, the personal computer would always run on some version of Windows.

After the speech, interviewer Gantz asked Gates to comment on the idea that some kind of Internet appliance would threaten the personal computer. Gates swatted the question away. He said, "People who think we're going to have dumb terminals in the world of the Internet," then stopped. How was he going to finish the sentence? *Will be kidnapped to Redmond and deprogrammed? Will be flushed out of their dark corners and squished like bugs?* No, Gates was too nice a guy to say any such thing. Of the idea that diskless terminals would someday outsell PCs, he said, "I just don't agree with that. It's the old X-terminal argument over and over again." X-terminal was a nicer name for dumb terminal. Gates never mentioned Ellison's name.

Yet the day had been a success for Ellison. At a time when a lot of people were sick and tired of hearing about Microsoft, he had offered a compelling and not implausible vision of a future in which

Microsoft did not rule the world. Having gotten a reaction from Bill Gates—even if the reaction was brief and contemptuous—was in itself a victory. Before the Paris conference Ellison was at best a marginal player in personal computing. But when Gates disputed Ellison's vision of the computing future, Gates unintentionally gave legitimacy to what he had said.

Still, Ellison's speech did not exactly cripple Microsoft. A month after the Paris conference Microsoft reported revenues of $2.02 billion for the three months ending September 30, 1995—a 62 percent increase over the same quarter in the previous year. The company also announced that it was selling copies of Windows 95 at a rate of one million a week.

An interviewer for *Upside* magazine later asked Ellison to talk about the origins of the network computer idea, and with his answer Ellison once again rewrote history. He told interviewer Richard Brandt, "The first time I used the term 'network computer' and talked about a $500 computer was in a conversation with William Jefferson Clinton, the President of the United States, when he was out here last August," meaning August 1995. Ellison was referring to a meeting at San Francisco's Palace of Fine Arts at which Clinton, Vice President Al Gore, and a group of high-tech executives discussed how technology could be used to improve education. Among the Silicon Valley leaders present were Michael Spindler, then the chief executive of Apple Computer, and Sun Microsystems chief Scott McNealy. During the meeting Clinton, Gore, and the executives discussed the importance of wiring every classroom so students could use the Internet.

"And while that's great and we were all excited about that, wiring the schools to the Internet is simply wiring buildings to the Internet," Ellison told Brandt. "What you have to do is attach human beings, students, kids, to the Internet. How do you do that? . . . I proposed that the President challenge our industry to build a machine for $500 that could be used in all the schools. That would be affordable and very, very easy to use."

"When was this?" Brandt said.

"This was August of last year [1995]," Ellison said. "He could be like John Kennedy, saying we're going to put a man on the moon by 19—choose any year. Bill Clinton could say we could put a computer in every student's desk by a certain year."

In fact Ellison's meeting with Clinton and Gore did not take place in August 1995. It happened on September 21, more than two weeks after Ellison had introduced the idea of the network computer in Paris. It didn't matter; nobody really cared whether Ellison first mentioned the idea in San Francisco or Paris. So why did he change the facts in the retelling? Maybe he got mixed up; his interview with Brandt took place long after the Paris speech. Or maybe he simply liked the Clinton story better. The truth about the origins of the NC—that the idea sprouted organically from other projects, was nurtured among mid-level managers at Oracle, and flowered at a computer-industry conference—was itself fairly interesting. But it wasn't memorable, wasn't *historic*. If Ellison claimed that he first mentioned the idea in the presence of the President of the United States ("William Jefferson Clinton," in Ellison's grand locution)—now *that* would be something. Ellison altered the network computer story in the same way that he often changed the facts of his childhood: He took a good story and made it better.

There is a telling postscript to Ellison's September 21 meeting with the President. This was how Ellison told the story to *Upside*'s Brandt: "It was funny, I was on a tight schedule that day, and the President said, 'Meet me at my limo, I want to talk to you more about [the network computer], and some other things.' Then he dove into the crowd, shaking hands, and I was left on my own devices to make it to the presidential limousine, which was an exciting experience. You normally don't see these guys with machine guns when you see the President, but as I went down the corridor I saw all these guys with the machine guns guarding the limousine.

"I finally got to a line of first defense, and I said, 'The President told me that I'm supposed to meet him by the limousine.' The guard said, 'I'm sorry, sir, you cannot come down here.' 'Well, they told me the limousine was down here and the President told me to meet him here.' He said, 'Yes, sir, I believe the President told you that. You may not come in here.' 'Well, what do you suggest I do? The President told me to be here. I could try to knock you down and make a dash for the limousine.' "

Ellison went on, "You're not supposed to joke like that with the Secret Service, right? He gave me a quizzical look and I said, 'But that looks like a bad idea.' He said, 'Good choice, Mr. Ellison.' I said, 'What would you do?' He said, 'Well, I'd find a staffer and have him come over here and tell me you're supposed to ride over to the Fairmont with the President.' I said, 'I'll go with that approach.' " Eventually Ellison found a staffer who escorted him to Clinton's car.

Clinton and Ellison had a lot in common. Both were born near the end of World War II, both were raised at least partly by surrogates, and neither ever knew his birth father. Both were exceptionally bright and curious adolescents who wowed people with their charm but always remained focused on their own needs. As adults both were ambitious to the point of obsession, and as a result, both were extremely successful. Both were immensely likable yet morally and intellectually flexible, capable of saying and doing whatever seemed most expedient. People loved and hated both of them. And both were motivated by ego, insecurity, and an innate ability to do great things—in Clinton's case, to be President of the United States and in Ellison's, to rule the world of high technology.

It was not surprising, then, that the conversation in Clinton's limo quickly turned to size. "I ended up riding over to the Fairmont with the President and the Vice-President," Ellison told *Upside*. "It's a very small limo, and I joked that the limo I arrived in was larger than his limo.

"He assured me that that might be the case, but his limo was safer."

* * *

As the computer press began exploring and writing about the network computer, there was, shall we say, some skepticism.

"I think it's a rip-off, a pipe dream," wrote Christopher Barr, a columnist for the on-line news service c/net. He wondered how anyone could build a computer for five hundred dollars when a decent screen alone would cost three hundred dollars. "I don't think it can be done. Hey, Larry, the emperor has no clothes!"

Even if it could be done, who would want one? Network computers would not have hard or floppy disks, so users would have to send their files across the network to a server maintained by somebody else. What if the server crashed? What would NC users do then? "[Do you really want to] do without a floppy, hard, or CD-ROM drive? Be unable to compute—or even access your data—when the server goes down? Watch performance slow to a crawl during peak hours?" Eric Knorr wrote in *Multimedia World*. "An Internet appliance has everyman appeal at first glance; but on closer inspection, it's two steps back toward those bad old mainframe days when Big Brother owned the computer, not you."

In an interview with one industry trade publication Nathan Myhrvold, in charge of advanced technology at Microsoft, described the NC with a single word: "Dorky."

Not everyone agreed. *Business Week* published an article titled "Internet Lite: Who Needs a PC?" It quoted an industry researcher as saying that banks would someday give away network computers to every new customer who signed up for home banking. "Once that happens, it becomes a completely different PC industry," the researcher was quoted as saying. "If I were [the PC manufacturer] Compaq, IBM, or Dell, I'd be thinking hard about that." Another influential analyst, Bruce Lupatkin of Hambrecht and Quist LLC, allowed that "the availability of an inexpensive dedicated Internet browser box would probably create a very exciting incremental market among consumers."

Best of all for Ellison, *Newsweek* ran an article titled "Is Your PC Too Complex? Get Ready for the 'NC.' " The author, Michael Meyer, wrote, "No one actually needs a complicated 'mini-mainframe' on their desktop anymore, the thinking goes. Instead, people might prefer something easier and more friendly, a sort of information appliance that . . . is as nonthreatening as a television." The article quoted a high-tech pundit, Paul Saffo: "The PC is dead. It's the horse and carriage of the Information Revolution." Along with the article was a picture of Larry Ellison sitting on the floor of a gym in shorts and a tank top, the new muscle man of Silicon Valley.

Ellison referred to the *Newsweek* article many times, and each time he did, he misquoted it to his own advantage. In several interviews and public appearances he mentioned a quotation that Microsoft's Myhrvold (the NC is "dorky") supposedly gave to *Newsweek*. According to Ellison, "He said about Oracle's plans to build an NC, 'They're lying. L-Y-I-N-G.' That's what he said."[1] Myhrvold never said those words, but Ellison always repeated this quotation with wide-eyed indignation, as if he were astonished that someone had dared question his integrity. In doing this, he portrayed Microsoft as a snarling corporate ogre willing to do anything to trounce its competition.

Ellison's portrayal of Microsoft was not entirely unfair, but never mind. Like the tale about his meeting with Bill Clinton, his story was compelling but untrue. This was what the *Newsweek* article said: "[Does the NC] sound too good to be true? You bet, say the folks who make those vanilla-colored boxes on sale at CompUSA. 'It's fantasy,' says Bob Stearns, a senior executive at Compaq Computer who professes to be 'shocked' by pie-in-the-sky talk of cheapo PCs. 'It will never happen. They're lying.' Nathan Myhrvold, head of advanced technology at mighty Microsoft, the biggest software company, is no less skeptical. 'People want more from their computers, not less. Sorry, I just don't see it.' "

Ellison had taken the words of a Compaq executive and attributed them to Myhrvold, whose reaction to the NC was merely skeptical, not slanderous. Maybe it was an innocent mistake on Ellison's part: Someone skimming the article easily could have thought that Myhrvold had called Ellison a liar. Mistake or not, the story reached its intended audience, and to Ellison, that was what mattered. He was a masterful storyteller, a peerless creator of images. As always, some of the stories he told were undeniably true; many were hard to check.

Larry Ellison spent his 1995 Christmas vacation on a boat, vomiting. But he didn't mind getting sick because the boat belonged to him, he was competing in a major race, and he was winning.

Ellison had not sailed much since he drove his first wife into counseling by buying a thirty-four-foot sailboat they couldn't afford. Then one day he was at the gym, climbing the stair machine to nowhere, when an acquaintance approached him and asked him if he sailed. Not recently, Ellison said. The guy told him about a new class of yachts called maxis that would soon be entering races around the world. "There'll probably be a half dozen of these things in the world, and I wondered if you'd be interested in building one," the acquaintance said. Why not? After a few months and a lot of phone calls, Ellison hired Bruce Farr ("the world's best designer," said the billionaire) to build him a plaything. The result was *Sayonara,* a seventy-eight-foot carbon fiber rocket, the best boat money could buy.

In December 1995 Ellison's crew ("the best crew ever assembled," he said) sailed *Sayonara* down under for the annual Sydney, Australia, to Hobart, Tasmania, yacht race, a 630-nautical mile dash across the underbelly of the planet. To prepare for the big event, Ellison and company entered the Sydney Big Boat Challenge, a 12-nautical-mile sprint in Sydney Harbor. Ellison shared the helm with Australian media magnate Rupert Murdoch, one of his many wealthy acquaintances. After the race, which *Sayonara* won, Murdoch carelessly placed his hand near one of the heavy, fast-moving Kevlar lines that hold the mast in place, and it ripped off the tip of his finger.

"I was standing right next to him. I didn't realize what he had done," Ellison said. "It was silly; you don't ever touch those lines. But it was after the race was over, so you feel real casual. But you feel dumb—you get your finger ripped off and you feel dumb." No problem. Murdoch was loaded into a chase boat, rushed to Sydney for some microsurgery, and delivered on time to the crew party that evening.

Murdoch, his hand wrapped in a waterproof cast, also crewed for Ellison in the Sydney to Hobart race. That was when the vomiting started. The boats in the race trace the coast of Australia, then break free across the Southern Ocean toward Tasmania. Out in the open sea, the waves circle like sharks, forever growing and gaining strength. "There's nothing to stop the wind, and there's nothing to stop the waves. So they just get bigger and bigger and bigger," Ellison said. In the Bass Strait the water is shallow, and the waves seem to ride one another piggyback.

"So you have these very steep waves, very close together," Ellison said. "They just bash into the boat. Crash, boom! Crash, boom! There's seventy knots of apparent wind and water flying. You've got double lifelines on; you can't move on the deck without clipping into one thing, unclipping from something else, and clipping into something else. . . . If you have to go down below to go to the bathroom, you have to take all your clothes off. And you can't take your clothes off without being thrown around the boat because you're hitting these waves. So getting undressed to go to the bathroom is a life-threatening activity. You just hope to be able to go to the bathroom and not break an arm or a leg or be knocked unconscious into a coma."

The crew members were supposed to work for four hours, then sleep for four hours. "But you can't fall asleep. Two thirds of the people are throwing up. This, the world's best crew, the best crew ever assembled, on this boat. And everybody was sick," Ellison said. "We smelled pretty bad and looked worse when we finally got into Hobart, but we got into Hobart before everybody else, so we won the race. We were the first American boat to win the race in seventeen years.

"In fact we were going to break the record, but just after going through a storm, and really going for the record, we got becalmed. The wind just died. And we sat there and watched the record we'd worked so hard for just disappear as the sails quietly flopped. But it was great watching the sun rise. We got a great film of it."

* * *

One day that same December Ellison and his lawyers met with San Mateo County District Attorney James P. Fox and Fox's chief deputy, Steve Wagstaffe. Ellison wanted Fox to do something for him: prosecute Adelyn Lee for shaking him down for money.

Ellison may have settled the case with Adelyn Lee months earlier, but he definitely had not dropped it. For several months Oracle's lawyers had been gathering the technical evidence necessary to show that Lee had broken into the Oracle computer system from her home computer, forged an E-mail note in Craig Ramsey's name, and used the message to defame Ellison and extort money from the company. Ellison had been looking forward to this day for months. He had settled the civil lawsuit only after his lawyers had assured him that doing so would not prevent Lee from being charged criminally for what she had done. Why, given all the other things he had to do, would Ellison spend time getting even with Adelyn Lee?

"It's really easy," his friend Steve Jobs said. "When you're defamed, it's on the front page, and when you're vindicated, it's on page five hundred twenty-three, if at all. I think that Larry felt that he'd been really wronged by this woman and just wanted someone to take notice of it."

The district attorney's office took notice. After the meeting with Ellison, Wagstaffe approached a young prosecutor named Paul Wasserman and asked him to look into the case. Never mind that Larry Ellison and Oracle want this woman prosecuted, Wagstaffe told him. Just evaluate it the way you would any other case.

Wasserman, thirty-one years old at the time, was the obvious choice to prosecute a computer-related crime. As a boy growing up in the Southern California town of Tustin, he had spent countless hours playing Star Trek games on an Apple II, the first truly functional personal computer. He later hung around the local computer store so long that the manager hired him to do odd jobs. Even after he finished law school and became a felony prosecutor (after a brief and unhappy stint in civil practice), he retained an abiding interest in computers. His home machine, which he put together himself, had a 133-megahertz Pentium chip, a 3-gigabyte hard drive, 32 megs of RAM, and an ISDN modem for connecting to the Internet. Of *course* he got the Adelyn Lee case. "I had the dubious distinction of being the office computer geek," he said.

Most of Wasserman's cases were everyday violent crimes: domestic beatings, brutal gang assaults, and so on. The job wasn't always pretty, but he enjoyed it immensely, and it showed. Wasserman was a great schmoozer; he liked prosecutors, defense lawyers, cops, and courtroom staff all the same. He had the ability, important in a trial attorney, to make the transition from adversary to friend as soon as a case was over. He also had a habit, probably unconscious, of ending telephone conversations by saying, "Be good"—helpful advice, coming from a prosecutor. All in all, if you had to go to prison, he was the sort of person you'd want to send you there.

Wasserman knew that some people would think he was just doing Larry Ellison's bidding if he prosecuted Adelyn Lee. There wasn't much he could do about that; if she had committed a crime, he would have to prosecute it. The fact was, he was about the last person to do favors for Ellison. Years earlier Wasserman's wife, Sutten, had worked as a sales representative for Oracle. Things didn't go well for her: She was among the four hundred people Ellison laid off when the company self-destructed in the fall of 1990. When Oracle was slow to forward her last commission check, Wasserman, full of young lawyer bravado, called to find out why. He was told that the sales-tracking system was so fouled up that Oracle didn't know how much it owed. This, sadly, was true, but Wasserman did not know that at the time.

"Come on, you're a freaking database company!" he remembered yelling into the phone. "You're telling me you can't figure this out? Under California law you have seventy-two hours to pay her. Pay her! We need the money!" Now, several years later, Wasserman wasn't upset with Ellison anymore, but he wasn't going to carry water for him either.

Wasserman began his investigation by requesting everything Oracle had on the case. He got thousands of pages of paper in response. Oracle wanted Wasserman to prosecute Lee for committing perjury in her civil deposition. The company believed that she lied when she said she did not use Craig Ramsey's E-mail account to send the incriminating message to Ellison. It offered a lot of documentation to support that charge.

But Ramsey had testified in his deposition that he changed his E-mail password after he fired Lee. She always knew his password while she was working for him. But if he had changed it, how could she have used the account to send a message under his name? District Attorney Fox was so concerned about this that he had made a note in the Oracle files: "This could be reasonable doubt."

Wasserman talked to the people who ran Oracle's internal computer system. Like everyone else at the company, Craig Ramsey actually had two computer accounts. One was his Unix account, which gave him access to the entire Oracle system. The other was the E-mail account, which allowed him to send and receive messages. He had a separate password for each account. Adelyn Lee knew both passwords while she was working for him.

Oracle believed that Lee had signed on to the company computer system from home and sent E-mail to Ellison in Ramsey's name. In order to do this, Lee first would have had to log on to his Unix account from home. That would have meant phoning the Oracle computer system and entering his name when prompted. The system would have asked her to enter Ramsey's Unix password, which, significantly, he had not changed after firing her. Wasserman understood all that. But what good would the Unix password have done her? If Lee had wanted to send electronic mail, wouldn't she also have needed Ramsey's E-mail password, the one he had just changed?

No. That was the key. Oracle's electronic mail system assumed that anyone who logged on to a Unix account was also permitted to use the accompanying E-mail account. In other words, if you logged on to Unix as Craig Ramsey, the E-mail system didn't ask you for ID. Lee would have needed only one password to send the message, the one Craig Ramsey had never changed.

Every time someone logged on to a Unix account, the Oracle computers made a record of it. By looking at these accounts, Wasserman could clearly see what Adelyn Lee had done. At 8:17 the morning she was fired, she logged on to Ramsey's account and typed an

E-mail message to Larry Ellison: "I have terminated adelyn per your request. cdr"

Ellison's computer records showed that he got the message at 8:19 A.M. At the same time Lee made two copies of the message. She sent one to the E-mail account of Chad Beecher, a friend who worked a couple of buildings away from Ramsey on the Oracle campus. She sent the other copy to Beecher's printer. She logged off Ramsey's Unix account at 8:24.

At 8:25 she logged on to Beecher's Unix account (she knew his password too), presumably to see if the incriminating message was there. She then logged on and off the Ramsey and Beecher accounts a couple more times before calling it a day.

In her deposition Lee had testified that she called Beecher that morning and asked him to check the printer for E-mail messages from Ramsey. She said she was hoping to find some explanation for her firing. Why would Ramsey send his messages to a printer two buildings away? Because the printer nearby wasn't working, Lee said. Lee was never able to explain why Ramsey would have sent a copy of the incriminating message to Beecher, whom he barely knew.

Wasserman had seen just about all he needed to see. The evidence may have been circumstantial, but it all pointed in one direction. Still, before he filed charges against Adelyn Lee, he wanted to speak to Ellison, who would have to testify if the case went to trial. Everything Wasserman knew about Ellison was secondhand, and not much of it was flattering. He had seen Ellison's face on a billboard near the courthouse and thought, *Sheesh, what an ego.* He also knew that Ellison's legendary arrogance might not play too well before a jury. "I knew going into this thing that my biggest liability was Larry Ellison," he said. So he arranged a meeting to find out if he was credible.

The meeting took place in Ellison's office at 500 Oracle Parkway. Wasserman was accompanied by an investigator, a former Internal Revenue Service agent named Randy Curtis. Ellison began with a speech about how the rich were often victimized, and there was no justice anymore, and so on. Wasserman nodded at the billionaire and thought, *Absolutely, the injustice you've suffered is just horrible.* He wasn't sure he was going to like this guy.

Randy Curtis wasn't either. During the meeting Ellison continually sipped from a tall glass of fruit juice but didn't offer his visitors anything to drink. This almost drove Curtis crazy, but it wasn't unusual. Other visitors had had similar experiences. Once a dozen or so Oracle customers were in Ellison's office for a meeting. After about thirty minutes an assistant strolled in and handed Ellison a dish of sorbet, which he ate as his guests watched.[2]

Then Wasserman did something that changed the tone of the meeting. He wanted to know exactly what Ellison and Lee had done in bed during their last date. Had she "manually stimulated" him, as she claimed? Ellison said he didn't remember.

Wasserman leaned forward and got as close to Ellison as he could. He said, "Larry"—he wasn't about to call him Mr. Ellison—"I find it really hard to believe that you don't remember whether you got jerked off that night."

According to Wasserman, Ellison's whole demeanor changed. He was no longer self-righteous or bombastic. Suddenly he was earnest and forthright and humble. Likable even. He said he honestly did not remember what he and Lee had done. They had had a lot of sex during their relationship (that was the point, wasn't it?), and whatever it was that they did on April 16, 1993, did not stand out in his mind. Wasserman was pleased. He had given Ellison an idea of what it would be like to be cross-examined, and Ellison had done beautifully. So far as Wasserman could tell, he had simply told the truth.

The State of California charged Adelyn Lee with two counts of perjury (one for lying about using Ramsey's account and one for lying about sending the message), one count of manufacturing false evidence, and one count of computer intrusion.

One of the many paradoxes of Silicon Valley life was that companies were often great pals and snarling enemies all at the same time. It was not uncommon for a high-tech executive to ridicule a company one day and share a stage with its representatives the next. This paradox could be seen live onstage at a hotel in downtown San Francisco on May 20, 1996.

Oracle held a press conference that day to make a couple of announcements about the network computer. First it was spinning off its NC development group as a wholly owned subsidiary, Network Computer Inc. The new company would develop and license software for the new devices. The NC was becoming ever more real; Ellison had given a public demonstration of a prototype only a couple of months earlier.

The other news was that four major high-tech companies were joining Oracle to develop open technological standards for NCs. The idea was to make sure that all NCs worked roughly the same way. The companies—IBM, Apple, Sun Microsystems, and Netscape Communications—all had sent high-ranking representatives to say that the network computer was a good and righteous thing and that they supported it. Moderator Richard Brandt, of *Upside* magazine, referred to this group as "the anti-Microsoft coalition," a description that was right on the mark. These companies wanted to make money at the expense of Bill Gates. They were all in it together.

At least they were for today. The truth was, Oracle's new "partners" were also some of its bitterest rivals. IBM was an example. The rivalry between Oracle and Big Blue went back to the days when Ellison took IBM's idea for a relational database and used it to take market share from IBM. In the early days IBM's lawyers had sent Oracle a letter complaining about the remarkable similarity between IBM's SQL documentation and Oracle's. It was no wonder the two documents looked alike: "We took the [IBM] book, and we typed it in, and we shipped it," Oracle's Kirk Bradley said. In more recent years Ellison had often said that IBM made the "single worst mistake in the history of enterprise on earth"[3] when it licensed its first PC operating system from Microsoft. Even so, a high-ranking IBM software executive was here to pledge allegiance to the NC.

Then there was Apple. Yes, Oracle and Apple had collaborated on set-top boxes, and Oracle had some products that ran on Macintosh computers. But as for Dr. Gil Amelio, Apple's chief executive—well, according to Ellison, he was no Steve Jobs. "Anyone who insists on being called doctor—it's truly bizarre," Ellison once said. "I make a point of always calling him Dr. Amelio. I feel like saying, 'I've got this cold. Can you prescribe something?' " Still, here was Dr. Amelio (he had a Ph.D. in something), plugging the network computer.

Oracle's strangest bedfellow that day had to be Netscape, maker of the software that enabled people to get around the World Wide Web. Ellison believed Netscape was overrated, both as a stock and as a Web browser. In an interview published a month before the press conference, he said, "I think their stuff is very, very thin," and urged the company to use its cash to buy "real estate [and] tankers of oil at sea."[4] Oracle, which had just developed its own Web browser, was a direct competitor of Netscape.

Netscape's chief technologist, twenty-five-year-old Marc Andreessen, was no devotee of Larry Ellison's either. A couple of months before the May press conference, he had given an interview to *Red Herring* magazine[5] in which he was asked about the NC.

HERRING: Can you envision a diskless network computer that runs Netscape?

ANDREESSEN: That's [Ellison's] game. Those are set-top boxes. The big question that nobody's been willing to answer is: what are you willing to strip out of the PC? How can you get it down to $500? The answer is you don't—for the same reason that you can't get a set-top box for less than $3,000. You can't give up the RAM, the hard drive, the video, the graphics, the sound, the CD, the keyboard, or the mouse. People need all that stuff. That's what a PC is. People love it. The PC is it for the next five years. This $500 terminal thing is nonsense.

HERRING: As PCs get more powerful and more complex, won't there be extra functionality that some people won't need?

ANDREESSEN: The more power you give people, the more ways they find to use it. End of story. There's no question that a lot more interesting stuff will be happening on the network, but it's a matter of time scales. For the next five years, I expect PCs to increase in performance at their current rate, and people will keep buying them at the same price point.

HERRING: Whether they do or not is really immaterial to your business plan.

ANDREESSEN: That's true. Unless, of course, the $500 network system is so incredibly compelling that people buy tens of millions of them and we're not on that platform. And I'll look really stupid, because I think it's nonsense.

HERRING: Mr. Ellison has said: "We don't expect Netscape to make any money at all." . . .

ANDREESSEN: I can't believe people still listen to him after the whole interactive TV thing. As Tevye said in *Fiddler on the Roof,* "When you're rich, they think you know."

But now here was Andreessen, sitting next to Ellison on the San Francisco hotel stage, shilling for the network computer. Apparently his thinking had changed. Where once he had believed "the PC is it for the next five years," he was now enthusing about "a whole new class of devices that we think can be very successful very quickly." He predicted that hundreds of millions, or even billions, of NC-like devices could be sold in the next decade or two. Where once he had called the network computer "nonsense," he now viewed it as "a pretty major new business opportunity" for Netscape.

After the press conference I happened to see Andreessen standing on the sidewalk, waiting for a car to pick him up. He looked different now. Onstage he had seemed impressive, even commanding, a real Silicon Valley player. But now, among the everyday people bustling about, he looked like what he was: a likable twenty-five-year-old midwestern kid in blue jeans.

I introduced myself and told him I had seen the *Red Herring* interview, and he actually blushed. "I was wondering if anybody was going to mention that," he said.

I reminded him of something he had said: that he couldn't believe people listened to Larry Ellison anymore. Just then his car pulled up to the curb. I had time for only one question. "You were up there with him today, talking about the NC," I said. "Does that mean you're listening to him now?"

This was Marc Andreessen's answer: "Ummmmmmmmm."

Then he climbed into the car and drove away.

Jenny Overstreet, Larry Ellison's longtime assistant, was thinking about quitting her job. It was not because of anything Ellison had done. Yes, he was often blunt or abrupt or harsh with her. People were always asking Overstreet's companion Kirk Bradley, "Why does Jenny stay? Why is she still putting up with this shit?"[6] After all, she did not need the job. Because of all the stock options Ellison had given her, she had enough money to retire and live comfortably for the rest of her life. But she wasn't thinking of quitting because of a few bad moments with Ellison; she was used to bad moments. No, Overstreet had other reasons for wanting to leave. A relentless organizer, she divided her reasons into two categories: positive and negative. The positive ones were things that really had nothing to do with Oracle, and the negative ones were things she did not like about her job.

Overstreet's principal positive reason for resigning had to do with family. Her mother, who lived a few miles down Highway 101, was not well, and Overstreet wanted to spend time with her while she still could. There were other positive reasons. Several years earlier Overstreet had bought a house on a hill overlooking San Francisco Bay, but she had always been so busy at Oracle that she had never had time even to paint the bedroom. "It still looked like the other people lived there," she said. Finally, at age thirty-five, Overstreet became uncomfortable with the fact that she had never done anything except work for Oracle Corporation. Ever since she finished college, every bead of her sweat, every watt of her considerable energy had been expended in the service of Larry Ellison's vision. She was beginning to wonder what else she could do, what else she could achieve. As she put it, "I had not taken a hard look at what kind of person I wanted to be since I was twenty-one." It was time to take a look.

Then there were the negative reasons. The main one was that the job just was not fun anymore. There was a time when Overstreet had done all sorts of enjoyable research for Ellison, whether that meant taking part in the design meetings for his homes or arranging a meeting with some high-tech mogul. But the company was gigantic now, and so was her job; she could no longer set aside the time to do such things. More and more Overstreet was delegating all the really fun tasks to her assistants.

The shiba story was a good example. For a long time Overstreet kept telling Ellison he should get a dog. He loved animals and had plenty of people to care for them while he was away. But Ellison kept brushing her off; getting a dog was not one of the three or four things he was concentrating on.

Then Ellison made one of his frequent trips to Japan. When he returned, he was bursting with enthusiasm for the dog idea; he had found just the breed he wanted: a shiba. Ellison had chosen the breed because of a statue he had seen at a train station. He told Overstreet the story printed on the base of the statue: Every morning a Japanese businessman walked to the train station, accompanied by his loyal shiba. Every afternoon the dog returned to the station to meet its master. One day, after leaving the dog and riding the train to work, the dog's master had a heart attack and died. The dog was so faithful, according to what Ellison had read, that it waited at the station for nine days for his return. The dog died waiting, Ellison told Overstreet. By the time he finished telling the story, Overstreet was in tears, and Ellison was misting up. That cinched it: Ellison had to have a shiba.

If Overstreet had heard the dog story a few years earlier, she would have devoted a lot of her time to finding the perfect dog for Ellison. She would have researched the breed to make sure it was right for him, then worked the phones until she got the name of the best shiba breeder in the world. In the end she would have given him a puppy, a perfect puppy, one befitting a billionaire. It would have been great fun. But now, in 1996, she couldn't possibly take on such a job. There were too many other things to watch over—as she put it, too many "homes and companies and entities and boats and cars and staff people." In the end she let one of her assistants, Cynthia Turner, do the dog research. Ellison never got a shiba, but that did not lessen Overstreet's frustration with what her job had become. A perfectionist, she wanted to be able to handle everything, but that wasn't possible anymore. By July 1996 she had made up her mind to quit.

She chose a week in late July to break the news to Ellison. Finding the right moment was difficult. The first couple of days passed without their having more than a brief exchange. Wednesday and Thursday would be her last chances that week; she was taking Friday off. And she wasn't sure she could still do it after the weekend.

When she arrived at the office that Wednesday morning, she told Carolyn Balkenhol and Cynthia Turner, her assistants in Ellison's office, what she had decided. She thought they had a right to know about anything that would affect them so profoundly. But Overstreet had another reason to tell them: She knew there was a chance that Ellison would be so upset with her that he would call security and bar her from the building. It did not matter that she had served him faithfully for her entire adult life. She knew he might see her decision as a betrayal. In her worst moments she could imagine him exploding at her, shouting that she should have left years ago, that she had always been grossly overpaid, that she was never right for that job anyway. Overstreet had seen Ellison do that to other people too many times to think it couldn't happen to her. "Larry doesn't take abandonment very well," was the way she put it. If he banned her from the building, she wanted Balkenhol and Turner to pack up her office for her.

Overstreet's other fear was that Ellison would, in her words, "disengage," that he would refuse to acknowledge how he really felt about her decision. She had seen him disengage when Bob Miner died—"I haven't seen the body"—and didn't want to see it again.

She left the Oracle campus after lunch that Wednesday and drove to Ellison's house in Atherton, where he was working. She did not call to say she was coming; if he thought she was going to deliver bad news, he might have tried to avoid her. She arrived about the time he finished a long meeting. As soon as she saw him, she started to consider backing out.

Overstreet and Ellison sat down in the breakfast room. "We have to do four things," she said. That was Overstreet: businesslike, efficient, exact. They spent a few minutes going over the first three things. Ellison seemed anxious to get up and get going.

"There's a fourth thing," Overstreet said.

"OK," Ellison said.

"I'm resigning," she said, almost in a whisper.

"What?" He hadn't heard her.

"I'm resigning."

He did not say anything at first. She immediately began filling the silence with her reasons; as she spoke, Ellison just nodded blankly, occasionally mumbling "hmm." When she gave him her "negative reasons"—the job wasn't fun, all she did was make snap decisions—he began to nod as if he understood. At first he tried to persuade her to stay; he offered her a couple of other jobs within the company, at least one of which already belonged to somebody else. No, she said, she didn't want a different job at Oracle. Then he told her she could take a leave of absence and come back in a few months. No, she said. She wasn't coming back.

Ellison was out of options. But he did not explode. He did not disengage. "Well," he said in his hyperbolic way, "everything in my life just changed forever."

They sat and talked for a few more minutes. Overstreet thought that Ellison was listening, really listening, to her; this was exactly what she had hoped for. They started to talk about ways that Ellison could improve the executive offices, but both soon realized they did not have the stomach to talk about that now. "I hope we'll keep in touch," Overstreet said. Ellison stood and opened his arms, and she stepped into his hug. They walked out of the house holding hands. Then Ellison got into his car and left, off to his next appointment.

Overstreet later sent an E-mail message about her departure. (She had written it the day before because she was afraid she would be "such a wreck" after resigning that she would not be able to announce the news in the proper way.) She sent copies to about two dozen people inside and outside Oracle, including Ellison and me.

> Subject: Jumping into the unknown . . .
>
> . . . was what I did the day I accepted my job at Oracle 13½ years ago. I was 22, knew nothing about anything except for the French classical poets and a little Ibsen, and started work on a national holiday for a boss I'd never even met, in the field—computer science—in which I'd gotten the lowest grade of my academic career. Today I'm jumping again. Today, with sadness, certainly great fear, but also excitement and anticipation, I am resigning from Oracle Corporation, to get my house in order—figuratively and literally—and to figure out what might be next for me, which, given the way I like to work and my affection for and allegiance to this place, has been impossible for me to even begin to consider while here.
>
> . . . I see greatness and the promise of greatness all around me here. . . . I have always been proud, and will be proud forever, to be able to say that I worked at Oracle Corporation, and I wish you all of the best that I always had here.
>
> Jenny

Jenny Overstreet's last day at Oracle was August 8, 1996. Two months later I asked Ellison about her decision to leave. We were sitting on the deck at the Atherton house, looking out over the ponds, pool, and the redwoods. Ellison had his customary glass of carrot juice. The breakfast room, where he had received the news, was just inside the door from where we were sitting. How, I wanted to know, was he getting along without her? I wanted to know how *he* was doing, but his answer was all about her.

"I think she's absolutely wonderful. I think she did the right thing. She was so devoted to her profession that it was distracting from other things in her life. I think she did the right thing by resigning," he said. He told me that he had since hired Rick Moore, a former aide in the Clinton White House, as his—these were his words—"chief of staff." With Ellison, it was always hard to talk about what happened in the past; he only wanted to discuss what would happen next. "Jenny loved to do everything herself. And the job got beyond a threshold where it was simply impossible, regardless of how Herculean you were individually and regardless of how enormous your memory was."

Part of Overstreet's frustration was that she could not do everything, Ellison said. "Everyone believes—and perhaps Jenny believed it too—that Jenny knew everything that was going on," he said. But she didn't, and couldn't have. Still, the idea that she knew everything was precious to her. And when our images of ourselves begin to conflict with reality, he said, "the reality ceases to be fun."

But what about him? What effect did her departure have on him?

"I didn't feel betrayed," he said, apparently aware of what Overstreet had been afraid of. "These are kind of awkward questions. This is why politicians never answer—why some people don't answer some questions honestly. I thought it was good for her to leave, and I thought it was good for me that she left. The job was overwhelming her, and things weren't getting done. But for someone to read this and interpret this that either I'm not grateful for everything she did or don't think she did an extraordinary job—nothing could be further from the truth. But it was time."

I told him nobody had said he was not grateful to Jenny. But he was sensitive to the accusation anyway. Sometimes, he said, "these people who were so wonderful in building Oracle to a certain level are not the right people to carry on." After they have left, "those people have felt abandoned. They've felt I've given up on them. They've been very angry about that. As Oracle has grown very rapidly, I've joked, I've said, 'I'm the only survivor.' This is not a good feeling, you know, to have no survivors." Ellison went on. "A lot of these people have been good friends. But they were good friends because I saw them every day at the office. Or if I saw them socially, it was always associated with business. And then they felt that I abandoned them as friends. That's been very—I don't know what to think about that. I don't want to make anyone feel bad."

I told him that Overstreet was happy with the way he reacted to her decision to leave.

"I love her. I love her dearly. I wanted her to be happy, and just because this is not exactly working out for both of us right now and I'm frustrated doesn't mean she hasn't been great and contributed a lot. . . . I was very proud of her."

I told him that she was afraid he would "disengage," and he said, "That's interesting. I guess that kind of hurts my feelings. Someone who knows me so well, to think that I would just do that to her."

Ellison laughed, but it was plain that he didn't think the idea was funny. He didn't like to think of himself as someone who was unkind to people just because they left his company. But there seemed to be part of him that couldn't help himself, a part that really *was* hurt that Overstreet had resigned. He said he loved her dearly and didn't feel betrayed, but he also said that he was frustrated with her and that "things weren't getting done." Overstreet was right: He didn't take abandonment very well. He also didn't like to have that pointed out to him.

Ellison then went on one of his rhetorical adventures. "Life is the enlightened pursuit of happiness, not the unenlightened pursuit of as much money as you can accumulate. And the only things that are important in our lives are love and work—not necessarily in that order, but they're both important," he said, and laughed a genuine laugh. The phrase sounded familiar to me; days later I remembered that he had said the same thing at Bob Miner's memorial service.

"We work because work is an act of creation. We identify with it. . . . I look at the company and I think, *This is me.* But that's not my whole life," he said. "People I care about, people I love, are essential for my sanity to make it through every day. So both of those things, work and love, conspire to deliver some kind of happiness. And if we get reasonably good at both of them, we're in really great shape."

Fifteen

IN LATE SEPTEMBER 1996 LARRY ELLISON MADE A VISIT TO HIS HOMEtown, Chicago. He went there not to visit relatives or meet with an Oracle customer but to prostrate himself before the body politic, to take part in a uniquely American ritual of self-promotion, self-revelation, and self-humiliation. This was without question one of the strangest episodes in the history of high-tech marketing. How much did Larry Ellison want to be the richest, most famous, and most beloved man in computing? This much: He went to Chicago to do *Oprah*.

The taped program was broadcast around the country on October 8. Though its ostensible subject was the homes of wealthy people, Ellison was not there to discuss real estate but to promote the network computer and of course himself. He succeeded, though weirdly. The program reduced him from a complex and unpredictable figure to a sort of goody-two-shoes cartoon character, a grown-up Richie Rich. *Oprah*'s portrayal of Ellison was clean, crisp, cloyingly positive—and almost completely false. Such was the magic of television.

Ellison was the last guest to appear. After a commercial break the camera focused on host Oprah Winfrey, who was sitting on the floor, wearing a brown bodysuit and holding a black wireless microphone. "Whoa!" she shouted, an expression of mock enthusiasm for the audience's applause. As the noise subsided, Winfrey exclaimed, "The life of a computer mogul! Larry Ellison is a self-made multi*billionaire,* that's with a *B,* who founded Oracle, that's a computer company. And here's an inside look at Larry's life."

At that point the producer rolled a videotape, a brief, *Lifestyles of the Rich and Famous*–type profile of Ellison. Winfrey and Ellison watched the tape along with the studio audience. The profile began with a shot of Ellison driving his convertible turbo Bentley. When Winfrey saw the car, she exclaimed to Ellison, "I have one of those too!"[1] She was, like Ellison, on *Forbes* magazine's list of the four hundred richest Americans, though, with a measly $415 million, she ranked dead last.

As the camera led a tour of Ellison's house in San Francisco, Ellison could be heard saying that the wall of windows in the living room was forty-four feet wide and thirteen feet high. "When you walk in, the view just assaults you," he said. The camera then moved down the peninsula to what Winfrey called Ellison's "fifteen-million-dollar Japanese retreat." In real life Ellison almost never answered his own door—his staff did it for him—but this was not real life. "Hi. Come on in," Ellison, the perfect host, told the camera, swinging open the front gates.

The videotape then portrayed Ellison in a way that, at long last, seemed genuine. Ellison was standing in the house, talking about his Japanese statues. "These are temple guardians," he said. "They're four hundred years old. These came to me because a museum wanted to get them but couldn't afford them. So sometimes when a museum can't buy something, they'll notify me that a piece of art is available, and I buy it. And they just keep calling my doctor and asking about my health." Finally Ellison was shown being truly clever and ironic and interesting.

Then the videotape altered reality again. The Winfrey voice-over said, "The man who has it all says he wants to get married again." Ellison was shown sitting in the cockpit of his jet, talking about Ms. Right. "I'm looking for a woman who is smart and funny and compassionate—and great-looking," he said. The fourth Mrs. Ellison would also need to overlook the fact that Ellison already had a steady girlfriend, one who frequently accompanied him on trips in that very plane.

After a commercial break Winfrey reintroduced Ellison with these words: "He grew up in one of the roughest neighborhoods on the South Side of Chicago." There it was again. Clearly the fiction about Ellison's youth in South Shore had been repeated so many times that it had a life of its own.

"Larry Ellison is obviously a self-made man. He's a self-made *billionaire,* who hit gold in starting his own computer company," Winfrey went on. She turned to Ellison. "One of the things I like about you—I mean, you have lots of nice houses and beautiful cars and planes, but you're one of those people who's interested in giving back." Ellison looked back at her blankly, as if he thought she was talking to somebody else. "That's what I hear about you," Winfrey said, pressing on. "Is that true?"

Ellison's eyebrows jumped, and he said, "Uh, yes!"

Then the show got to what was, for Ellison, the point. When Winfrey spoke of Ellison "giving back," she was referring to Oracle's promise to give a network computer to each of the 294 children at the James Flood Science and Technology Magnet School in Menlo Park, down Highway 101 from Oracle. The gift of the NCs was a story in itself.

The James Flood School, which comprised grades kindergarten through eight, was in one of the poorest school districts in the San Francisco Bay Area. Ninety-eight percent of its students qualified for a free or reduced-price lunch. Four years earlier Oracle had given the school some computers—Apple computers, not Windows-based PCs—and sponsored classes on how to use them. Oracle employees also volunteered to mentor some of the kids. Ellison knew of the company's beneficence but had not found time to visit the school himself.

In September 1996, about the time the *Oprah* people asked Ellison to be on the show, Ellison went to the school. He told principal Ellen Spencer that he wanted Oracle to give an NC to each of the 294 students and then announce the gift on *Oprah.* That was fine with Spencer, who told me, "His bottom line is to sell NCs. My bottom line is to have them for my students. So it works out." Ellison may have wanted to market network computers, but Spencer also thought that he genuinely "wanted to make a difference."

Now, on *Oprah,* all America was going to hear about Ellison's generosity. "The whole idea is to give each and every one of those students a computer to use at school," Ellison said.

"Really?" Winfrey said. "So your goal is to get every student a computer?" Wait, hadn't Ellison just said that? Winfrey was like a kindergarten teacher, and twenty million Americans constituted her not too sophisticated class.

Ellison pitched on. "The computers we currently make, personal computers, are rather expensive and very, very complex. So we've introduced this new class of computer—I have a copy of it with me—for students or for normal human beings—"

"That's pretty!" Winfrey said, except she pronounced it *purty,* as if she were a poor Appalachian girl seeing a frilly dress for the first time.

Ellison took no notice of Winfrey's dopey locutions; he just kept talking. "Very lightweight, very low-cost," Ellison said, bouncing the sleek silver network computer in his hands to show its lightness. (The computer was just a mock-up.) "And the kids can use it at school, and they can pick it up and unplug it and take it home with them and then use them at home."

"So you're giving each of the students in the school a computer?" Winfrey was hopelessly stuck on this idea.

"Every student, ah, in the school, every teacher in the school, has a computer." Well, not yet they didn't. Ellison had promised the computers, but the students were not scheduled to get them until the following spring, after the classrooms were wired and after the Oracle-based NC became commercially available. It wasn't on the market yet.

With that Winfrey reinvented herself again. She spoke to Ellison in baby talk, as if he were a cocker spaniel puppy that she had just trained to beg. "What a nice guy you are! That's *very nice,"* she said. The audience applauded the puppy's niceness. The show never made it clear that the NCs would be donated by Oracle Corporation, not by Ellison personally.

Winfrey turned to Spencer, principal of the James Flood School, who was sitting next to Ellison, wearing pearls and beaming. Winfrey said, "Every kid's going to get their own computer? I could weep over this!"

"Wonderful," Spencer said. She explained that there were only twenty-three computers in the entire school before Ellison offered to provide the NCs. Now—in case anybody missed the point—every child would have one. Spencer later said that she wished Winfrey had spent more time talking about education and less time cooing over cars and planes. But at least the school got *some* attention.

What Winfrey's viewers did not see was that Ellison was a tough-minded, calculating, egotistical man who succeeded in business at least partly through intimidation, exaggeration, and Machiavellian attention to his own needs. On *Oprah,* he was bland as cardboard: sincere, kindhearted, altruistic, almost Franciscan in his goodness. The show was superficial by design, but even given its limitations, Ellison got short shrift. To Oprah's audience, he was just a lonely heart, a misunderstood rich guy in search of that special someone.

Yet his appearance had to be seen as a resounding marketing success. Ellison's discussion of the NC was simply brilliant, another example of his ability to define reality on his own terms. As he portrayed the NC, it was not an expensive toy, like a multimedia personal computer, but a learning device—a serious, staid electronic learning aid. Ellison was hardly the first to try to market computers in schools—Apple Computer had done the same thing years earlier—but that did not make the strategy any less clever. Ellison's pitch for the NC was so compelling that a lot of people were under the impression that the product actually existed (Oracle's customers had been making the same mistake for years). "I had parents asking me, 'Where can I buy this computer?' " Principal Spencer said.

It wasn't just what Ellison said on *Oprah* that was so smart but how he said it. Not once did he wield a scary technical term like "cache," "server," or "application." He did not even utter the word "network," even though he had a network computer in his hands. He referred to the computer only as "a new class of computer" that was "very low-cost and very easy to use." Ellison, the image maker, talked about the NC in a way that was certain to give at least a few people the nerve to go computer shopping.

Winfrey's hour was up now; it was time to go. Referring to Ellison's gift to the school, she shouted, "I'm inspired by that!" Then, taking on one final persona, that of an inarticulate street person, she said, "*I'm gonna find me a school and I'm gonna give some computers too!* I'm going to do that." The audience clapped louder than ever. "I'm inspired by you, Larry. That's a brilliant idea! It really is! Thank you, thank you very much.

"Thanks, everybody! Don't write to me about marrying Larry!"

They didn't. They wrote to Ellen Spencer at the James Flood School instead. "We got some scented stationery," she said.

Ellison and I were sitting on the deck in Atherton when Klaus, his houseman, interrupted us. There was a phone call; did he want to take it? When he was told who was calling, Ellison sprang out of his chair. "My son is on the line," he said, excusing himself. "My son wants to talk about the air show again."

Some dads take their sons fishing; others teach them to play golf or throw a baseball. What Larry Ellison did with his thirteen-year-old son was fly. "Larry hasn't been real involved in the kids' lives as far as going to baseball games and school events and things like that. He makes one or two, but he doesn't make them all the time," Barb Ellison, Larry's former wife, had told me. "Now he and David have their flying, and it's pretty cool that they have this thing that is theirs to do together." The day after I spoke with Ellison at his home, he and David flew one of Ellison's planes to Salinas for a big air show. David had been looking forward to it all week.

When Ellison finished with his phone call, he told a story about the time he took Barbara up in his twin-engine Cessna Citation jet so she could see David fly. David and his instructor were flying the Lancair plane called the Dreamcatcher. "It's a really tiny, tiny, tiny airplane," Ellison said. "And it's painted like a kid's plane: it's white with beautiful turquoise blue and purple stripes. They look like they're dripping, like ice cream streaks along the side of the plane." David was cruising along in the Dreamcatcher when his parents' jet pulled up next to him. Barb was already nervous; she was about to be more so.

"She said, 'You never should have brought me out there. I'm better off not knowing he's doing this,' " Ellison remembered. "And when we pulled up next to him, I told her, 'Get ready for this.' I said, 'OK, David, do a peel-off.' And he just kind of rolled the plane over and dropped away. It was great." The maneuver left Barb Ellison breathless.

For Ellison, flying with David was more than just a fun weekend outing. Showing his son how to fly was a way of showing him how to live. "Larry decided that teaching his son to fly was a very good way to teach him a lot of things about responsibility," Steve Jobs said. Jobs pointed out that the Ellisons often flew together in the same plane. "I mean, he's risking his life with his son. There's not a lot of fathers who would risk their life, would put their life in their sons' hands."

I told Ellison what his friend had said. "The fact of the matter is that if David does something stupid, I'll grab the stick from him. I'm not going to let him kill himself," he said. "But what a great feeling for a thirteen-year-old to go out and fly this airplane and make all of these adult decisions. It's given him a level of confidence he otherwise would have no way of achieving. I think he feels very good about the fact that he does this. He makes the decisions on his own; he lands the plane beautifully; his aerobatics are getting very, very good right now. He's learning almost too fast. Have you seen the commercial with Kareem Abdul-Jabbar? He's teaching this kid to play basketball, and this young kid's doing too well. So he says, 'OK, kid, that's enough self-esteem for the day.' And he takes the ball and leaves." Ellison jokingly said he sometimes has the same thought about David: *That's enough self-esteem for the day.*

I had met David months earlier, while interviewing his mother at her home. He was a pleasant kid, shy but polite, with a bright, freshly upholstered look about him. He and his father had gone flying that day; David's instructor had dropped him off while Barb and I were talking. David had no trouble finding things to do: He rode his dirt bike, played a game on his computer, and watched the film *Jumanji,* which was just out on video. Seeing him, I wondered what his life must have been like. How does a kid maintain any sense of values when his dad is a multibillionaire who gives him everything he wants, including light aircraft? Barb Ellison had thought a lot about this issue. She said she tried hard to—well, to keep David's feet on the ground. Once he asked if he could fly to Seattle with Larry to try out a flight simulator at Boeing. Barb said no because David had not finished some of his schoolwork.

Instilling any sense of humility was difficult when raising children in a moneyed community such as Woodside, Barb Ellison said. "You've got a lot of big names and big money around here. I mean, nobody in Larry's stratosphere. But you've got the money here."

I asked Ellison if he worried about spoiling David. How do you teach values to someone when everything is available to him?

"But everything isn't available. Money is available, and that's nice. But you know, Freud talked about work and love. Freud didn't mention money. Money didn't make the list," he said. "David is going to have to create something that will make him feel good. The old saying, which is absolutely wrong, is, 'Why did you climb the mountain?' 'Because it was there.' That's not why people climb mountains. They climb mountains because *they're* there. And they're curious to know if they can do it. We test ourselves all the time. We're endlessly curious about ourselves. What can I do? Can I do this? How will I react in this situation?

"Flying an airplane is very complicated," he continued. "And it can be very dangerous if you're not competent. If you're competent, it isn't that dangerous. And he's perfectly capable of learning. This is not a seven-year-old girl, that poor little girl that died." Ellison was talking about Jessica Dubroff, a California girl who had died earlier that year while trying to set a cross-country flying record. "David isn't a child. This is a big kid. He's very coordinated and very strong and very smart, and he can be very disciplined about something he likes. And it has been wonderful watching him rise to the challenge and seeing how good he feels about being able to do this."

But if he never has to work—and surely he won't have to—what will motivate him to do anything besides fly planes and play video games? What challenges will he face?

"There are all different kinds of challenges. Maybe he won't be challenged to make money. But I'm not sure that not having to make money is a bad thing," Ellison said. "He's got the same problem all the rest of us have. He has to engage in an enlightened pursuit of happiness to figure out what makes him happy. That's all about how you feel about yourself, how you relate to other people, what your work is, what you create, what you make. We are builders. Human beings are builders. He's going to have to find something he really wants to build. He's going to have to have some idea and create something out of that idea. Maybe he can spend his life giving away money rather than making it. That's not a bad thing. Giving it away in the right ways is very difficult to do. He's got a very different situation from the one I grew up in, but he has to do something himself to feel right, to feel complete."

Ellison had his own way of feeling complete. During our conversation he mentioned that he had been flying the Citation jet earlier that day. When he landed, he said, he stopped the jet in twelve hundred feet, leaving fourteen hundred feet of runway to spare. "Actually stopping it in twelve hundred—that's the shortest anyone's ever stopped a jet and turned off," he said. "The tower was very impressed. It was a good moment for me today."

Coincidentally, about the same time Ellison was landing his jet, I was meeting with a source who told a joke that had gone around Oracle. This was the joke:

Q. What's the difference between God and Larry Ellison?

A. God doesn't think he's Larry Ellison.

* * *

The paperback edition of *The Road Ahead* was published in November 1996, about a year after the hardcover. The difference between the two editions was stark—and comical.

The reason was timing. Bill Gates finished writing *The Road Ahead* sometime in mid-1995, when people in the industry were still talking about set-top boxes and interactive television. So Gates wrote about whether the set-top box would ever usurp the personal computer (he didn't think so). Then, just before *The Road Ahead* reached the bookstores, Ellison introduced the idea of the network computer. Suddenly nobody was talking about set-top boxes anymore. At least part of *The Road Ahead* was already in the rearview mirror on the day the book was published.

A year later came the paperback edition. Its cover said it was "COMPLETELY REVISED AND UP-TO-DATE." In fact, as the hardcover had proved, it was impossible to publish a book about the computer industry that was anywhere near "up-to-date." By the time you felled the trees to make the paper, the industry would have gone through a half dozen major paradigm shifts. (This problem haunts almost every computer industry book project, including this one.) There was no way to keep up with such a hyperkinetic industry.

Still, Gates tried. In the hardcover he had used the term "information highway" dozens of times. A year later the term was hopelessly outdated; anyone who used it in the fall of 1996 would have been ridiculed as a greenhorn. Thus the term was all but banished from the paperback edition, the index had only nine references to the phrase, compared with twenty-four in the hardcover. In the "UP-TO-DATE" edition, the words "information highway" were changed to "communications revolution," "global interactive network," and a host of other phrases. At one point in the paperback Gates remarked on "how quaint . . . the term 'information highway' is beginning to sound."

The changes in *The Road Ahead* went far beyond the rewriting of some phrases. In the paperback edition—a sort of debugged Version 2 of the book—Gates paid a lot more attention to the World Wide Web than he had a year earlier; the number of references in the index boomed to twenty-seven, from a not very visionary four in the hardcover. Even the chapter titles changed. "Applications and Appliances" became "Information Appliances and Applications," "Implications for Business" became "Business on the Internet," and "Race for the Gold" became "The Internet Gold Rush." Yes, Microsoft had finally become aware of the Internet.

Gates had little choice but to address the idea of the network computer in the paperback version. To do otherwise would have been disingenuous and disappointing; all his readers wanted to know what he had to say. "Another competition engendered by the popularity of the Internet is determining how PCs and terminals should change to be cheaper and more suitable for browsing," he said. He mentioned Oracle and Sun Microsystems by name, then outlined their positions; basically, that users should have a cheap computer terminal connected to a powerful (and therefore not cheap) server. Gates quickly laid bare the agendas of the two companies: Sun Microsystems sold servers, and Oracle peddled server software. Then he took a swipe at his rivals. "Sun and Oracle contend that the Internet will finally do for them what they've failed to do in the past—reverse the movement toward powerful personal machines and recentralize computing." In the narrow-shouldered world of high tech, this amounted to a vicious right hook.

Gates briefly described the rationale for the NC, then not so briefly argued the other side. "Sun and Oracle used to promote diskless 'X-terminals' for corporate local-area networks. These so-called 'dumb terminals' never caught on, in part because there wasn't much of a savings compared to the cost of a PC," Gates wrote. The future lay not in the use of skin-and-bones terminals and muscle-bound servers but in a "balanced" model in which the terminal—"usually a PC"—and the server both are "quite capable." "I don't see much reason to dumb down the desktop machine," Gates wrote.

When it came to marketing, Larry Ellison had met his match. If Bill Gates could convince the market that Ellison wanted to "dumb down" computers, Ellison didn't stand a chance.

Gates continued. "The supposed advantages of dumb terminals"—the term was now conspicuously out of quotation marks—will be equally available to people who invest a little more and get a personal computer that has a disk drive and can run any of the thousands of Windows-based applications," he wrote, in one of the book's many advertisements for the Microsoft operating system. "Entry level PCs are now about $1,000 or $1,200 in the United States, but the price could approach $600 as manufacturers direct some of their innovative energy into lowering prices." He predicted PCs and "lots of other information appliances" would get a share of the consumer market. He concluded, "I'm unenthusiastic about the prospects of incompatible dumb terminals that are almost, but not quite, real PCs."

This was where that Gates really drew blood. By raising the possibility of inexpensive Windows-based computers, he neutralized one of Ellison's most powerful weapons: price. Why would anybody buy a bony little network computer—a dumb terminal!—when he or she could get a "real" personal computer for about the same money? Ellison had other weapons, of course: PCs were expensive to maintain, hard disks often failed, and the Windows operating system was a mess, a great bulging bag of code. But computer users were accustomed to Windows and were unlikely to give it up without a compelling reason. Price had been a reason. But now the natural forces of the computer market were taking that away.

"Now some people wonder if Microsoft has met its match in the Internet," Gates says in the last chapter. That idea had scarcely been addressed in the hardcover; now it formed part of Gates's conclusion. Larry Ellison and the NC were at least partly responsible.

One thing about *The Road Ahead* didn't change from one edition to the next. Lawrence J. Ellison's name had not appeared in the hardcover book and didn't make it into the paperback either.

Adelyn J. Lee was scheduled to stand trial for her alleged crimes against Larry Ellison in the first week of January 1997. That Lee was going to trial at all could be interpreted as proof of her arrogance and poor judgment. Months earlier prosecutor Paul Wasserman had offered a deal: If she pleaded guilty to one count of perjury, she would not have to pay back the hundred thousand dollars she and her lawyer had received from Oracle and would face no more than one year in county jail. The judge told Lee he would sentence her to no more than ninety days and might not give her any time at all. Lee refused. She wanted a trial, and she was going to get one.

Ellison could hardly wait. Court TV had contacted the judge about televising the proceedings, and though Ellison predicted that cameras would be excluded (he was right), he said he wouldn't mind if the trial were televised. "I'd love for the truth to be out. But I don't think that it can ever be fully out," he said. "There's really no way to wash away the accusation [she made] or stop people from believing it's true if they want to. But you know, the more graphically what she did is portrayed, the happier I would be."

He seemed upset that he had misjudged someone so completely. He had always believed that there was "something off" about Lee, some vague, disturbing thing he couldn't put into words. Well, he could put it into words now. "My opinion—and I don't know what I can get sued for—is that she is an evil person, and I really mean evil. . . . She takes pride in being evil. And this is the thing that I couldn't get," he said. Ellison pointed out that Lee's password for the Oracle E-mail system was "Darkside."

"I think she actually sees herself as a cunning, ruthless woman to be reckoned with. . . . She's smart enough and devious enough to always get her way. And no one had better cross her," he said.

Adelyn Lee's character and her actions would be the subject of the trial. But for now, there was another issue worth considering—namely, whether Ellison should have been dating her in the first place. He was, after all, the fabulously rich and powerful man at the top of the company, and she was a basically powerless woman many levels beneath him. To look at the situation strictly from his point of view (her turn is coming), wasn't it unwise for Ellison to date a female employee? Didn't he leave himself vulnerable to precisely the kind of suit she filed? During the trial Wasserman asked Ellison the question this way. "On a scale of one to ten, with ten being not at all bright, don't you think you score a twenty for dating an employee?"

Ellison didn't think so. "It has nothing to do with working at Oracle. I think working at Oracle is a minor issue. . . . It could be anyone. It makes no difference. Anyone could say, 'This is what he did.' Anyone," Ellison said. "Anyone who says anything about me—you know, it's going to get in the newspaper. 'Larry Ellison set my cat on fire. Sadistic Larry Ellison.' They don't have to work at Oracle." He missed the point. It was precisely because Adelyn Lee worked at Oracle that she might have had any basis for a claim against him. She didn't sue him because (according to her) she refused to have intercourse with him. She sued him because *he fired her* after she refused to have intercourse with him. (Again, that was her claim.) Ellison was right. Conceivably, anyone could accuse him of anything at any time, even feline immolation. But it was because Lee worked for him that her allegations were plausible.

If dating a female employee could be harmful to Ellison, wasn't it just as bad for his company? Craig Ramsey testified that he was "uncomfortable" when he learned that Lee was the CEO's girlfriend, and everyone involved in her firing trod lightly. Ellison called their behavior "bizarre," but was it really? Maybe it had been so long since Ellison answered to a boss that he forgot how it felt. Maybe Craig Ramsey was not in an awful predicament with Adelyn Lee, but he must have felt that he was. Didn't that bother Ellison?

He answered with a hypothetical situation. Suppose, he said, that he had a male friend at Oracle—say, a programmer—who was doing a poor job. "Should they be afraid to fire him too?" Ellison asked. If so, he said, "I shouldn't date anybody, and I shouldn't have any friends in the company. And I certainly shouldn't have any relatives in the company. Ridiculous."

Was it? Plenty of companies had strict rules against nepotism. And Ellison's hypothetical example about the male programmer wasn't really apt. Firing somebody who plays basketball with the CEO presumably would be a lot easier than getting rid of somebody who sleeps with him. Even so, lots of executives maintain a certain distance from their employees because they don't want friendship to become a factor in professional decisions.

What about the women at Oracle? How did they feel knowing that the CEO was on the lookout for dates when riding the elevators? To some, it must have been exciting. Ellison was attractive and entertaining and (if you believed what he said on *Oprah*) marriageable. Surely lots of women would have liked to catch his eye. Ellison could see no reason why he should not ask them to dinner or why, for that matter, any manager should not date an employee. "This happens a lot. You know, Ray Lane ended up marrying his secretary. I have no problem with that. Bill Gates ended up marrying somebody who worked at Microsoft. I have no problem with that," he said.

He rejected the idea that Oracle should have a policy forbidding executives to date those who reported directly to them. "I'm just not convinced you can legislate [that sort of thing]. You know, people are working very closely together, and if a relationship happens, it happens. Do you think people are going to adhere to the policy? Let's imagine you were managing someone and you fell madly in love with them. Would you say, 'Oh golly, I would love to go out with you on a date and kiss you, but this policy says we're not allowed to do that, so I'm going to stop talking to you'? I don't think so," he said. "I'm not sure I have fully baked thoughts on all of this, but in general I think people should be free to date within the company and managers should be able to date people who are working for them, within the constraints that you can't force someone to go out with you, and the person shouldn't feel that they're being coerced in any way."

That last idea was important. One could argue (and plenty of people did) that anytime a supervisor expressed romantic feelings toward an employee, there was coercion. Let's say, since we're talking about Ellison, that a male supervisor asked a female employee for a date. She might react in any number of ways. She might be delighted and say yes. She might be uninterested and say no and never think about it again. There were lots of other possibilities, and these were the troublesome ones. She might say yes out of fear that her boss would retaliate if she said no. She might say no and then have to live with the gut-wrenching knowledge that things at work might never be the same. She might wish that her boss had appreciated and respected her for her work and left it at that. She might have a complicated and unsettling combination of feelings.

All these things seemed magnified when the boss was the tremendously powerful chief executive officer. Ellison was presented with a hypothetical situation. Suppose he dated a female employee once, had a nice time, and asked her for a second date. Suppose that she thought he was a nice enough guy but didn't want to go out with him again. Would it be so unreasonable for her to think, *What's going to happen if I say no?* And wouldn't it be best if no one were ever put in that position?

Answer: Things would never get to that point because Ellison would always be certain the woman was interested before he asked. He would *already know* how she felt.

"I'm extremely careful," he said. "Every time I've gone out with someone at Oracle, I've been very careful not to initiate the relationship, to constantly test and see if there's any reluctance whatsoever. I think it's incumbent upon the senior person in the company to constantly be doing these tests. The women almost have to be the aggressor, or you're foolish to pursue it."

There was little doubt that Ellison was a gallant and charming suitor. Once, after a first date in a San Mateo restaurant, he drove his companion (not an Oracle employee) back to Atherton for a tour of his house. He pulled up to the front gate, reached out the window to punch in the security code—and drew a blank. He had forgotten the code. Unembarrassed, he climbed out of his seventy-thousand-dollar sports car, stripped off his tailor-made suit jacket, and shed his shoes. Then, like an eighth grader trying to impress the new girl in class, the middle-aged billionaire vaulted the fence and ran stocking-footed to the house to open the gate from inside. "It was funny to see the richest man in California climbing his own fence," the young woman said.[2]

While Ellison may have been kind to the women he dated, he was not often willing to make women his partners at his company. Though many women considered Oracle the most exciting and challenging place they had ever worked, they also knew its history: At a company with more than twenty thousand employees only a couple of women had ever risen to the level of senior vice-president. No woman had ever served on Oracle's board of directors.

At Oracle's annual shareholders' meeting in 1996 a self-assured young woman stepped to the microphone to grill Ellison on this point. Why were there so few women in positions of real authority, she asked him, and what was Oracle doing to recruit more? This would have been a fairly tense and interesting moment no matter who was asking the question; after all, this person was raising a highly charged issue in a meeting that was normally dedicated strictly to the bottom line. But Ellison's inquisitor wasn't just anyone. She was Nicola Miner, the daughter of the late Bob Miner.

Nicola Miner, now twenty-six, had long been discouraged about the prospects for women at the company where she and her family were still major stockholders. She herself had worked there over the years, mostly in marketing, and had been frustrated that women never had a say when the most important decisions were made. "We have plenty of women vice-presidents," she said later. "But at Oracle being a vice-president is not really the top echelon, or even close."

She knew who was at least partly responsible too: her late father. For all his charms, Bob Miner was something of a Neanderthal when it came to women at work. He and Nicola argued about the issue all the time. Miner thought women didn't know how to program computers and weren't dependable anyway because they were prone to having babies in mid-career. "It's very hard to find a woman who's capable," he used to tell Nicola. That proved to be a self-fulfilling prophecy. "It's frustrating to me that the company my father founded—and he had two daughters—wasn't a little more progressive," Nicola Miner said.

Oracle was never the sort of place where women were prevented from doing interesting work. Usually women (and men) had the opposite problem: They couldn't work fast enough, or hard enough, to keep up with Ellison's demands. Kate Mitchell, a marketing vice-president in the early 1990s, said she had "complete and total respect" when she worked at Oracle and "was given tremendous leeway and authority." Marcia Wells-Lawson, the company's first female employee, said working at Ellison's company was "the most perfect job I could have asked for." Plenty of women retired young because of Oracle's success and Ellison's generosity.

Yet even Wells-Lawson, a true Oracle loyalist, thought that

"there was a glass ceiling." As corporate secretary she was for many years the only woman present for board meetings. And she didn't expect to see another anytime soon. Wells-Lawson blamed other board members for turning the board into a men's club. Still, Ellison had the clout to bring women in but never used it for that purpose.

Now he had to explain why. Now he had to answer Nicola Miner's questions: Why weren't there more women in upper management, and why weren't there any women on the board? He answered the second question by saying that he had identified two women who possessed the qualities needed for board membership, and neither was available. "I found that to be a pretty stock, irritating answer," Nicola Miner said later. Jack Kemp was on the Oracle board, for God's sake. Ellison considered him "an unbelievably bright guy" and "a leader," but others rolled their eyes. "The whole board thought that was crap," former board member Joe Costello said. "I mean, *Jack Kemp*? What value added is that guy going to have on the board of directors of Oracle, really?" Certainly Kemp was no high-tech genius; his experience was in government (not coincidentally, a huge market for Oracle). Nicola Miner wondered why Ellison couldn't have found an "unbelievably bright" female with similar skills and connections.

As for Nicola Miner's first question—the one about women in upper management—Ellison dodged it completely. "I knew I wasn't going to get the answer I wanted," she said.

The night before Ellison testified in the trial of Adelyn J. Lee, prosecutor Paul Wasserman drove to Atherton to prepare him. Wasserman was met at Ellison's door by a woman he described as "a gorgeous blonde"; he was under the impression that she worked for Ellison but didn't really know and didn't ask. When Wasserman entered the house and saw the beautifully burnished white oak floors, he was so impressed that he was tempted to shout, "Cool floors!" Larry Ellison greeted him warmly and said, "Want to see the scene of the alleged crime?" Ellison led him back to the master bedroom, where Adelyn Lee said he had got rough with her almost four years earlier, only to withdraw the allegation later. Wasserman thought, *I feel like I'm playing the DA in some weird movie.*

Wasserman, who had talked to Ellison several times while preparing for the trial, had grown to like the boyish billionaire and told him so. But the prosecutor was afraid that the jury's first impression of Ellison would be as unfavorable as his had been. Wasserman was blunt. "You come off like a prick," he said.

"I changed your mind, didn't I?" Ellison said.

"Yes. But you and I have been talking for a long time, and I think that over time I have come to understand you," Wasserman said. He explained that the jury might see only "Larry Ellison, the playboy billionaire." So it was important for Ellison to be pleasant and solicitous and unflappable. No matter how confrontational Lee's attorney might become, Wasserman told Ellison, "Don't argue things with him. Just get up and tell the truth. Don't guess, and don't soft-pedal things."

"I hear you. I hear you," Ellison said. He promised to be good.

Accompanied by a couple of Oracle lawyers, Ellison arrived for court on time the next day. He wore a blue-gray suit with a red tie and white shirt. He and Adelyn Lee would have made a lovely couple if he had not been trying to send her to prison. "You should have seen [Adelyn], with her sweet-looking white cardigan sweater on, buttoned demurely at the throat," one observer[3] told a friend. "All she was missing was a curved bamboo staff and a couple of lambs milling around her to complete the picture."

Ellison testified for more than two hours. At one point Judge Carl Holm asked him to speak more slowly and made a very apt joke about salesmen who talk too fast. Ellison occasionally licked his lips or glanced at the wall clock above the jurors' heads, but mostly he remained calm and composed. One of the newspaper reports about his testimony said that he was "flustered" during cross-examination, but others in the courtroom disagreed. When Wasserman polled the court staff about Ellison's performance, they pronounced him "charming."

Gordon Rockhill, Lee's clever and well-respected defense lawyer, scored at least one point on cross-examination. First, he got Ellison to repeat the story of how he had told his managers to treat Adelyn Lee the way they would treat any other employee. "You didn't want to mix business with pleasure, did you?" Rockhill asked, and Ellison said that was correct.

Then Rockhill got Ellison to admit that he had once intervened on Lee's behalf when something unfair happened to her at Oracle (she wasn't given a chance to move into a new position because of a bureaucratic error). Ellison said he acted only because Lee raised "a very serious issue." Rockhill asked Ellison if he had intervened on behalf of three other people who were in the same situation, and Ellison said, "I wasn't asked to." Rockhill's point was clear: If Ellison would intervene to help Lee when he was happy with her, he might also intervene to hurt her when he wasn't.

At other times Ellison fended off Rockhill's attacks with ease and grace. When Rockhill asked Ellison if he was trying to "save face" when Lee sued him, he said no, he wanted only to "get the truth out."

"You had some animosity toward that woman, didn't you?" Rockhill said.

"I was very angry that she lied, yes," Ellison shot back.

There wasn't much that Rockhill, having failed to undermine Ellison's credibility, could do for Adelyn Lee. The technical evidence was straightforward and convincing, as was the testimony of the Oracle people involved in Lee's firing. Perhaps worst of all for the defense, Judge Holm had ruled that the jury could not be told about Oracle's payment to Lee of the hundred thousand dollars to settle the civil lawsuit. The ruling destroyed what might have been Lee's best defense: that Ellison never would have settled with her if he thought she had done something criminal and that he had sicced the district attorney on her purely out of vengeance. Now that Rockhill could not play the spite card, he had to try the case on the evidence, unfortunate for his client indeed.

In his closing argument Wasserman asked the jury of eight women and four men to concentrate on Lee's crimes, not on her relationship with Ellison and certainly not on Ellison's lifestyle. "This is not about attacking a rich guy and making him look sleazy so you'll hate him," he said. Rockhill, for his part, argued that the technical evidence against Lee was insufficient and certainly could have been doctored. He also suggested that the whole prosecution had been initiated by a vindictive billionaire. "What you've heard is Oracle's case," he said.

Rockhill was right, but so was Ellison. After deliberating for most of a day, the jury convicted Lee of two counts of perjury and one of creating false evidence. (The computer intrusion charge had been dropped mid-trial because the statute of limitations on that crime had run out.) Lee was sentenced to a year in jail and was ordered to repay the one hundred thousand dollars Oracle had paid her and her lawyer.

When Wasserman got back to his office after the verdict, he called Ellison to give him the news. Wasserman was too late; Ellison had already heard it from some Oracle people who attended the trial. Still, Ellison was grateful. "Thanks for looking at the facts the way they really are," Ellison told Wasserman on the phone. "I knew that it was going to be an uphill battle because of my reputation, and I appreciate everything you've done."

By the spring of 1997 the future of the network computer was still unclear. As usual in Silicon Valley, a lot had not happened. Larry Ellison had long ago promised that a device with an Intel chip would be available in the fall of 1996, but fall had turned to winter and winter to spring, and the thing still was not out. Network Computer Inc., the company Oracle had created to market the devices, listed several manufacturing partners on its Web site, including Zenith Electronics and IDEA, but none of the partners actually had an NC for sale. This nonexistence of the NC was beginning to bother the industry press. In the beginning even those who hated the idea of the network computer wrote about it with a sense of immediacy. Now the press was beginning to get impatient. "For Larry Ellison," *Information Week* said, "it's time to deliver the goods."[4]

There was still no consensus on the issue of the network computer. Nicholas Negroponte, the founder of the Media Lab at the Massachusetts Institute of Technology and a sort of dial-a-quote for journalists covering technology, liked the idea. But he thought the NC might not be accepted because Ellison was, in his considered opinion, "something of a nutcake." The Gartner Group, an industry research firm, worried that Ellison's enthusiasm for the NC "bordered on obsession."[5] The firm wrote, "If the NC comes to fruition, it will happen with or without Oracle. So why does Mr. Ellison spend so much time on it? He wants to keep the vision alive. . . . Apart from his media contest with Bill Gates, promoting centralization . . . can only help server-dependent companies such as Oracle. The focus on the NC distracts attention from the lack of exciting news from his own factory. If the NC fails to take off, it will be forgotten in two years' time." The bottom line, according to Gartner: "Larry Ellison has built a great company, but his visions have been wrong as frequently as they have been right." Even members of Ellison's own management team understood that the NC was far from a sure thing. "He's gotten a tremendous amount of notoriety out of this thing, which has been good for Oracle," chief financial officer and board member Jeff Henley said. "If nothing happens, we're going to look kind of stupid in a year or two. But that's Larry. Larry's got balls, and he'll take a risk."

Maybe the network computer was not going to cause the revolution that Ellison had hoped for. But even Ellison's critics believed that it would serve a purpose. Over time Ellison had spoken more and more about what everyone believed was an important issue for computer users: cost of ownership. A first-rate Windows-based personal computer cost more than $2,000 in 1996, and that was just for starters. The Gartner Group estimated that it cost corporations $13,200 a year to run each PC they owned.[6] That figure included the cost of buying the machine, upgrading the hardware and software, fixing the things that broke, and the work time lost while people fiddled with their machines. Ellison argued that the NC would cost much less to buy and thousands less to administer. That may or may not have been true; getting all those NCs to work properly within a network was going to be complicated. But the argument certainly got the attention of a lot of corporate information managers, who were always trying to make their budgets go as far as possible.

By the spring of 1997 there seemed to be a consensus that the NC would eventually find a niche in big corporations. "The NC is perfect for a reservations clerk who only has to fill out a standard form or a bank teller who wants access to an account. These workers do not need a PC because the tasks they perform are predetermined, limited and take place in a fixed location. If the NC can do this more efficiently and cost effectively than existing terminals, it will be adopted for those tasks," wrote Eric Nee, editor of *Upside* magazine.[7] IDC Research went even further, saying the NC could take 7 to 10 percent of the market for new corporate desktop machines by 2001. The research firm said 1998 "will be the make-it-or-break-it-year" for the NC. Even if the NC never earned a penny for Oracle Corporation and Larry Ellison, it seemed that the device—when and if it ever appeared—would be at least a small success.

As a weapon in Ellison's private war against Microsoft, it already was. When Ellison introduced the idea of the NC in Paris in 1995, Bill Gates had essentially pooh-poohed the idea. But as time went by, the cost of ownership issue had forced a change, albeit a minor one, in his thinking. First Gates introduced a concept he called the Simply Interactive Personal Computer, or SIPC. This device, really just a twinkle in Gates's eye, sounded a lot like a network computer, though there were substantial differences. "A SIPC system will be quite easy to use," he said.[8] "It will turn on instantly, like most other consumer appliances. It will interconnect with VCRs, stereos, and TVs." Here is where it would differ from an NC: "Every SIPC will run thousands of Windows applications, including web browsers and software for faxing, voice messaging, conferencing, and exchanging E-mail." If Gates had anything to say about it, you were never going to get a PC without paying Microsoft for it. Even as he reacted to the threat of the NC, Gates dismissed it: "The Internet terminal is too close to being a PC, without really being one. It loses the advantage of being a general-purpose computer able to run off-the-shelf software, yet offers little in return other than a somewhat lower price tag."

Later in 1996 Microsoft and Intel announced a new kind of machine called the NetPC, which was even more like an NC than the SIPC. The companies said the NetPC would be aimed at "task-oriented users that do not require the flexibility and expandability of the traditional PC"[9]—in other words, reservations clerks and bank tellers. The machines would run Windows software (of course) but would be cheaper to use and easier to maintain. At the same time Microsoft announced a broad effort to reduce the cost of computing within corporations. Wrote the on-line news service c/net: "The NetPC platform materialized in response to pressure from other vendors proposing low-cost network computer initiatives. Those vendors include IBM, Oracle, and Sun Microsystems."

As 1996 gave way to 1997, Gates remained insistent that the PC—and therefore Bill Gates—would continue to rule the world. "We see the PC going forward," he said. "The network computer would throw away all software. Clearly Microsoft doesn't think people want to do that."[10]

He was probably right. But Larry Ellison, Mr. Unconventional Thinker, would never admit it. "The battle of ideas is over," he said, "and the battle of markets has begun."

Epilogue

I HAD MY LAST INTERVIEW WITH LAWRENCE JOSEPH ELLISON ABOARD the 192-foot motor yacht *October Rose,* anchored in a cove off the Caribbean island of Antigua. Ellison, who was relaxing on the yacht with his steady girlfriend Christine Petermeier, had told me a few days earlier that he wouldn't mind taking a vacation from his vacation so we could talk. I boarded an American Airlines flight to Antigua on a Tuesday night and met him on the boat the next morning. When Ellison greeted me, he was wearing a pair of black swimming trunks and nothing else, making it clearer than ever that he was in excellent physical condition. We spent seven hours together, during which time he added only one piece of clothing—a lightweight tank top shirt.

As always, he was a fast and ebullient talker; sometimes he answered questions directly, and other times he moved from subject to subject like a rock skipping across the water. He talked about network computers, flying, his affection for Steve Jobs, Adelyn Lee, and O. J. Simpson's acquittal on murder charges ("Oh, my God!"). He talked about the time he bought a supermarket tabloid because he couldn't resist the headline WOMAN GIVES BIRTH TO FROG. He talked about his dear departed cat, Clio, who died while he was out of town and whom he ordered exhumed and then buried under her favorite tree. While making a point about effective advertising, he said the best ad he ever saw was the one in which Crest toothpaste said it was certified by the American Dental Association to fight cavities. He quoted Clarence Darrow and Joe Montana and Bill Gates. And he talked a lot about the *October Rose,* which had a crew of fourteen and was capable of crossing the Atlantic twice without refueling. Ellison said he owned the yacht and told an elaborate story of having bought it out from under the nose of a fellow who wanted it but moved too slowly. At one point we took a break so Ellison could introduce me to Christine, a couple of her young friends, and her parents, who were also visiting. Then Ellison and I moved below-decks to the dining room to talk some more.

Ellison was most articulate and seemingly the most sincere when he talked about Japan. He had made his first trip there in the mid-1970s, when he was working as a programmer for Ampex. During the visit he made a side trip to Kyoto, where he first became enchanted with Japanese gardens. "They were just astounding. They were designed to mimic nature, as opposed to Western gardens, which are designed to show man's ability to dominate nature," he said.

"Look at the archetypical Western garden, Versailles, where man subdues nature and turns hedges into elaborate geometric forms. The scale of the garden is not natural. Everything is kind of off; it's all designed to impress your friends. That's what Versailles was designed to do. Foreign kings who arrived in France were supposed to be cowed by the grandeur of Versailles. Not so the Japanese gardens and Japanese palaces. They were designed to make people feel comfortable, not uncomfortable. It's interesting that we design our gardens arguably to make people feel uncomfortable. They design their gardens to make people feel part of the garden, feel comfortable inside the garden.

"We were evolved in the forest. The sounds of the forest, whether it's the wind in the bamboo or water running over rocks in a stream, or the smells of pine or the damp earth after a rain are all things that millions of years of evolution tell us are a natural state of affairs. We should feel comfortable here. We should feel at home here."

Finally it was time to go. As I was leaving, Christine and her friends were preparing to play canasta; Ellison was going to play too. So much for his image as a playboy. He escorted me to the stern, where a crew member was waiting in the skiff to carry me to shore.

As I was climbing in, Ellison noticed a barracuda and then a sea snake in the clear blue water; he seemed as excited as a child to see them. As the boat carried me off, I could still see him standing there, gazing into the water in hopes of seeing them again.

We had talked about everything, yet the thing I thought about most later was the yacht, *October Rose*. Had he really bought it? Or had he simply chartered it and said he was the owner because it sounded better? Was he telling the truth? Or was this another example of Ellison's self-promotion?

I might not even have been curious if not for a conversation I'd had with a crew member on the way to the interview. This fellow, an Englishman named Simon, used a skiff to pick me up at a seaside hotel not far from where the boat was anchored. As I climbed aboard, I asked him, "Is this Larry's yacht we're going to?"

"Well," he said, smiling, "I don't know what he's going to tell you in your interview."

Now I wanted to know: Did the *October Rose* really belong to Ellison? I called Bermuda, where the ship was registered, and got the basics on the vessel: gross tonnage 819, net tonnage 245, engines by Caterpillar, estimated speed 15 knots, built in Germany in 1986, registered in Bermuda in 1992 by October Rose Limited of Great Britain. And who was the beneficial owner? The authorities in Bermuda hadn't the foggiest.

After searching the World Wide Web for information about yachts (the sort of thing I might have done on a network computer had one existed) and after placing a few calls, I was directed to a yacht broker in Fort Lauderdale who was said to have the listing on the *October Rose*. His name was Merle Wood. Yes, he said, he had the listing, and yes, he knew Mr. Ellison. Was it true? Had Larry Ellison bought the boat? Wood said he could not tell me without first speaking to Ellison. He promised to call me back, but he never did, and he never took my calls after that either.

Did Ellison really own a motor yacht? Did he really steal it away from another interested but hesitant buyer? The answer, I knew, was the story of his life: Maybe. And maybe not. Maybe that was Larry Ellison's boat, and maybe it wasn't. For that matter, maybe it wasn't even a boat. It looked like a boat to me, but it also looked like a hotel. Maybe it was just really good beachfront property.

Maybe I wasn't even in Antigua. How would I know? I didn't fly the plane.

Epilogue to the 2003 Paperback Edition

WELL, IT WAS HIS BOAT, ALL RIGHT.

The motor yacht *October Rose,* on which I interviewed a tanned and shirtless Larry Ellison around Christmas 1996, was neither a charter boat nor expensive oceanfront property—nor a figment of my imagination. It belonged to Ellison, all sixty-four yards of it, from its engines generating thirty-one hundred horses to its gleaming teak decks to its on-board fleet of runabouts and personal watercraft. Ellison's weekend plaything had five staterooms (including VIP quarters reserved for Steve Jobs), a pool, and an entertainment center featuring a half-dozen audiovisual systems. Ellison, it turned out, had bought the yacht for $10 million only weeks before our interview, and would soon repaint her stern with a new name: *Sakura,* which is Japanese for "stock option." Just kidding. It means "cherry blossom."

The yacht had been built in Germany in 1986 for Kirk Kerkorian, who had become rich buying and selling hotels, casinos, movie studios, you name it. In the months before I spoke to Ellison, Kerkorian had often used the *October Rose* as a floating office as he and former Chrysler chairman Lee Iacocca put together their bid to buy the automaker. In 1996 the takeover attempt failed, and Kerkorian, perhaps short on cash after loading up on Chrysler stock in preparation for the coup, sold the boat to Ellison. (According to a brochure sent to Ellison during the negotiations, "The sundeck is given totally to hedonistic pleasure"—just what he was looking for.) Ellison ranked about a dozen notches higher than Kerkorian on the Forbes 400, so you could say the boat moved up in the world.

The details of the purchase might never have become public if not for a civil lawsuit that came to trial in a San Mateo County, California, court in 2000. A Florida yacht salesman named Michael Rafferty claimed he had brokered the sale of the *October Rose* to Ellison, only to have the Oracle chairman buy the yacht from another broker at the last minute. (That broker was Merle Wood, who had declined to discuss the sale as it was happening when I called him in 1996.) Rafferty said Ellison's double-dealing had cost him a $700,000 commission. His suit rested on the claim that an Oracle executive, negotiating on Ellison's behalf, had entered into an oral contract to buy the boat through him. Rafferty had never met Ellison, but insisted he had been assured by Ellison's front man, an Oracle speechwriter named Richard "Dickie" Brass, that a deal was in the offing. (Rafferty said a friend of a friend of his wife had introduced him to Brass. You make your connections where you can.) The trial made for a good show; when potential jurors were asked if they had ever owned or worked on a yacht, they burst out laughing. But Ellison, who was as glib and engaging as ever, delivered the best performance.

Dickie Brass, who had since left Oracle to work for Microsoft in Washington State, did not appear in the California courtroom, but instead gave a deposition. It was entered into the record during the trial. Rafferty would later sorely regret this; he felt it hurt his case that he couldn't get Brass to California to testify. Brass acknowledged that he had corresponded with Rafferty about the possibility of Ellison's buying a boat from him, but said he warned the yacht salesman that Ellison was "a force of nature" and hard to do business with. Brass also admitted to having arranged a finder's fee of up to 20 percent of Rafferty's commission, should a deal be completed. Why did he do that? "I thought there was a possibility Larry would let me keep it," Brass said. Well, fine. If you had to come up with a single phrase to express the business ethic of Silicon Valley at the end of the twentieth century, that one—*Maybe they'll let me keep it*—would do pretty well.

The thing was, if Dickie Brass was negotiating to buy Ellison a boat, Ellison didn't know anything about it. That was the story presented by Ellison's legal team, anyway. His witnesses insisted that he never saw any of the brochures Rafferty sent by Federal Express, never saw the contract Rafferty overnighted, never knew anything about the deal. Ellison said the same things himself when he testified. He said he became interested in yachts when a girlfriend (he didn't say her name) turned up her nose when he referred to his racing vessel *Sayonara* as a yacht. "That," she said, pointing to a much larger craft, "is a yacht." "She made me feel terribly inadequate," Ellison testified. "As a result, I went on a search to reclaim my adequacy." I remembered what Ellison's boyhood friend Dennis Coleman had said about him: "On some sort of fundamental level he felt inadequate." In those days Ellison had produced questionable medical-school acceptance letters. Now he bought yachts.

Ellison's search, he said, led him to Merle Wood, who showed him a number of yachts, including some real doozies. Ellison spoke of seeing a boat in San Diego that had constellations painted on the ceiling, slot machines, and a clear Plexiglas staircase. "I don't think staircases should be transparent," he said, issuing one of his frequent, spur-of-the-moment statements of principle. "It was the tackiest thing I'd ever seen in my life." In fall 1996 Kerkorian flew Ellison to Antibes, France, to see the *October Rose*. Ellison and Wood took the yacht to Monaco, where Ellison fell in love. "It was bathed in light. I felt like I was in a James Bond movie," he said. "I didn't know life could be this cool." (When I spoke to Rafferty about the trial, he referred to this testimony as "that whole line of crap.") Ellison testified that the boat suited him because it was larger than the others he'd seen, and less formal; he wouldn't mind if his kids plopped down on the seats in their wet bathing suits. The *San Jose Mercury News* said he "behaved like a choir boy" and "played the courtroom like a seasoned performer"—which would sound like a contradiction if you were talking about anyone but Ellison. His testimony was the talk of the courthouse that day; people poked their heads into the courtroom just to have a look at him.

"Larry Ellison just absolutely snowed everybody on that jury," Rafferty said later.

The jurors deliberated for less than two hours before deciding that Rafferty wasn't owed a penny. He was able to document a long correspondence with Brass, but he had never met Ellison and couldn't produce any proof that Brass told Ellison what was going on. Rafferty's stack of FedEx receipts signed by Oracle minions failed to impress the jury. "It was a very clear and clean deal," the foreman said. The newspapers were unable to reach Ellison for comment. The week the verdict was reached, he was on *Sayonara,* competing in the Chicago-Mackinac sailboat race. He won.

When I talked to Rafferty about the case a year after it ended, I could still hear the anger and frustration in his voice. "He tried to deny even knowing who I was," he said. A lot of people who did business with Ellison ended up feeling the way Rafferty did—that Ellison rewrote history according to his needs, that he couldn't be held accountable for anything, and that he didn't care. Rafferty would have liked to forget all about Larry Ellison, but it wasn't easy, given the business he was in. He was a regular reader of the boating press, and the boating press was full of stories about Ellison. After years of racing in *Sayonara,* he was assembling a team to compete in the America's Cup races. When he sold *Sakura* he bought another 192-foot luxury yacht and named it *Izanami.* Its interior was entirely Japanese in design. Then he bought an even larger boat—the 244-foot *Katana*, which had a pyramidlike superstructure, a basketball court on the aft deck, and three engines, including a General Electric turbine. The sale price was rumored to be $25 million. The boat was mentioned in *Power and Motor Yacht* magazine about the time I talked to Rafferty.

"I'm sure this upsets him quite a bit. It says here he's only twenty-sixth on the list of the one hundred largest yachts," Rafferty said.

Between boat rides, Ellison had a company to run.

When I last spoke to him, he was evangelizing the Network Computer, the inexpensive device he said would make the PC and Microsoft Windows obsolete and bring about a new era in communications. More than a few writers were saying he had "bet the company" on the NC. Business writers in Silicon Valley were forever saying executives had "bet their companies" on new ideas, yet over and over again the ideas fell short and the companies kept bouncing along. This proved to be the case with Oracle and the NC. I am writing this epilogue (just as I wrote the book) on a PC equipped with Windows software—not because I am crazy about Windows but because the NC, at least as Ellison envisioned it, hasn't yet caught on. When he was pitching the NC, Ellison behaved the way he did when he was wooing corporate information technology guys in the 1980s: He made extravagant promises about what his software could do and then he went back to the office and told his team to build it.

Ellison certainly gave it the old college try. He spun off the NC development group into a separate company, Network Computer Inc., or

NCI. For a year or so its engineers tried to write software for the device—an operating system, a browser, an application or two—but never came up with anything workable. And though Ellison had promised to ship one million units by the end of 1997, not many people found one under the tree that Christmas. The manufacturers NCI teamed up with, RCA and IBM, shipped only ten thousand each—and RCA later recalled its entire shipment. Ellison was not entirely to blame for this inauspicious start. Soon after he developed the NC, which sold for around $500, the price of a fully equipped personal computer fell below $1,000, including the monitor. Before long, consumers could get a PC with everything the NC lacked—a hard drive, speakers, and loads of memory—for $799 or less. Consumers may have been frustrated with the complexity of PCs, as Ellison said, but PCs did so much more for close to the same price that they didn't care.

Somehow things worked out anyway. In 1998, NCI brought in a new chief executive—Mitchell Kertzman, the former CEO of Oracle's database rival, Sybase. NCI had recently purchased Navio, a Netscape spinoff that was developing a way to get to the Internet through a gizmo that would sit on top of the television set. Kertzman shifted NCI's efforts away from Ellison's grand vision and toward Navio and interactive TV. Here, at last, was something consumers wanted. NCI changed its name to Liberate—as in, let's Liberate this company from any association with the NC—and started selling interactive TV software to everybody with a dog in the home entertainment fight, from cable operators to satellite companies. Liberate went public in 1999 and at this writing has a market capitalization of $1.1 billion. In 2001 Oracle owned 33 percent of the company.

Ellison was wrong about the Network Computer . . . but not dead wrong. His prediction that corporate tech people would like the Network Computer concept proved true; they liked the simplicity and low cost of these so-called "thin clients," which they could add to their networks and replace when they needed to. Information technology managers bought 700,000 of the devices in 1999. Unfortunately for Ellison, they bought most of them from companies aligned with Microsoft, which had developed its own cheap Internet appliance, the NetPC. (The NetPC had Windows software.) Ellison's plot to destroy Microsoft had one major beneficiary—Microsoft.

Even so, he continued to believe in the NC. One market research company predicted sales of slimmed-down computers would reach 2.7 million units a year by 2005. Ellison wanted part of that market. In 2000 he created yet another company to build and sell the devices. This one was called the New Internet Computer Company, or NICC. To run it he drafted Gina Smith, a tech journalist in her thirties who had (it was noted again and again in the press) no experience running a company. (Ellison had often been criticized for putting inexperienced people in charge of major divisions of Oracle, which after all was a publicly traded company. Hiring Smith may have been a gamble, but what could anyone say? NICC was a private company. It was Ellison's money and he could do what he wanted with it.) What Smith had going for her were strong connections in the schmoozy world of high tech, some ideas about what consumers wanted, and a track record of supporting Ellison on the NC. The company introduced its first product, the NIC, with a publicity-generating auction on Amazon.com. Though the machines would retail for only $199 (with a monitor costing $129 more), the first ten sold on Amazon fetched $1,650 each. The buyers said they were enticed partly by the certificates of authenticity signed by Ellison. Smith was quoted as saying she was delighted with the auction results: She just wanted the machines "to get to the retail price."

Ellison said he wanted to sell 5 to 10 million NICs in the first eighteen months; Smith, hovering closer to the home planet, said she hoped for a million. The idea was to get NICs into schools and small businesses, where Ellison and Smith saw a need for cheap, easy-to-use Internet boxes. By the middle of 2001 Smith was saying she hoped to have shipped 100,000 by the end of the year. Ellison's early estimates were off by a few zeros. Still, who could say he wouldn't be right about the NC in the long run? He wasn't the only one making the bet; lots of companies, from WebTV to Netpliance to Gateway, were developing similar products. If Ellison failed, a lot of other people probably would too. But of course Ellison would do it the loudest.

"Larry certainly has a style that lends itself to lampooning," Liberate

CEO Kertzman told Gary Rivlin of *The Industry Standard*. "But I don't know if I'd bet against him."

The part of Larry Ellison's business that had nothing to do with the Network Computer—which is to say, virtually all of it—continued to grow as the new century began. In 2000, Oracle, which by then employed forty-one thousand people, had sales of more than $10 billion and an operating profit of $3 billion. It faced some challenges—most big businesses already had Oracle databases, so it was hard to sell any new ones, and the company was still struggling for a larger share of the market for software applications. Still, stock in Oracle was soaring at the end of that year, making shareholders happy—and raising Ellison's net worth to $58 billion, according to *Forbes*. That made him the second-richest man in the world, behind Bill Gates.

To reward Ellison, the Oracle board of directors came up with a novel scheme: it decided not to pay him any salary or bonus at all over the next four years. It granted him an option to buy twenty million shares of Oracle at the fair market value of the stock at the time of the grant. The options would vest at a rate of five million a year and would expire in 2009. These changes in his compensation, the board explained, "more closely align his compensation with the Company's stock performance." The board estimated the value of the options at $191 million, and added, "It will be the Chief Executive Officer's only option grant for the four-year period." Somehow he'd get by. At first glance, Ellison's willingness to take his compensation this way was admirable; he'd make money if Oracle's stock performed well. But in reality he wasn't making much of a sacrifice. The chances were good that the stock price would exceed the option price sometime in the next ten years, no matter how well or badly he ran the company. When it did, he'd cash in, and $191 million (if that was all he got) wasn't bad pay for four years' work, to say the least. In 2001 *Fortune* magazine cited Ellison's deal as an example in a story titled "The Great CEO Pay Heist."

Some in Silicon Valley groused that Ellison wouldn't have deserved that fat package if not for Ray Lane, who ran the company day to day and held the title of president. Since arriving at the company in the midst of a crisis in 1992, Lane had helped raise its annual revenues from $1.5 billion to $10 billion, and had grown its consulting business into a $5.6-billion-a-year juggernaut. (Lane had come to Oracle from a consulting company, Booz•Allen & Hamilton.) More important, his presence reassured Wall Street and Oracle's customers that the company had a grownup making the big everyday decisions. It was his very success, the cynic might say, that finally cost him his job.

Lane was on vacation on his farm in Oregon on June 30, 2000, when Ellison called him on the phone. He said he wanted to take back the title of president. Ellison was reengineering the company around applications and the Internet and he wanted everyone at Oracle to understand that he was solely in charge. He likened himself and Lane to a married couple trying to raise difficult children. "The company operates with a kind of two-parent philosophy," Lane recalled Ellison saying. "If they don't like what Mom said they go to Dad, and if they don't like what Dad said they go to Mom."

So was Ellison firing him? Not exactly. "Larry can't fire people. He doesn't know how. And he hates conflict," Lane said. "And secondly, I had about $80 million worth of options vesting within two and a half weeks. [Options worth $200 million more would vest in a few months.] If he had fired me, that probably would have brought the options into play. So clearly the idea was do something dramatic enough to make me resign. Which I did. I just said, 'Look, if you want to do that, it's over.' " According to Lane, the conversation then turned to Lane's severance and to the press release Oracle would make when he left. He said Ellison promised to look into those things and call him back, but Ellison never did. The next morning Lane's mother, in Pittsburgh, read about Lane's departure in the newspaper. The company's announcement included no comment from him. The word Lane used to describe this was "chickenshit." He never returned to the office, but had his things sent to him.

Lane had worked for Oracle for eight years, and during that time saw less of Ellison than one might think. They met once a week with other top executives and did their own things after that. Ellison ran engineering, Jeff Henley handled finance, and Lane oversaw what he called "the field"

—sales, consulting, day-to-day operations. "Even when I wanted him to, he wouldn't intervene," Lane said. That formula worked well.

But in the mid to late 1990s, Ellison became obsessed with the idea of building his company around the Internet, and he wanted to do it personally. As a result, Lane said, his last year at Oracle "was hell." Ellison started taking over everything. He insisted that some of Lane's operations people report directly to him instead. He met with Lane's European management team without inviting Lane to attend. "This," Lane said, "drove me crazy." In February 2000, and again in April, Lane said, he had "very private" meetings with Ellison in which he asked if Ellison wanted him to leave. Ellison insisted he didn't; he just wanted Lane to travel less, to communicate with Ellison more, to make more public appearances with him. Lane said he tried to do all of that, but sensed the end was near.

When Ellison made it impossible for him to stay, Lane thought a lot about his motives. He believed that Ellison had cut him loose partly because he was getting too much credit for Oracle's success. "A year ago we had a market cap that put us in the top five in the world," Lane told me in August 2001. "We were starting to have discussions about catching Microsoft in market cap, passing General Electric . . . We were starting to get full of ourselves. And so afterwards, I was thinking, you know, when the most valuable company has been built and Larry is richer than Bill Gates, he doesn't want to have this guy Ray getting the credit. There's just too much being written about Ray doing the turnaround from near-bankruptcy. When we are a near-trillion-dollar company, who the hell is going to get the credit?"

Lane would eventually arrive at other theories for why Ellison squeezed him out. For years Lane had been complaining about Ron Wohl, the young lieutenant Ellison had chosen to run Oracle's applications division. Nobody questioned Wohl's talent and intelligence, but a lot of people doubted he had the skill and experience needed to lead Oracle to victory in the applications market. Joe Costello, the former Oracle board member, had told me in 1996 that putting Wohl up against the brain trust at archrival SAP was "like asking Mister Rogers to fight Mike Tyson." Lane felt the same way. "Ron Wohl was the single worst decision Larry's ever made, and he's never backed away from it," Lane said. He said he nagged Ellison about Wohl repeatedly, telling him, "Ron is not delivering, he is in over his head." But Ellison stuck with him. Lane said he even confronted Wohl once and asked him to resign for the good of the company. No way, Wohl told him.

Finally, with Oracle's applications still struggling and some of Oracle's top sales guys ready to revolt, according to Lane, Ellison acknowledged the company had a problem. He took over Oracle's applications himself and started building server-based applications that would run off a browser.

Once during that time, Lane was at a large meeting during which Oracle salespeople were complaining about the company's applications. He called Ellison and asked him to come and answer some questions. Ellison did—and was immediately subjected to a grilling by Lane's people. "This group just pelted him with questions about when are we going to fix this, when are we going to fix that," Lane said. He would later learn from someone else that Ellison thought Lane was staging a palace coup. "He thought I was setting him up. I don't know how I can do that. He's my boss. I'd have to be really stupid," Lane said.

Lane also concluded that Ellison was angry with him because of an episode a couple of years earlier. Lane was being pursued by the software company Novell, which wanted him to be its CEO. Lane said he resisted at first, but when the company offered $15 million in cash and four million stock options—"In those days it was a lot of money to me"—he couldn't say no. He went to Ellison's house in Atherton to break the news. Here is how Lane remembers the conversation:

"Larry, I'm going to Novell."

"Over my dead body."

"Larry, I've already accepted."

"No, you can't go. How about two million options? Look, Ray, you know Novell's not going anywhere."

So Lane stayed. And Ellison delivered not 2 million options, but 2.5 million. It seemed a happy ending, but others would later tell Lane that Ellison was anything but happy.

"He had a feeling of vulnerability that I could walk into his house, resign, and leave him in that position," Lane said. Lane believed he never got over the resentment. "It's a shame that he couldn't come to me and say, 'Ray, if you ever put me in that position again' . . . whatever. Just say it. It was like a bad marriage in that he kept it to himself, allowed it to boil, and decided, 'Somehow I've got to get Ray out of the company.' "

In the summer of 2000 he got him out. Even though Lane left a lot of options on the table, he still managed to depart with $800 million worth of Oracle stock. A year later it was worth about $200 million. I asked Lane, who had gone to work for a venture capitalist firm after Oracle, how much the dip in Oracle's stock could be attributed to a weak economy, and how much to the way the company was being run. "That is a great question," he said. "It's easy to blame the economy right now. That said, I would probably say 50 percent is due to the economy. It takes a while for lack of management attention or bad management decisions to show an effect. We'll see the difference when the economy picks up again, and we see the relative performance differences between SAP and Oracle and Microsoft and Oracle." After he left Oracle, Lane said, he received a call from an Oracle account manager who wanted him to help sell a database system to General Motors. The GM people trust you, the account manager told Lane. The implication, Lane believed, was that they didn't trust anyone else at the top of Oracle.

Even though he was long gone from the company, Lane helped make the sale.

"I want the company to do well," he told me. "I own twenty million shares."

After this book was first published in 1997, I became sort of a dial-a-quote resource for journalists writing about Larry Ellison. A lot of the calls came from people who were following Oracle much more closely than I was, frankly, which made me a little uneasy. But if their questions were about Ellison as a person and not about whatever new business strategy he had announced that day, I did what I could to help.

People writing profiles or producing TV segments about Ellison usually began with the same question: Did he really hate Bill Gates? The conventional wisdom was that he did. In a 2000 cover story about Ellison,

Fortune magazine said he was "sick of playing second banana" to Gates and was going to become first banana by making Oracle "the Microsoft of the Internet." About the same time, *The Washington Post* quoted Ellison referring to the Microsoft founder derisively as "William Gates the Third."

Of course, it was often hard to know how Ellison really felt about anything because you could never get past the wisecracks. Still, I told other journalists I was skeptical about the Larry-hates-Bill thing. Certainly Ellison wanted to be number one in the world of high tech, and he definitely didn't like Windows software, which he complained was bloated and untrustworthy. (Anybody who ever tried to decode a Windows error message knew he had a point.) But hate Bill Gates? I didn't see it. Nothing in Ellison's description of their first meeting suggested he hated him; on the contrary, he seemed to admire his intelligence and business savvy. Once, when I arrived at Ellison's home in Atherton for an appointment, he said he had just finished a phone conversation with Gates. This was at a time when the two were jabbing at each other in the press over the NC, and yet Ellison seemed perfectly relaxed after their chat. If they had been screaming at each other, it didn't show on Ellison's face.

Ray Lane said Ellison considered Gates "a geek," and that was probably true. But I sensed something calculating in the way Ellison chided Gates in the press. Microsoft was (and is, as I write) a vastly larger, stronger and better known company than Oracle. And though Ellison's net worth briefly rivaled Gates's in 2000, Gates was far more famous than Larry Ellison. By encouraging the idea that Oracle and Microsoft, and he and Gates, were somehow competing as equals, Ellison did great things for his own PR, and absolutely nothing for Gates's. Their back-and-forth over the Network Computer was a good example. This was a war in which Ellison was firing a popgun at his enemy's rumbling tanks, and yet it got Ellison and his company more attention than they had ever received before. Brilliant. By pretending to have it in for Gates, Ellison generated attention for himself and Oracle and created an irritating distraction for his rival. Hate Bill Gates? I didn't think so. I told anyone who called that the whole thing was, in fact, mostly shtick.

Then I got some calls that started to make me think I was wrong.

In the summer of 2000 came the news that private detectives employed by Investigative Group International had been spying on a proMicrosoft trade group in Washington, D.C. There was even a suggestion that the detectives had offered to pay for a look at the trade group's garbage. And who had put them up to this Dumpster-diving? Ellison. When questioned, he immediately served up a justification. The trade groups IGI was investigating (there were several) weren't the independent bodies they pretended to be. According to an Oracle press release, they "were misrepresenting themselves as independent advocacy groups, when in fact their work was funded by Microsoft for the express purpose of influencing public opinion in favor of Microsoft during its antitrust trial." Larry Ellison certainly couldn't let people go around misrepresenting things!

When the reporters called, I laughed off the whole thing. This, I said, was exactly the kind of cockamamie, Little Rascals–type scheme Ellison would dream up. But when I thought about it later, I had to admit that this was different from Ellison's usual Gates-baiting. Hiring private eyes to go through Microsoft's garbage bespoke a level of weird obsession I had somehow missed in Ellison. He was quoted saying the snooping was his "civic duty," and a reasonable person could see why he might think so. Microsoft was, after all, fighting for its life (the government was pursuing the mother of all antitrust cases against it), and the public had a right to know what it was up to. But did Ellison really have to get involved? Who had appointed *him* to the Justice Department legal staff? It was creepy to think of Ellison getting so worked up that he called in the private dicks. And the only plausible explanation was that he couldn't stand his nemesis getting away with something. That's not calculating or clever. It's pathological.

Ellison is about to celebrate his fifty-seventh birthday as I write this, so it's natural to speculate on what he'll do with the rest of his life. Maybe he'll ease out of Oracle and pour his money and energy into education and medicine, two areas to which he has already given (though compared with Gates he hasn't given much). Maybe he'll get married again; he sometimes referred to his girlfriend, romance writer Melanie Craft, as his fianceé (and Ray Lane's impression was that he was indeed in love with her). Maybe he'll win the America's Cup. Maybe he'll spend his days tooling around in his cars and boats and planes.

Maybe he'll find satisfaction in what he accomplished. Starting with three other guys and a good idea, he built a company that employed legions of people and created astonishing wealth for himself and thousands of shareholders. His product, for all its early bugs, was a godsend to businesspeople all around the planet; with it, they made more money and did a better job for their customers. If history is fair to Larry Ellison, it will acknowledge that he helped change the world. Maybe he'll appreciate that.

Or maybe he'll worry about this status right up to the end. Maybe he'll smear charcoal on his face, don a black pullover and black slacks, and head up to Redmond, where he can go through Bill Gates's garbage personally, in hopes that whatever he finds there will somehow make him number one. Or at least make him adequate.

Acknowledgments

I INTERVIEWED LARRY ELLISON FOUR TIMES, FOR A TOTAL OF SIXTEEN hours. The first two conversations took place in October 1996, the last two in December of that year. Ellison is a forward-looking person, and talking about the past was sometimes hard for him. I appreciate the help he gave me in telling this story.

I spoke with dozens of people who have been associated with Oracle Corporation over the years; save for those who asked to remain anonymous, their names appear in the long list below. A few sources were remarkably generous with their time. Jenny Overstreet, Ellison's longtime personal assistant, met with me a half dozen times. We also kept up a running conversation by E-mail during the months I was working on the book. She is a true professional, and I am lucky to know her. Carolyn Balkenhol and Cynthia Turner in Ellison's office were always gracious and accommodating. Ed Oates, an Oracle cofounder, gave me two long interviews and was patient when I was checking facts, and Stuart Feigin was a wonderful source and dinner companion.

Another Oracle person was so helpful in getting me in touch with sources that I came to refer to him as "the source of all wisdom." He knows who he is.

Oracle engineer Roger Bamford and his wife, Ginger, sat for an interview in April 1996. Soon after, they invited me to dinner at their home, something they did again and again during my year of research. The Bamfords are great admirers of Larry Ellison. I'm grateful for their insights into Silicon Valley and its unique culture and for their overwhelming generosity. The reader should know that when I quote the Bamfords, I am quoting friends.

The following people sat for interviews or helped in other ways: Steve Abramowitz, William Aspray, Rick Bennett, Mike Blasgen, Andre Boisvert, Michael Boskin, Kirk Bradley, Roy Bukstein, Katherine Bull, Martin Campbell-Kelly, Warren Capps, Ken Cohen, Dennis Coleman, Harold Coleman, Richard Coleman, Craig Conway, Kevin D. Corbitt, Joe Costello, Sarah Crawford, Paul Cubbage, Kitty Cullen, Katherine Daugherty, Dwight B. Davis, Don Deutsch, Farzad Dibachi, Barbara Ellison, Mike Feibus, Donald Feinberg, Richard Finkelstein, Tom Ford, Bill Friend, Len Gallagher, Errol Getner, Joe Gillach, Evan Goldberg, Stacey Griffin, Al Guibord, Sheila Maydet Gutterman, Jeff Henley, James G. Hix, Bob Howie, Stephen Imbler, Steve Jobs, John Kemp, Gary Kennedy, Jane Kennedy, Frank King, Anne King, George Koch, Andy Laursen, Kelly Lawson, Joshua Lederberg, Don Lucas, Ken Marshall, Charles Masters, Brian Mattal, Stephen McClellan, Tom McGowan, Sam Miller, Nicola Miner, Kate Mitchell, Bob Ney, Dale Osborn, Chuck Phillips, Bob Preger, Habeeb Qadri, Adda Quinn, Dave Roberts, Rick Rosenfield, Erik Salbu, John Schill, George Schussel, Bruce Scott, Mike Seashols, Phil Shaw, Jean Shildneck, Tom Siebel, Scott Smith, Mike Sottak, Ellen Spencer, David Steinberg, Mike Stonebraker, Jeff Tarter, Jeffrey Tash, Les Tepper, Irv Tjomsland, Danny Turano, Dennis Voorhees, Jeff Walker, Paul Wasserman, Chuck Weiss, Marcia Wells-Lawson, Doran Wilde, Richard Winter, Ron Wohl, and Tony Ziemba.

I was fortunate during this project to have the help of many fine journalists. Joanne Huddleston covered the Adelyn Lee trial in my stead; the chapter about the trial is based largely on her excellent reporting. I am also grateful to Don Clark, Bart Ziegler, and Stefan Fatsis at the *Wall Street Journal* and to Scott Thurm and S. L. Wykes at the San Jose *Mercury News.* Nora Paul and Bill Boyd of the Poynter Institute for Media Studies offered ideas and research help. Thanks also to Julian Guthrie at the San Francisco *Examiner,* Richard Brandt at *Upside* magazine, Anthony Perkins at the *Red Herring* magazine, and Marc Ferranti of the IDG News Service.

The St. Petersburg *Times,* where I am a staff writer, graciously granted me a leave of absence to work on the book. My thanks to Paul Tash, Neil Brown, Nancy Waclawek, Gretchen Letterman, and the talented writing and editing staff at that newspaper.

Henry Ferris, my editor at William Morrow and Company, was a good and encouraging friend from beginning to weary end. Ann Treistman provided help in many ways. I'm grateful, as always, to Esther Newberg at International Creative Management and to her patient and hardworking assistant, Jack Horner.

Phil and Joanne Taylor opened their home to me the many weeks I was in California, a kindness I cannot hope to repay. This book could not have been written without them.

Notes

MOST OF THE FACTS AND QUOTATIONS IN THIS BOOK ARE THE PRODucts of original research. Sources not identified in the text are listed here. The book also includes information first published by other authors and journalists. The sources are also listed here.

Chapter One

1. The designer was Olle Lundberg.
2. This statistic was provided by Oracle chief financial officer Jeff Henley.
3. Katherine Daugherty was the source of this quotation.
4. The man was Stephen Imbler, later Oracle's vice-president of corporate finance.
5. The executive was George Koch.
6. The associate was advertising man Rick Bennett.
7. Martin Campbell-Kelley and William Aspray, *Computer: A History of the Information Machine* (New York: Basic Books, 1996).
8. Martin Campbell-Kelly, "Development and Structure of the International Software Industry, 1950-1990," *Business and Economic History,* vol. 24, no. 2 (Winter 1995).
9. *The HP Way: How Bill Hewlett and I Built Our Company* (New York: HarperBusiness, 1995).
10. The board member was Joe Costello.
11. David Coursey, editor of *PC Letter,* wrote this passage in *Windows Sources* magazine, May 20, 1996.
12. Ellison made this remark in a television interview with Charlie Rose on April 10, 1995.
13. Oracle engineer Roger Bamford was the source of this comment.
14. Ellison made this remark in an interview with *Forbes ASAP* (April 8, 1996).
15. The source was Rick Bennett.

16. Ellison was quoted in the San Francisco *Examiner,* December 11, 1994.
17. Oracle executive Raymond Lane was quoted in *PC Week* on February 26, 1996.
18. This column by Ziff-Davis columnist Woody Leonhard was dated December 8, 1995.

Chapter Two

1. Ellison made this remark to Errol Getner, a boyhood friend.
2. Ellison was quoted in Brenton R. Schendler, "Software Tiger: Oracle Corp.'s Ellison Spurs Its Fast Growth with Aggressive Style," *Wall Street Journal,* May 31, 1989.
3. The word "rough" appeared in Louise Kehoe, "Software Billionaire's New Ambitions," *Financial Times,* March 1, 1995. "Notoriously rough" was used by Kathleen Doler in a February 22, 1995, article in *Investor's Business Daily.* The adjective "tough" appeared in Rich Karlgaard, "Larry Ellison," *Forbes ASAP* (June 7, 1993). The *USA Today* story, "Ellison: Survivor of Corporate Storms," appeared on February 3, 1995, and was written by James Kim.
4. Richard Speck murdered eight student nurses in an apartment building in a neighborhood near Ellison's in July 1966, only a few weeks before Ellison left Chicago for good. A ninth woman survived Speck's attack.
5. Ellison was the source of the story about his sister's joke.
6. The friend was Dennis Coleman.
7. Dennis Coleman sent me a copy of Larry Ellison's E-mail message to him.
8. Ellison said this in a television interview with Charlie Rose, April 10, 1995.
9. Jenny Overstreet was the source of this quotation.
10. The Churchill letter was quoted in Martin Gilbert, *Churchill: A Life* (New York: Henry Holt, 1991)
11. Ellison's friend and employee Chuck Weiss was the source of this information.

Chapter Three

1. Ellen Ullman's essay on programming, "Getting Close to the Machine," was excerpted in the June 1995 issue of *Harper's* magazine. It was also published as part of the book *Resisting the Virtual Life: The Culture and Politics of Information,* ed. James Brook and Iain A. Boal (San Francisco: City Lights Books, 1995).
2. Bob Miner made this remark in a 1993 interview with the Bay Area cable television show *High-Tech Heroes.*
3. Former Oracle employee Brian Cassidy told this story at a 1994 memorial service for Bob Miner.
4. Oracle engineer Andy Laursen said this about Bob Miner.

5. Brian Cassidy.
6. Craig Conway, a former Oracle vice-president, was the source of this quotation.

Chapter Four

1. This story was first told in the excellent book *Computer: A History of the Information Machine,* loc. cit.
2. IBM's Irv Traiger was quoted in the System R reunion papers posted on the World Wide Web in 1996.
3. This point is made in a slightly different way in George Koch and Kevin Loney, *Oracle: The Complete Reference,* 3d ed. (New York: McGraw-Hill, 1995).
4. Bob Miner said this during the *High-Tech Heroes* interview.
5. This prediction appeared in David Kroenke, (SRA, 1977).
6. *High-Tech Heroes.*
7. Ibid.
8. Ibid.
9. Bruce Scott was the source of this quotation.
10. *High-Tech Heroes.*
11. This statistic was taken from Paul Carroll, *Big Blues: The Unmaking of IBM* (New York: Crown, 1993).
12. This quotation from IBM's Rich Seidner is taken from the public TV program *Triumph of the Nerds,* produced by Oregon Public Broadcasting and John Gau Productions.
13. Steve Jobs made this remark in an interview for *Triumph of the Nerds.*
14. The employee was Dave Roberts.

Chapter Five

1. The source of this quotation was Roy Bukstein, Oracle's first chief financial officer.
2. Ed Oates said he was one of the employees who got a security clearance. Accountant Roy Bukstein was the source of the anecdote about plain white envelopes coming from the CIA.
3. *High-Tech Heroes.*
4. Oracle engineer Bob Brandt said this about the IBM 4331.
5. Roy Bukstein made this comment.
6. *High-Tech Heroes.*
7. Larry Ellison was quoted to that effect in the August 1989 issue of *Software* magazine.

8. This detail was taken from Glenn Rifkin and George Harrar, *The Ultimate Entrepreneur: The Story of Ken Olsen and Digital Equipment Corporation* (Chicago: Contemporary Books, 1988). I am indebted to the authors for much of the material about the VAX.
9. Ibid.
10. The statistics in this paragraph are taken from Oracle's 10-K annual reports, filed with the Securities and Exchange Commission.
11. This term appeared in Jerry Kaplan, *Startup: A Silicon Valley Adventure* (Boston: Houghton Mifflin, 1995).
12. These statistics were published in Oracle's red herring, titled "Preliminary Prospectus Dated February 4, 1986."
13. The source of this quotation was Oracle sales manager Bob Preger.
14. Roger Bamford remembered Bob Miner's saying this.
15. Oracle employee Katherine Daugherty was the source.
16. Ellison was quoted to this effect in the February 28, 1988, *New York Times.*
17. Roy Bukstein heard Ellison make this point.
18. Roy Bukstein was the Miner family accountant when I interviewed him. He later left his accounting firm to become the family's full-time financial adviser.

Chapter Six

1. Brian Cassidy made this remark, and the ones that follow, at a memorial service for Bob Miner in November 1994.
2. Ellison said this in a speech to the Commonwealth Club of California on March 8, 1996. I am indebted to the club for providing the audiotape.
3. These ideals appeared in Thomas J. Watson, Jr., and Peter Petre, *Father, Son & Co.: My Life at IBM and Beyond* (New York: Bantam Books, 1990).
4. This quotation and the ones that follow are from the Bob Miner interview on *High-Tech Heroes.*
5. Barbara Ellison did not say much about their wedding day either. The account of that day is drawn from a lawsuit she filed, years later, challenging the prenuptial agreement.
6. The source of this quotation was Dave Roberts, an early Oracle salesman.

Chapter Seven

1. Oracle provided a list of the operating systems in early 1986 in the prospectus for its initial public offering.
2. Richard Finkelstein was quoted to this effect in *Lotus* magazine in October 1989.

3. The company was Timeplex, whose story is told in more detail later in this chapter.
4. This appeared in Oracle form 10-K for fiscal year 1986.
5. The source of this anecdote did not want to be identified.

Chapter Eight

1. This phrase appeared in the *Business Week* article "Cullinet Struggles to Get Out of the Doldrums," March 31, 1986.
2. Craig Conway was quoted in the Los Angeles *Times* on March 26, 1995.
3. This figure appeared in the November 15, 1988, issue of *Datamation* magazine; the magazine quoted market researcher InfoCorp.
4. This source asked to remain anonymous.
5. Ibid.
6. Ibid.

Chapter Nine

1. The engineer was Dennis Voorhees.
2. Larry Ellison was the source of the "un-sales" anecdote. Stephen Imbler was the source of the "sales prevention" quotation.
3. Oracle always finished the fiscal year with high receivables because it took some time to get paid for deals made in the always hectic last days of the final quarter.
4. The information about Oracle's initial public offering comes from a press release issued by Oracle chief financial officer Jeffrey O. Henley in March 1996. The information about Microsoft's IPO was taken from Stephen Manes and Scott Andrews, *Gates: How Microsoft's Mogul Reinvented an Industry—and Made Himself the Richest Man in America* (New York: Doubleday, 1993).
5. The article was Myriam Weisang Misrach, "Eastern Exposure." It appeared the February 1993 issue of *Northern California Home & Garden.*
6. The source was Paul Cubbage of Dataquest.
7. The Oracle users—Joe Sparks, Dave Neimeyer, and Dan Haden, in that order—were quoted in the February 1990 issue of *NY Oracle User.*
8. Kabcenell's talk was tape-recorded on September 7, 1989, by a member of the St. Louis Oracle User Group and reprinted in the February 1990 edition of the *NY Oracle User* newsletter.
9. Rick Bennett was the source.
10. According to Imbler, an exception was made only once, in the case of a job candidate who had been a partner at the consulting firm Arthur Andersen.

Chapter Ten

1. The information about Oracle's stock price and trading volume is taken from "Oracle Stock Dives; Dow Advances 6.75," San Jose *Mercury News*, March 29, 1990.
2. This analyst report was attributed, in the lawsuit, to Paine Webber Mitchell Hutchins, Inc.
3. This quotation was taken from Oracle's March 27, 1990, announcement of its Q3 results.
4. The source was Oracle's Roger Bamford.
5. SEC Form 10-K for the fiscal year ended May 31, 1990.
6. The quotation and the facts about Tom Siebel's career were taken from an internal Oracle publication called *Milestones.*
7. *Milestones.*
8. The piece was published on September 8, 1990, and was written by Lee Gomes.
9. The first quotation was taken from "CEO Blames Managers for Oracle Woes," San Francisco *Chronicle*, September 26, 1990. The source of the second was "Oracle 'Fiasco' Results in $36 Million Loss," *PC Week* (October 1, 1990).
10. *Oracle User Resource* (November 1990).
11. Ellison remembered having this conversation with Ken Cohen but did not recall the details.
12. Tony Ziemba's article appeared in the November 1990 issue of *Oracle User Resource.*
13. This letter, dated January 21, 1991, was reprinted in the February 1991 issue of *Oracle User Resource.*

Chapter Eleven

1. The Oracle stock prices were taken from the *New York Times* stock tables for the day after each of the dates in question.
2. The article appeared on November 2, 1990, under the headline THE ACTION DRIES UP FOR SMALLER STOCKS.
3. The value of Ellison's stock during 1990 was calculated using information provided by CDA Investment Technologies, Inc, which tracks insider trades. (The author is indebted to Stacey Griffin for this information.) For example, the $954 million figure was calculated by multiplying the number of shares Ellison owned (33,646,864) by the stock price on March 19, 1990 ($28.38).
4. San Jose *Mercury News*, June 10, 1991.
5. The source was Bill Friend.
6. The source was chief financial officer Jeff Henley.
7. The information about Nippon Steel was taken from "Oracle Fulfills Cash Yen," *Information Week* (June 10, 1991).

8. The source was Jeff Henley.
9. These figures were taken from the Securities and Exchange Commission's Complaint for Permanent Injunction and Other Relief, filed September 24, 1993, in U.S. District Court in San Francisco.

Chapter Twelve

1. Alan Deutschman, "Software's Other Billionaire," *Fortune* (November 29, 1993).
2. "Debt Restructuring Agreement," filed in San Mateo County Superior Court as part of one of the nCube-related lawsuits against Ellison.
3. "Memorandum of Points and Authorities in Opposition to Motion to Tax Costs," by Melissa A. Finocchio, filed in San Mateo County Superior Court, March 19, 1993.
4. The Oracle user who tested Oracle7 was Rob Watson of Motorola, Inc. He wrote about the database in the December 1991 issue of *Azora,* the newsletter of the Arizona Oracle Users Group. The article was reprinted in the March 1992 issue of *Oracle User Resource.*
5. *Computer Reseller News* (November 18, 1996).
6. Ellison told the story about selling the car to Ray Lane.
7. "Can Larry Beat Bill?," *Business Week* (May 15, 1995).
8. This person asked to remain anonymous.
9. Deposition of Craig Ramsey.
10. Deposition of Oracle lawyer Juana Schurman.
11. The source for Adelyn Lee's behavior when she was fired was prosecutor Paul Wasserman.
12. Nicola Miner.

Chapter Thirteen

1. Alan Deutschman, "Software's Other Billionaire," *Fortune* (November 29, 1993).
2. Ellison was interviewed by *Red Herring* magazine (June 1994).
3. "The Oracle of Silicon Valley," Los Angeles *Times,* March 26, 1995. The article notes that Ellison made the remark about not settling only a few weeks before he settled.
4. Prosecutor Paul Wasserman was the source of this figure.
5. The information about Don Lucas's board service is taken from Cadence Design Systems' 1995 and 1996 proxy statements filed with the SEC.
6. *Datamation* (February 1997).

7. Ellison made these remarks in a speech to the Commonwealth Club of California on March 8, 1996.
8. Livingstone gave this interview to nCube's Paul Godfrey. nCube published the interview on the World Wide Web in 1996.
9. This quotation from Steve Jobs and the one below from Apple scientist Larry Tesler were taken from *Triumph of the Nerds.*
10. "Bell Atlantic Video Services Unveils Programming Sources for its Stargazer Interactive Television Market Trial," Bell Atlantic press release, March 8, 1995.
11. *Red Herring* (June 1994).
12. Ellison spoke about the phone companies on September 4, 1995, at the European IT Forum in Paris.
13. "Larry Ellison Is Captain Ahab and Bill Gates Is Moby Dick," *Fortune* (October 28, 1996).
14. "What It's Really Like to Be Marc Andreessen," *Fortune* (December 9, 1996).
15. Lee Gomes "And Gates Said . . . Let There Be Hype," San Jose *Mercury News,* August 24, 1995. For information about the launch of Windows 95, I relied on that article, the TV program *Triumph of the Nerds,* and Dan Gillmor, "Is Microsoft Taking Us All for a Ride?," *Mercury News,* August 25, 1995.

Chapter Fourteen

1. *Forbes ASAP* (April 8, 1996).
2. The source was Oracle user Warren Capps, who was among those who went without sorbet.
3. Ellison was quoted in *Triumph of the Nerds.*
4. "Larry Ellison: Samurai Interview," *Forbes ASAP* (April 8, 1996).
5. "Bill Gates with a College Degree," *Red Herring* (January 1996).
6. Kirk Bradley was the source for this.

Chapter Fifteen

1. School principal Ellen Spencer heard Winfrey's remark.
2. Ellison's date requested anonymity.
3. The person who made this comment requested anonymity.
4. John Foley, "Time to Deliver," *Information Week* (February 24, 1997).
5. "Larry Ellison's Vision for the Network Computer," taken from the Gartner Group Web site. The site listed the source as "Gartner Group's Systems Software Architecture Research Note E-401-138, September 27, 1996."
6. The Gartner Group estimate was published in "Weighing the Case for the Network Computer," *Economist* (January 18, 1997).

7. Erik Nee, "Give It Up, Scott and Larry," *Upside* (December 1996).
8. Bill Gates, "The Internet PC," posted on the Microsoft Web site in April 1996.
9. Microsoft and Intel were quoted in Brooke Crothers, "Grove, Gates Gaga over NetPC," c/net, October 28, 1996.
10. Bill Gates's remarks were reported by Reuters, on February 6, 1997. He made the comments at a conference of Internet users in Madrid.

Notes to the Epilogue for Paperback Edition

Some of the details about *Sakura* first emerged in David A. Kaplan's very funny 1999 book, *The Silicon Boys* (William Morrow, New York). Bill Vlasic and Bradley A. Stertz wrote about Kirk Kerkorian in "Taken for a Ride," their 2000 book about his Chrysler takeover attempt (William Morrow, New York). Some of the details of the *Sakura* trial were reported by Tracy Seipel and Camille Mojica Rey in the *San Jose Mercury News,* and by Michele R. Marcucci and Millicent Mayfield in the *San Mateo County Times.* Some of the information for my section about the Network Computer was first reported by Gary Rivlin in an excellent article in the *Industry Standard,* "The Network Computer Strikes Again!" (March 27, 2000). I also benefited from the work of Ian Fried of CNET News.com and Jeff St. John of *Upside* magazine. Mark Leibovich's profile of Ellison, published on October 30, 2000, is probably the best newspaper piece ever done on the subject. In the section about Oracle's hiring of private detectives, I have made use of several details first reported by Joshua Micah Marshall on Salon.com on June 30, 2000.

Bibliography

Barry, Dave. *Dave Barry in Cyberspace.* New York: Crown, 1996.

Brand, Stewart. *The Media Lab: Inventing the Future at MIT.* New York: Viking Penguin, 1987.

Brooks, Frederick P., Jr. *The Mythical Man-Month: Essays on Software Engineering.* Reading, MA: Addison-Wesley, 1975.

Campbell-Kelly, Martin, and William Aspray. *Computer: A History of the Information Machine.* New York: Basic Books, 1996.

Carroll, Paul. *Big Blues: The Unmaking of IBM.* New York: Crown, 1993.

Dinerstein, Nelson T. *Database and File Management Systems for the Microcomputer.* Glenview, IL: Scott, Foresman and Company, 1985.

Gates, Bill, Nathan Myhrvold, and Peter Rinearson. *The Road Ahead.* New York: Viking, 1995.

Gilbert, Martin. *Churchill: A Life.* New York: Henry Holt, 1991.

Kaplan, Jerry. *Startup: A Silicon Valley Adventure.* Boston: Houghton Mifflin, 1995.

Kidder, Tracy. *The Soul of a New Machine.* Boston: Little, Brown, 1981.

Koch, George, and Kevin Loney. *Oracle: The Complete Reference,* 3d ed. New York: McGraw-Hill, 1995.

Kroenke, David. *Database Processing: Fundamentals, Modeling, Applications.* Worthington, OH: SRA, 1977.

Manes, Stephen, and Scott Andrews. *Gates: How Microsoft's Mogul Reinvented an Industry—and Made Himself the Richest Man in America.* New York: Touchstone/Doubleday, 1993.

McClellan, Stephen T. *The Coming Computer Industry Shakeout: Winners, Losers & Survivors.* New York: John Wiley & Sons, 1984.

Packard, David. *The HP Way: How Bill Hewlett and I Built Our Company.* New York: HarperBusiness, 1995.

Pacyga, Dominic A., and Ellen Skerrett. *Chicago, City of Neighborhoods: Histories & Tours.* Chicago: Loyola University Press, 1986.

Rifkin, Glenn, and George Harrar. *The Ultimate Entrepreneur: The Story of Ken Olsen and Digital Equipment Corporation.* Chicago: Contemporary Books, 1988.

Sobel, Robert. *IBM: Colossus in Transition.* New York: Times Books, 1981.

Wallace, James, and Jim Erickson. *Hard Drive: Bill Gates and the Making of the Microsoft Empire.* New York: John Wiley & Sons, 1992.

Watson, Thomas J., Jr., and Peter Petre. *Father, Son & Co.: My Life at IBM and Beyond.* New York: Bantam Books, 1990.

Index